# GENERALIZED
# HYPERGEOMETRIC FUNCTIONS

T0276134

TO THE MEMORY OF
W. N. BAILEY

# GENERALIZED HYPERGEOMETRIC FUNCTIONS

BY

LUCY JOAN SLATER

D.LIT., PH.D. (London), PH.D. (Cambridge)

*Special Appointment*
*Department of Applied Economics*
*University of Cambridge*

CAMBRIDGE

AT THE UNIVERSITY PRESS

1966

CAMBRIDGE UNIVERSITY PRESS
Cambridge, New York, Melbourne, Madrid, Cape Town, Singapore, São Paulo, Delhi

Cambridge University Press
The Edinburgh Building, Cambridge CB2 8RU, UK

Published in the United States of America by Cambridge University Press, New York

www.cambridge.org
Information on this title: www.cambridge.org/9780521064835

First published 1966
This digitally printed version 2008

*A catalogue record for this publication is available from the British Library*

*Library of Congress Catalogue Card Number: 66–10050*

ISBN 978-0-521-06483-5 hardback
ISBN 978-0-521-09061-2 paperback

# CONTENTS

# 3 Basic Hypergeometric Functions

# 4 Hypergeometric Integrals

# 9 Basic Appell Series

# Appendices

# PREFACE

This book should really be attributed to Bailey and Slater. It was Professor Bailey's intention to write a comprehensive work on hypergeometric functions, with my assistance. This present work is based in part on notes for a series of lectures which he gave in 1947–50 at Bedford College, London University. The rest of the book contains the results of my own researches into the general theory. It also covers the great advances made in the subject since 1936 when W. N. Bailey's Cambridge Tract 'Generalized Hypergeometric Series' was first published.

The theory of generalized hypergeometric functions is fundamental in the field of Mathematical Physics, since the general functions studied here contain as special cases all the commonly used functions of analysis. The present work should prove of use and interest to all Mathematical Analysts and Theoretical Physicists. The generalized Gauss function is also used increasingly in Mathematical Statistics, and the basic analogues of the Gauss functions have many interesting applications in the field of Number Theory.

I should like to thank Dr Theo Chaundy for a very careful reading of the manuscript, and several helpful comments.

L. J. S.

*Cambridge*
*June, 1964*

# 1

## THE GAUSS FUNCTION

### 1.1 Historical introduction

The series

$$1+\frac{ab}{c}\frac{z}{1!}+\frac{a(a+1)\,b(b+1)}{c(c+1)}\frac{z^2}{2!}+\frac{a(a+1)\,(a+2)\,b(b+1)\,(b+2)}{c(c+1)\,(c+2)}\frac{z^3}{3!}+\cdots,$$

$$(1.1.1)$$

is called the Gauss series or the ordinary hypergeometric series. It is usually represented by the symbol

$$_2F_1[a,b;\,c;\,z].$$

The variable is $z$, and $a$, $b$ and $c$ are called the parameters of the function. If either of the quantities $a$ or $b$ is a negative integer $-n$, the series has only a finite number of terms and becomes in fact a polynomial $\qquad _2F_1[-n,b;\,c;\,z].$

For example, suppose that $a = -2$, then the series becomes

$$_2F_1[-2,b;\,c;\,z]=1+\frac{(-2)\,b}{c}\frac{z}{1!}+\frac{(-2)\,(-1)\,b(b+1)}{c(c+1)}\frac{z^2}{2!}+0,$$

that is $\qquad _2F_1[-2,b;\,c;\,z]=1-\frac{2bz}{c}+\frac{b(b+1)\,z^2}{c(c+1)},$  $\qquad(1.1.2)$

since all the later terms are zero.

In his work *Arithmetica Infinitorum* (1655), the Oxford professor John Wallis (1616–1703) first used the term 'hypergeometric' (from the Greek ὑπερ, above or beyond) to denote any series which was beyond the ordinary geometric series

$$1+x+x^2+x^3+\cdots.$$

In particular, he studied the series

$$1+a+a(a+1)+a(a+1)\,(a+2)+\cdots.$$

During the next one hundred and fifty years many other mathematicians studied similar series, notably the Swiss L. Euler (1707–1783)[†] who gave amongst many other results, the famous relation

$$_2F_1[-n,b;\,c;\,z]=(1-z)^{c+n-b}\,_2F_1[c+n,c-b;\,c;\,z],\qquad(1.1.3)$$

† Euler (1748). Full details of all references are to be found in the bibliography.

In 1770, the Frenchman, A. T. Vandermonde (1735–1796) stated his theorem, an extension of the binomial theorem, in the form

$$_2F_1[-n, b;\; c;\; 1] = \frac{(c-b)(c-b+1)(c-b+2)\ldots(c-b+n-1)}{c(c+1)(c+2)(c+3)\ldots(c+n-1)},$$

(1.1.4)

but during the next forty years the Göttingen school under C. F. Hindenberg (1741–1808) wasted much effort on various complicated extensions of the binomial and multinomial theorems. All this was changed dramatically, when on 20th January, 1812, C. F. Gauss (1777–1855) delivered his famous thesis 'Disquisitiones generales circa seriem infinitam'† before the Royal Society in Göttingen. In it, this brilliant mathematician defined the modern infinite series of (1.1.1) above and introduced the notation $F[a, b;\; c;\; z]$ for it. He also proved his famous summation theorem

$$_2F_1[a, b;\; c;\; 1] = \frac{\Gamma(c)\,\Gamma(c-a-b)}{\Gamma(c-a)\,\Gamma(c-b)},$$

(1.1.5)

and he gave many relations between two or more of these series. He showed clearly that he was already regarding $_2F_1[a, b;\; c;\; z]$ as a function in four variables, rather than as a series in $z$, and in a note added 10 February, 1812, he gave a remarkably full discussion of the convergence of such series.

The next major advance was made in 1836 by E. E. Kummer (1810–93), who first used the term 'hypergeometric' for series of the type (1.1.1) only. He showed that the differential equation

$$z(1-z)\frac{d^2y}{dz^2} + \{c - (1+a+b)z\}\frac{dy}{dz} - aby = 0,$$

(1.1.6)

is satisfied by the function

$$_2F_1[a, b;\; c;\; z],$$

and has in all twenty-four solutions in terms of similar Gauss functions.‡ In 1857§, G. F. B. Riemann (1826–66) extended this theory by the introduction of his $P$ functions, which in a way, are generalizations of the Gaussian $_2F_1[a, b;\; c;\; z].$

Riemann also discussed the general theory of the transformation of the variable in a differential equation and this theory was applied to Kummer's work by J. Thomae who, in 1879, worked out in detail the relationships between Kummer's twenty-four solutions.‖

---

† Gauss (1812).          ‡ Kummer (1836).
§ Riemann (1857).        ‖ Thomae (1879).

The first integral representation of the Gauss function goes back to Euler[†] who showed that

$$_2F_1[-n, b; c; z] = \frac{n!}{c(c+1)(c+2)...(c+n-1)}$$

$$\times \int_0^1 t^{-n-1}(1-t)^{c+n-1}(1-tz)^{-b}\,dt. \quad (1.1.7)$$

The basic idea of representing a function by a contour integral with gamma functions in the integrand seems to be due to S. Pincherle (1853–1936) who used contours of a type which stems from Riemann's work. This side of the subject was developed extensively by R. Mellin and E. W. Barnes.[‡] In 1907, Barnes published his contour integral representations of Kummer's twenty-four functions, and later, in 1910,[§] he proved the integral analogue of Gauss's theorem

$$\frac{1}{2\pi i}\int_{-\infty}^{\infty}\Gamma(a+s)\,\Gamma(b+s)\,\Gamma(c-s)\,\Gamma(d-s)\,ds$$

$$= \frac{\Gamma(a+c)\,\Gamma(a+d)\,\Gamma(b+c)\,\Gamma(b+d)}{\Gamma(a+b+c+d)}. \quad (1.1.8)$$

**1.1.1 The Gauss series and its convergence.** Let us write

$$(a)_n \equiv a(a+1)(a+2)(a+3)...(a+n-1), \quad (1.1.1.1)$$

and in particular, $(a)_0 \equiv 1$, so that, for example $(3)_5 = 3.4.5.6.7$, $= 2520$, and $(1)_n = n!$. Then

$$(a)_n = \frac{\Gamma(a+n)}{\Gamma(a)} \quad (1.1.1.2)$$

and

$$\lim_{n \to \infty}(a)_n = \frac{1}{\Gamma(a)}. \quad (1.1.1.3)$$

If $a$ is a negative integer $-m$, then

$$(a)_n = (-m)_n \quad \text{if} \quad m \geqslant n,$$

and

$$(a)_n = 0 \quad \text{if} \quad m < n,$$

so that $(-3)_3 = (-3)(-2)(-1) = -6$, but $(-3)_4 = 0$.

In this notation, the Gauss function becomes

$$_2F_1[a, b; c; z] = \sum_{n=1}^{\infty}\frac{(a)_n(b)_n z^n}{(c)_n n!}, \quad (1.1.1.4)$$

where $a$, $b$, $c$ and $z$ may be real or complex. From this, we see that if either of the numbers $a$ or $b$ is zero or a negative integer, the function

† Euler (1748).      ‡ Barnes (1907a).
§ Barnes (1910).

reduces to a polynomial, but if $c$ is zero or a negative integer, the function is not defined, since all but a finite number of the terms of the series become infinite. Also we have immediately

$$\frac{d}{dz}\left({}_2F_1[a,b;\ c;\ z]\right) = \frac{ab}{c}\ {}_2F_1[a+1,b+1;\ c+1;\ z].\quad (1.1.1.5)$$

Some alternative notations for the Gauss function, which are in common use, are:
Appell (1926) and Bailey (1935 a),

$$_2F_1\begin{bmatrix}a,b;\\ c;\end{bmatrix} z \end{bmatrix} = {}_2F_1[a,b;\ c;\ z],\quad (1.1.1.6)$$

$$F(a,b;\ c;\ z) = {}_2F_1[a,b;\ c;\ z],\quad (1.1.1.7)$$

Meijer (1953 c),

$$\Phi[a,b;\ c;\ z] = {}_2F_1[a,b;\ c;\ z]/\Gamma(c),\quad (1.1.1.8)$$

MacRobert (1947), p. 352,

$$E(2;\ a,\ b;\ 1;\ c;\ -1/z) = \frac{\Gamma(a)\,\Gamma(b)}{\Gamma(c)}\ {}_2F_1[a,b;\ c;\ z],\quad (1.1.1.9)$$

Meijer (1941 a),

$$G_{22}^{12}\left(-z\ \middle|\ \begin{matrix}-a,\ -b\\ -1,\ -c\end{matrix}\right) = -\frac{\Gamma(a)\,\Gamma(b)}{\Gamma(c)\,z}\ {}_2F_1[a,b;\ c;\ z],\quad (1.1.1.10)$$

Riemann (1857),

$$P\begin{Bmatrix}0 & \infty & 1 & \\ 0 & a & 0 & z\\ 1-c & b & c-a-b &\end{Bmatrix} = {}_2F_1[a,b;\ c;\ z].\quad (1.1.1.11)$$

Let $u_n = \dfrac{(a)_n\,(b)_n}{(c)_n\,(1)_n}$, then we have

$$(1+n)\,(c+n)\,u_{n+1} = (a+n)\,(b+n)\,u_n.\quad (1.1.1.12)$$

The ratio of the two successive terms $u_n$ and $u_{n+1}$ of the Gaussian series is

$$\frac{(a+n)\,(b+n)}{(c+n)\,(1+n)}\,z = \frac{(1+a/n)\,(1+b/n)}{(1+c/n)\,(1+1/n)}\,z,\quad (1.1.1.13)$$

so that as $n \to \infty$, the ratio

$$|u_{n+1}/u_n| \to |z|.$$

Hence, by D'Alembert's test[†], the series is convergent for all values of $z$, real or complex such that $|z| < 1$, and divergent for all values of $z$ real or complex, such that $|z| > 1$.

---

† Bromwich, *Infinite Series*, (1947), p. 39.

When $|z| = 1$,

$$|u_{n+1}/u_n| = \left| \left\{ 1 + \frac{a+b}{n} + O(1/n^2) \right\} \left\{ 1 - \frac{1+c}{n} + O(1/n^2) \right\} \right|,$$

$$= \left| 1 + \frac{a+b-c-1}{n} + O(1/n^2) \right|,$$

$$\leqslant 1 + \{ \mathrm{Rl}\,(a+b-c-1)/n \} + O(1/n^2). \qquad (1.1.1.14)$$

Thus, when $z = 1$, by Raabe's test†, the series is convergent if $\mathrm{Rl}\,(c-a-b) > 0$, and divergent if $\mathrm{Rl}\,(c-a-b) < 0$.

It is also divergent when $\mathrm{Rl}\,(c-a-b) = 0$, for in this case

$$|u_{n+1}/u_n| > 1 - \frac{1}{n} - \frac{C}{n^2},$$

where $C$ is a constant.

When $|z| = 1$, but $z \neq 1$, the series is absolutely convergent when $\mathrm{Rl}\,(c-a-b) > 0$, convergent but not absolutely so when

$$-1 < \mathrm{Rl}\,(c-a-b) \leqslant 0,$$

and divergent when $\mathrm{Rl}\,(c-a-b) < -1$. If $\mathrm{Rl}\,(c-a-b) = -1$, more delicate tests are needed. In this case, we find that

$$|u_{n+1}/u_n| = 1 - \frac{\mathrm{Rl}\,(a+b-ab+1)}{n^2} + O(1/n^3). \qquad (1.1.1.15)$$

Hence the series is convergent if $\mathrm{Rl}\,(a+b) > \mathrm{Rl}\,ab$, and divergent if $\mathrm{Rl}\,(a+b) \leqslant \mathrm{Rl}\,ab$.

For example, the series

$$1 - \tfrac{2}{3} + \tfrac{3}{4} - \tfrac{4}{5} + \tfrac{5}{6} - \tfrac{6}{7} + \dots = \tfrac{1}{2}\{1 + {}_2F_1[2,2;\ 3;\ -1]\}, \qquad (1.1.1.16)$$

is divergent.

We note also that

$$\frac{(a)_n\,(b)_n}{(c)_n\,n!} \to 0 \text{ as } n \to \infty, \text{ if } 0 < \mathrm{Rl}\,(1+c-a-b) < 1. \quad (1.1.1.17)$$

## 1.2 The Gauss equation

The differential equation

$$z(1-z)\frac{\mathrm{d}^2y}{\mathrm{d}z^2} + \{c - (1+a+b)z\}\frac{\mathrm{d}y}{\mathrm{d}z} - aby = 0, \qquad (1.2.1)$$

is called the Gauss equation or the hypergeometric equation. In the region $|z| < 1$, one solution is

$$y_1 = {}_2F_1[a,b;\ c;\ z]. \qquad (1.2.2)$$

† Bromwich (1947), p. 40.

This can be verified by direct differentiation of the series (1.1.1), and substitution in the above differential equation. But an alternative form of writing this equation is

$$\frac{d}{dz}\left(z\frac{d}{dz}+c-1\right)y = \left(z\frac{d}{dz}+a\right)\left(z\frac{d}{dz}+b\right)y, \qquad (1.2.3)$$

and this leads to an elegant proof, for

$$\left(z\frac{d}{dz}+a\right)y = \sum_{n=0}^{\infty}\frac{(a)_n(b)_n}{(c)_n n!}(n+a)z^n.$$

Hence

$$\left(z\frac{d}{dz}+a\right)\left(z\frac{d}{dz}+b\right)y = \sum_{n=0}^{\infty}\frac{(a)_{n+1}(b)_{n+1}}{(c)_n n!}z^n.$$

Similarly,

$$\left(z\frac{d}{dz}+c-1\right)y = \sum_{n=1}^{\infty}\frac{(a)_n(b)_n}{(c)_{n-1}n!}z^n.$$

Hence

$$\frac{d}{dz}\left(z\frac{d}{dz}+c-1\right)y = \sum_{n=1}^{\infty}\frac{(a)_n(b)_n}{(c)_{n-1}n!}nz^{n-1},$$

$$= \sum_{n=0}^{\infty}\frac{(a)_{n+1}(b)_{n+1}}{(c)_n n!}z^n.$$

The Gauss equation can be rewritten

$$\frac{d^2y}{dz^2}+\left\{\frac{c}{z(1-z)}-\frac{1+a+b}{1-z}\right\}\frac{dy}{dz}-\frac{ab}{z(1-z)}y = 0, \qquad (1.2.4)$$

from which 0 and 1 are seen to be regular singularities. If we write $1/z$ for $z$, we find that infinity is also a regular singularity of the Gauss equation.†

In the notation of operators, where $\Delta \equiv z\frac{d}{dz}$, the Gauss equation can also be written

$$\Delta(\Delta+c-1)y = z(\Delta+a)(\Delta+b)y. \qquad (1.2.5)$$

**1.2.1 The connexion with Riemann's equation.**  We shall now show that any equation of the general form

$$\frac{d^2y}{dz^2}+\left(\sum_{\nu=1}^{3}\frac{A_\nu}{z-z_\nu}\right)\frac{dy}{dz}+\left(\sum_{\nu=1}^{3}\frac{B_\nu}{z-z_\nu}\right)\frac{y}{(z-z_1)(z-z_2)(z-z_3)} = 0, \qquad (1.2.1.1)$$

where $A_\nu$ and $B_\nu$ are constants, can be reduced to a Gauss equation, provided that $A_1+A_2+A_3 = 2$, to ensure that the 'point at infinity'

† Whittaker & Watson (1947), § 10.3.

is an ordinary point of the equation. We shall also exhibit the inter-connexions between several well-known differential equations, as incidental to the proof given here.

First we note that in the equation (1.2.1.1) every point, including infinity, is an ordinary point of the equation, except the points $z = z_1$, $z = z_2$ and $z = z_3$. So let us write $\theta = z_2 - z_3$, $\phi = z_3 - z_1$ and $\psi = z_1 - z_2$, where $\theta + \phi + \psi = 0$. The indicial equation, formed for expansion about $z = z_1$, is

$$\rho(\rho - 1) + A_1\rho + \frac{B_1}{-\phi\psi} = 0,$$

with roots $\alpha$ and $\alpha'$ say. Then we can write

$$A_1 = 1 - \alpha - \alpha' \quad \text{and} \quad B_1 = -\phi\psi\alpha\alpha'.$$

Similarly, by considering the indicial equations formed for expansions about $z = z_2$ and $z = z_3$, respectively, we can write

$$A_2 = 1 - \beta - \beta', \quad B_2 = -\psi\theta\beta\beta',$$

and

$$A_3 = 1 - \gamma - \gamma', \quad B_3 = -\theta\phi\gamma\gamma',$$

where, since $A_1 + A_2 + A_3 = 2$, we must have

$$\alpha + \alpha' + \beta + \beta' + \gamma + \gamma' = 1.$$

The given equation then becomes Riemann's equation

$$\frac{d^2y}{dz^2} + \left(\frac{1-\alpha-\alpha'}{z-z_1} + \frac{1-\beta-\beta'}{z-z_2} + \frac{1-\gamma-\gamma'}{z-z_3}\right)\frac{dy}{dz}$$
$$= \left\{\frac{\alpha\alpha'}{(z-z_1)\,\theta} + \frac{\beta\beta'}{(z-z_2)\,\phi} + \frac{\gamma\gamma'}{(z-z_3)\,\psi}\right\}\frac{\theta\phi\psi y}{(z-z_1)\,(z-z_2)\,(z-z_3)}.$$

$$(1.2.1.2)$$

This equation is also known as Papperitz's equation.† Its solution is usually written in terms of Riemann's $P$ function as

$$u = P\begin{Bmatrix} z_1 & z_2 & z_3 & \\ \alpha & \beta & \gamma & z \\ \alpha' & \beta' & \gamma' & \end{Bmatrix}, \qquad (1.2.1.3)$$

or, in terms of the Gauss function, as

$$u = (z-z_1)^\alpha (z-z_2)^{-\alpha-\gamma} (z-z_3)^\gamma$$
$$\times {}_2F_1\left[\alpha+\beta+\gamma, \alpha+\beta'+\gamma; 1+\alpha-\alpha'; \frac{(z-z_1)\,(z_3-z_2)}{(z-z_2)\,(z_3-z_1)}\right].$$

$$(1.2.1.4)$$

† Papperitz (1885).

Twenty-four solutions of Riemann's equation can be written down immediately, simply by interchanging the triads $(z_1, \alpha, \alpha')$, $(z_2, \beta, \beta')$ and $(z_3, \gamma, \gamma')$, in a cyclic order.

If, in (1.2.1.2) we write $t = \{(z - z_1)\,\theta\}/\{(z - z_2)\,(-\phi)\}$, and divide by $(t\phi + \theta)^4/(\theta^2\phi^2\psi^2)$, the equation (1.2.1.2) becomes

$$\frac{d^2 y}{dt^2} + \left(\frac{A_1}{t} + \frac{A_3}{t-1}\right)\frac{dy}{dt} - \left(\frac{\alpha\alpha'}{t} - \beta\beta' - \frac{\gamma\gamma'}{t-1}\right)\frac{y}{t(t-1)} = 0.$$

(1.2.1.5)

If further we write $y = t^\alpha(t-1)^\gamma\, Y$, (1.2.1.5) reduces to

$$t(t-1)\frac{d^2 Y}{dt^2} + \{t(2 + \alpha - \alpha' + \gamma - \gamma') - (1 + \alpha - \alpha')\}\frac{dY}{dt}$$

$$+ \{(\alpha + \gamma)(1 - \alpha' - \gamma') + \beta\beta'\}\, Y = 0. \quad (1.2.1.6)$$

Finally, if we write $a + b$ for $1 + \alpha - \alpha' + \gamma - \gamma'$, $ab$ for

$$(\alpha + \gamma)(1 - \alpha' - \gamma') + \beta\beta'$$

and $c$ for $1 + \alpha - \alpha'$, (1.2.1.6) reduces to the ordinary Gauss equation (1.2.1). Thus we see that, in general, for any equation with three ordinary singularities at $z_1$, $z_2$ and $z_3$, these singularities can be transformed into the three singularities of the Gauss equation simply by writing $z$ for $\{(z - z_1)(z_3 - z_2)\}/\{(z - z_2)(z_3 - z_1)\}$.

## 1.3 Kummer's twenty-four solutions

Let us assume that $$y = z^g \sum_{n=0}^{\infty} u_n z^n,$$ (1.3.1)

(where $u_0 \neq 0$) is any solution of the Gauss equation (1.2.1). Then, by direct differentiation of this series, we find that

$$u_0 g(g + c - 1) z^{g-1} + \sum_{n=0}^{\infty} \{u_{n+1}(g + n + 1)$$

$$\times (g + n + c) - u_n(g + n + a)(g + n + b)\} z^{g+n} = 0. \quad (1.3.2)$$

Hence we must have as the indicial equation

$$g(g + c - 1) = 0,$$ (1.3.3)

and in general

$$(g + n + c)(g + n + 1)\, u_{n+1} = (g + n + a)(g + n + b)\, u_n.$$ (1.3.4)

The root $g = 0$ of the indicial equation (1.3.3) leads to the solution

$$y_1 = {}_2F_1[a, b;\, c;\, z],$$

provided that $c$ is not zero nor a negative integer, and the root $g = 1 - c$, gives a second solution in which

$$(1+n)(2-c+n)u_{n+1} = (a+1-c+n)(b+1-c+n)u_n. \quad (1.3.5)$$

This solution is

$$y_2 = z^{1-c}\,{}_2F_1[1+a-c, 1+b-c; 2-c; z], \quad (1.3.6)$$

provided that $c$ is not a positive integer $\geqslant 2$. Hence one complete solution of the Gauss equation (1.2.1) is

$$y = A\,{}_2F_1[a, b; c; z] + Bz^{1-c}\,{}_2F_1[1+a-c, 1+b-c; 2-c; z], \quad (1.3.7)$$

for $|z| < 1$, and for $c$ not an integer, where $A$ and $B$ are constants.

When $c = 1$, the two solutions are equivalent, and we have to follow the usual Frobenius process† in order to find that a second solution is now

$$y_2 = {}_2F_1[a, b; 1; z]\log z + \sum_{n=1}^{\infty}\left[\frac{\partial}{\partial g}\left\{\frac{(a+g)_n(b+g)_n}{(1+g)_n(1+g)_n}\right\}\right]_{g=0}z^n. \quad (1.3.8)$$

When $c = 0$, or a negative integer, the second solution (1.3.6) is still valid but the first solution has to be replaced by

$$y_1 = z^{1-c}\,{}_2F_1[1+a-c, 1+b-c; 2-c; z]\log z$$

$$+ \sum_{n=0}^{\infty}\left[\frac{\partial}{\partial g}\left\{\frac{(a+g)_n(b+g)_n}{(c+g)_n(1+g)_n}\right\}\right]_{g=1-c}z^n. \quad (1.3.9)$$

When $c$ is a positive integer $\geqslant 2$, the first solution is still valid but the second solution has to be replaced by

$$y_2 = {}_2F_1[a, b; c; z]\log z + \sum_{n=0}^{\infty}\left[\frac{\partial}{\partial g}\left\{\frac{(a+g)_n(b+g)_n}{(c+g)_n(1+g)_n}\right\}\right]_{g=0}z^n. \quad (1.3.10)$$

If $a$ or $b$ is a negative integer, as we have already seen, our first solution reduces to a polynomial in $z$, and if $1+a-c$ or $1+b-c$ is a negative integer, the second solution reduces to a polynomial in $z$.

When we are dealing with solutions of this type, it is useful to remember that if

$$u_n(g) = \frac{(a+g)_n(b+g)_n}{(c+g)_n(1+g)_n},$$

† Whittaker & Watson (1947), § 10.3.

then

$$\frac{1}{u_n(g)}\frac{\partial}{\partial g}\{u_n(g)\}+\frac{1}{g+n}+\frac{1}{g+c+n-1}-\frac{1}{g+a+n-1}-\frac{1}{g+b+n-1}$$

$$=\frac{1}{u_{n-1}(g)}\frac{\partial}{\partial g}\{u_{n-1}(g)\}, \quad (1.3.11)$$

(see Copson, *Functions of a complex variable* (1950), p. 248).

Next let us substitute $(1-z)^k w$ for $y$ in the Gauss equation. It becomes

$$z(1-z)\frac{d^2w}{dz^2}+\{c-(a+b+1+2k)z\}\frac{dw}{dz}$$

$$+\left[\frac{k(k-1)z-k\{c-(a+b+1)z\}}{1-z}-ab\right]w=0. \quad (1.3.12)$$

This equation is also of hypergeometric type if $1-z$ divides exactly into $k(k-1)z-k\{c-(a+b+1)z\}$, that is, if either $k=0$, or $k=c-a-b$. When $k=0$, the two solutions (1.2.2) and (1.3.6) are given, but when $k=c-a-b$, then two new solutions are given, valid in the region $|z|<1$. These are

$$y_3=(1-z)^{c-a-b}\,{}_2F_1[c-a,c-b;\,c;\,z] \quad (1.3.13)$$

and $\qquad y_4=z^{1-c}(1-z)^{c-a-b}\,{}_2F_1[1-a,1-b;\,2-c;\,z]. \quad (1.3.14)$

Since the Gauss equation is of order two, it can have only two linearly independent solutions. Hence there must exist constants $A$ and $B$ such that

$$(1-z)^{c-a-b}\,{}_2F_1[c-a,c-b;\,c;\,z]$$

$$=A\,{}_2F_1[a,b;\,c;\,z]+Bz^{1-c}\,{}_2F_1[1+a-c,1+b-c;\,2-c;\,z].$$

Now the left-hand side of this equation can be expanded in integral powers of $z$, but $z^{1-c}$ cannot, since $c$ is not an integer, by hypothesis. Hence $B=0$. If however we put $z=0$, we find that we must have $A=1$. Hence $y_1=y_3$, that is

$$_2F_1[a,b;\,c;\,z]=(1-z)^{c-a-b}\,{}_2F_1[c-a,c-b;\,c;\,z]. \quad (1.3.15)$$

This is the result usually known as Euler's identity. In a similar way we can show that $y_2=y_4$, that is

$$_2F_1[1+a-c,1+b-c;\,2-c;\,z]=(1-z)^{c-a-b}\,{}_2F_1[1-a,1-b;\,2-c;\,z].$$

$$(1.3.16)$$

**1.3.1 The region $|1-z| < 1$.** Next let us substitute $1-Z$ for $z$ in the Gauss equation (1.2.1). It becomes

$$Z(1-Z)\frac{\mathrm{d}^2y}{\mathrm{d}Z^2} - \{c-(a+b+1)(1-Z)\}\frac{\mathrm{d}y}{\mathrm{d}Z} - ab\,y = 0. \quad (1.3.1.1)$$

This is also a Gauss type equation satisfied by

$$y_5 = {}_2F_1[a,b;\; 1+a+b-c; Z]. \quad (1.3.1.2)$$

Hence four further solutions of the original Gauss equation (1.2.1) are

$$y_5 = {}_2F_1[a,b;\; 1+a+b-c;\; 1-z], \quad (1.3.1.3)$$

$$y_6 = (1-z)^{c-a-b}\,{}_2F_1[c-a,c-b;\; 1-a-b+c;\; 1-z], \quad (1.3.1.4)$$

$$y_7 = z^{1-c}\,{}_2F_1[1+a-c,1+b-c;\; 1+a+b-c;\; 1-z] \quad (1.3.1.5)$$

and

$$y_8 = (1-z)^{c-a-b}z^{1-c}\,{}_2F_1[1-a,1-b;\; 1-a-b+c;\; 1-z], \quad (1.3.1.6)$$

all valid in the region $|1-z| < 1$. These new solutions hold provided that $a$, $b$ and $c$ are not negative integers nor zero. If any of the coefficients are integers or zero special solutions can be found corresponding to those of the preceding section (1.3.8–1.3.10).

**1.3.2 The regions $|z| > 1$ and $|1-z| > 1$.** In the Gauss equation (1.2.1), next let us substitute $1/z$ for $z$, and $z^a Y$ for $y$. The equation becomes then

$$z(1-z)\frac{\mathrm{d}^2Y}{\mathrm{d}z^2} + \{1+a-b-(2a-c)z\}\frac{\mathrm{d}Y}{\mathrm{d}z} - a(1+a-c)\,Y = 0,$$

$$(1.3.2.1)$$

which is a Gauss type equation also, with the two solutions

$$Y = {}_2F_1[a,1+a-c;\; 1+a-b;\; z]$$

and

$$Y = z^{b-a}\,{}_2F_1[b,1+b-c;\; 1+b-a;\; z].$$

Hence by substitution, we can find four further solutions of our original equation,

$$y_9 = (-z)^{-a}\,{}_2F_1[a,1+a-c;\; 1+a-b;\; 1/z], \quad (1.3.2.2)$$

$$y_{10} = (-z)^{-b}\,{}_2F_1[b,1+b-c;\; 1+b-a;\; 1/z], \quad (1.3.2.3)$$

$$y_{11} = (-z)^{b-c}(1-z)^{c-a-b}\,{}_2F_1[1-b,c-b;\; 1+a-b;\; 1/z] \quad (1.3.2.4)$$

and

$$y_{12} = (-z)^{a-c}(1-z)^{c-a-b}\,{}_2F_1[1-a,c-a;\; 1+b-a;\; 1/z]. \quad (1.3.2.5)$$

All these solutions are valid in the region $|z| > 1$.

If we combine the above results in $1-z$ and $1/z$, we obtain a further four solutions again,

$$y_{13} = (1-z)^{-a}\,_2F_1[a, c-b;\ 1+a-b;\ 1/(1-z)], \tag{1.3.2.6}$$

$$y_{14} = (1-z)^{-b}\,_2F_1[b, c-a;\ 1+b-a;\ 1/(1-z)], \tag{1.3.2.7}$$

$$y_{15} = (-z)^{1-c}(1-z)^{c-a-1}\,_2F_1[1+a-c, 1-b;\ 1+a-b;\ 1/(1-z)] \tag{1.3.2.8}$$

and

$$y_{16} = (-z)^{1-c}(1-z)^{c-b-1}\,_2F_1[1+b-c, 1-a;\ 1+b-a;\ 1/(1-z)]. \tag{1.3.2.9}$$

These four solutions are valid in the region $|1-z| > 1$.

### 1.3.3 The regions $\mathrm{Rl}(z) > \tfrac{1}{2}$ and $\mathrm{Rl}(z) < \tfrac{1}{2}$. In the region

$$|(z-1)/z| < 1,$$

that is in the half plane $\mathrm{Rl}(z) > \tfrac{1}{2}$, we can now deduce the four solutions

$$y_{17} = z^{-a}\,_2F_1[a, 1+a-c;\ 1+a+b-c;\ 1-1/z], \tag{1.3.3.1}$$

$$y_{18} = z^{-b}\,_2F_1[b, 1+b-c;\ 1+a+b-c;\ 1-1/z], \tag{1.3.3.2}$$

$$y_{19} = (1-z)^{c-a-b}z^{b-c}\,_2F_1[1-b, c-b;\ 1-a-b+c;\ 1-1/z] \tag{1.3.3.3}$$

and

$$y_{20} = (1-z)^{c-a-b}z^{a-c}\,_2F_1[1-a, c-a;\ 1-a-b+c;\ 1-1/z], \tag{1.3.3.4}$$

by combining our results for $1-z$ and $1/z$. Finally from a further combination of the results in $1-z$ and $1/z$, we can deduce the last group of four solutions to the Gauss equation (1.2.1). These are

$$y_{21} = (1-z)^{-a}\,_2F_1[a, c-b;\ c;\ z/(z-1)], \tag{1.3.3.5}$$

$$y_{22} = (1-z)^{-b}\,_2F_1[b, c-a;\ c;\ z/(z-1)], \tag{1.3.3.6}$$

$$y_{23} = z^{1-c}(1-z)^{c-a-1}\,_2F_1[a-c+1, 1-b;\ 2-c;\ z/(z-1)] \tag{1.3.3.7}$$

and

$$y_{24} = z^{1-c}(1-z)^{c-b-1}\,_2F_1[b-c+1, 1-a;\ 2-c;\ z/(z-1)]. \tag{1.3.3.8}$$

These results are valid in the region $|z/(z-1)| < 1$, that is in the half plane $\mathrm{Rl}(z) < \tfrac{1}{2}$.

Thus, provided that $a$, $b$ and $c$ are not integers nor zero, we now have twenty-four solutions of the Gauss equation (1.2.1), which among them, provide complete solutions valid in any region of the complex $z$-plane. The existence of these twenty-four solutions was first discussed by Kummer (1836).

**1.3.4 Products of Gauss functions.** Later on, we shall find general theorems concerning products of generalized hypergeometric functions (see §2.6). For the moment we note that, from (1.3.15), we can deduce

$$_2F_1[x, y;\ u;\ \zeta]\, _2F_1[1 - n - v, 1 - n - w;\ 1 - n - z;\ \zeta]$$

$$= {}_2F_1[u - x, u - y;\ u;\ \zeta]\, _2F_1[v - z, w - z;\ 1 - n - z;\ \zeta], \quad (1.3.4.1)$$

provided that $\qquad (1 - \zeta)^{u-x-y} = (1 - \zeta)^{1+n+v+w-z}.$ $\qquad\qquad$ (1.3.4.2)

There are a number of similar results which can be deduced from relations like (1.3.15) under conditions similar to (1.3.4.2) above. This condition (1.3.4.2) is equivalent to

$$u - x - y = 1 + n + v + w - z \qquad\qquad (1.3.4.3)$$

provided that $\zeta \neq 1$.

## 1.4 Contiguous functions and recurrence relations

Two Gauss functions are said to be contiguous if they are alike except for one pair of parameters, and these differ by unity. Thus $_2F_1[a, b;\ c;\ z]$ is contiguous to the six functions

$$_2F_1[a \pm 1, b;\ c;\ z], \quad _2F_1[a, b \pm 1;\ c;\ z] \quad \text{and} \quad _2F_1[a, b;\ c \pm 1;\ z].$$

Any three of these functions can be connected by a linear relation in $z$. Such a relationship is called a recurrence relation. These relationships are of great use in extending numerical tables of the function, since for one fixed value of $z$, it is necessary only to calculate the values of the function over two units in $a$, $b$ and $c$, and apply some recurrence relations in order to find the function values over a large range of values of $a$, $b$ and $c$, in this particular $z$ plane. There are fifteen recurrence relations given by Gauss. If we write $F$ for

$$_2F_1[a, b;\ c;\ z], \quad \text{and} \quad F[a \pm 1], \quad F[b \pm 1] \quad \text{and} \quad F[c \pm 1]$$

for the six contiguous functions, these equations are

$$\{c - 2a + (a - b)z\}F + a(1 - z)F[a + 1] = (c - a)F[a - 1], \qquad (1.4.1)$$

$$(b - a)F + aF[a + 1] = bF[b + 1], \qquad\qquad (1.4.2)$$

$$(c - a - b)F + a(1 - z)F[a + 1] = (c - b)F[b - 1], \qquad\qquad (1.4.3)$$

$$c\{a + (b - c)z\}F + (c - a)(c - b)zF[c + 1] = ac(1 - z)F[a + 1], \qquad (1.4.4)$$

$$(c-a-1)F+aF[a+1] = (c-1)F[c-1], \tag{1.4.5}$$

$$(c-a-b)F+b(1-z)F[b+1] = (c-a)F[a-1], \tag{1.4.6}$$

$$(b-a)(1-z)F+(c-b)F[b-1] = (c-a)F[a-1], \tag{1.4.7}$$

$$c(1-z)F+(c-b)zF[c+1] = cF[a-1], \tag{1.4.8}$$

$$\{a-1+(1+b-c)z\}F+(c-a)F[a-1] = (c-1)(1-z)F[c-1], \tag{1.4.9}$$

$$\{c-2b+(b-a)z\}F+b(1-z)F[b+1] = (c-b)F[b-1], \tag{1.4.10}$$

$$c\{b+(a-c)z\}F+(c-a)(c-b)zF[c+1] = bc(1-z)F[b+1], \tag{1.4.11}$$

$$(c-b-1)F+bF[b+1] = (c-1)F[c-1], \tag{1.4.12}$$

$$c(1-z)F+(c-a)zF[c+1] = cF[b-1], \tag{1.4.13}$$

$$\{b-1+(1+a-c)z\}F+(c-b)F[b-1] = (c-1)(1-z)F[c-1], \tag{1.4.14}$$

$$c\{c-1+(1+a+b-2c)z\}F+(c-a)(c-b)zF[c+1]$$
$$= c(c-1)(1-z)F[c-1]. \tag{1.4.15}$$

In every case we assume that $a$, $b$ and $c$ are not zero nor integers such that the equations above would cease to have any real meaning. All these relations can be proved by the expansion of the various power series in $z$, so that we can equate the coefficients of $z^n$ throughout.

The series
$$_2F_1[a+l, b+m; c+n; z]$$

for $l, m, n$ integers, are called the associated series. These series can always be expressed in terms of a linear relation between $F$ and one of its contiguous functions. For example, we can show that

$$(c-a)(c-b)\,_2F_1[a,b; c+1; z]$$

$$= c(c-a-b)\,_2F_1[a,b;c;z]+ab(1-z)\,_2F_1[a+1,b+1; c+1; z], \tag{1.4.16}$$

for the coefficient of $z^n$ on the right is

$$c(c-a-b)\frac{(a)_n(b)_n}{(1)_n(c)_n}+ab\frac{(a+1)_n(b+1)_n}{(1)_n(c+1)_n}-ab\frac{(a+1)_{n-1}(b+1)_{n-1}}{(1)_{n-1}(c+1)_{n-1}}$$

$$= \frac{(a)_n(b)_n}{(1)_n(c+1)_n}\{(c-a-b)(c+n)+(a+n)(b+n)-n(c+n)\}$$

$$= \frac{(a)_n(b)_n}{(1)_n(c+1)_n}(c-a)(c-b)$$

which is the coefficient of $z^n$ on the left of (1.4.16).

The Gauss function can be represented as a continued fraction in some special cases. Thus, since

$$_2F_1[a, b+1; c+1; z] - {}_2F_1[a, b; c; z]$$

$$= \frac{a(c-b)}{c(c+1)} z {}_2F_1[a+1, b+1; c+2; z], \quad (1.4.17)$$

we can deduce that

$$\frac{_2F_1[a, b+1; c+1; z]}{_2F_1[a, b; c; z]} = 1 \bigg/ \left(1 - \frac{a(c-b)}{c(c+1)} z \bigg/ \left(1 - \frac{(b+1)(c-a+1)}{(c+1)(c+2)} z \bigg/ \left(1 \right.\right.\right.$$

$$\left. - \frac{(a+1)(c-b+1)}{(c+2)(c+3)} z \bigg/ \left(1 - \frac{(b+2)(c-a+2)}{(c+3)(c+4)} z \bigg/ \left(1 - \right.\right.\right.$$

$$\cdots \left. - \frac{(a+n-1)(c-b+n-1)}{(c+2n-2)(c+2n-1)} z \bigg/ \left(1 - \frac{(b+n)(c-a+n)z}{(c+2n-1)(c+2n)} \right.\right.$$

$$\left.\left.\left. \times \frac{_2F_1[a+n, b+n+1; c+2n+1; z]}{_2F_1[a+n, b+n; c+2n; z]} \right) \cdots \right) \right). \quad (1.4.18)$$

In particular, if $b = 0$, $_2F_1[a, b; c; z] = 1$, and we find† that

$$_2F_1[a, 1; c+1; z] = 1 \bigg/ \left(1 - \frac{ac}{c(c+1)} z \bigg/ \left(1 - \frac{1 \cdot (c-a+1)}{(c+1)(c+2)} z \bigg/ \left(1 - \right.\right.\right.$$

$$\cdots \left. - \frac{(a+n-1)(c+n-1)z}{(c+2n-2)(c+2n-1)} \bigg/ \left(1 - \right.\right.$$

$$\left.\left. \frac{n(c-a+n)z}{(c+2n-1)(c+2n)} \frac{_2F_1[a+n, n+1; c+2n+1; z]}{_2F_1[a+n, n; c+2n; z]} \right) \cdots \right) \right).$$

$$(1.4.19)$$

**1.4.1 Differential properties.**    If the Gauss series is differentiated term by term in $z$, it gives

$$\frac{\mathrm{d}}{\mathrm{d}z} \{{}_2F_1[a, b; c; z]\} = \frac{ab}{c} {}_2F_1[a+1, b+1; c+1; z]. \quad (1.4.1.1)$$

When this process is repeated, we find that in general

$$\frac{\mathrm{d}^n}{\mathrm{d}z^n} \{{}_2F_1[a, b; c; z]\} = \frac{(a)_n (b)_n}{(c)_n} {}_2F_1[a+n, b+n; c+n; z].$$

$$(1.4.1.2)$$

† Gauss (1866) p. 135.

There are seven other differential relations which are proved in a similar way. These are

$$\frac{d^n}{dz^n}\{z^{a+n-1}{}_2F_1[a,b;\,c;\,z]\}$$
$$= (a)_n z^{a-1}{}_2F_1[a+n,b;\,c;\,z], \tag{1.4.1.3}$$

$$\frac{d^n}{dz^n}\{z^{c-a+n-1}(1-z)^{a+b-c}{}_2F_1[a,b;\,c;\,z]\}$$
$$= (c-a)_n z^{c-a-1}(1-z)^{a+b-c-n}{}_2F_1[a-n,b;\,c;\,z], \tag{1.4.1.4}$$

$$\frac{d^n}{dz^n}\{(1-z)^{a+b-c}{}_2F_1[a,b;\,c;\,z]\}$$
$$= \frac{(c-a)_n(c-b)_n}{(c)_n}(1-z)^{a+b-c-n}{}_2F_1[a,b;\,c+n;\,z], \tag{1.4.1.5}$$

$$\frac{d^n}{dz^n}\{z^{c-1}{}_2F_1[a,b;\,c;\,z]\}$$
$$= (c-n)_n z^{c-n-1}{}_2F_1[a,b;\,c-n;\,z], \tag{1.4.1.6}$$

$$\frac{d^n}{dz^n}\{(1-z)^{a+n-1}{}_2F_1[a,b;\,c;\,z]\}$$
$$= \frac{(-1)^n(a)_n(c-b)_n(1-z)^{a-1}}{(c)_n}{}_2F_1[a+n,b;\,c+n;\,z], \tag{1.4.1.7}$$

$$\frac{d^n}{dz^n}\{z^{c-1}(1-z)^{b-c+n}{}_2F_1[a,b;\,c;\,z]\}$$
$$= (c-n)_n z^{c-1-n}(1-z)^{b-c}{}_2F_1[a-n,b;\,c-n;\,z], \tag{1.4.1.8}$$

$$\frac{d^n}{dz^n}\{z^{c-1}(1-z)^{a+b-c}{}_2F_1[a,b;\,c;\,z]\}$$
$$= (c-n)_n z^{c-1-n}(1-z)^{a+b-c-n}{}_2F_1[a-n,b-n;\,c-n;\,z]. \tag{1.4.1.9}$$

If we put $a+n$, $b+n$, $c+n$ for $a$, $b$ and $c$ in (1.4.1.3) and apply Euler's identity (1.3.15), we can deduce†

$$\frac{d^n}{dz^n}\{z^{n+c-1}(1-z)^{n+a+b-c}{}_2F_1[a+n,b+n;\,c+n;\,z]\}$$
$$= (c)_n z^{c-1}(1-z)^{a+b-c}{}_2F_1[a,b;\,c;\,z]. \tag{1.4.1.10}$$

In particular, if $a = -n$, we have

$$\frac{d^n}{dz^n}\{z^{n+c-1}(1-z)^{b-c}\}$$
$$= (c)_n z^{c-1}(1-z)^{b-c-n}{}_2F_1[-n,b;\,c;\,z]. \tag{1.4.1.11}$$

† Jacobi (1829).

## 1.5 Special cases of the Gauss function

Most of the more elementary functions which occur in Mathematical Physics, can be expressed in terms of the Gauss function. For example

$$_2F_1[a,b;\,b;\,z] = \sum_{n=0}^{\infty} \frac{(a)_n}{n!} z^n = \sum_{n=0}^{\infty} (-a)(-a-1)\ldots(-a-n+1)\frac{(-z)^n}{n!},$$

that is
$$_2F_1[a,b;\,b;\,z] = (1-z)^{-a}. \tag{1.5.1}$$

This is simply a statement of the Binomial theorem for $|z| < 1$. Similarly,

$$_2F_1[1,1;\,2;\,-z] = \sum_{n=0}^{\infty} \frac{(-z)^n}{1+n},$$

$$= \log(1+z), \tag{1.5.2}$$

for $|z| < 1$, and 
$$_2F_1[\tfrac{1}{2},1;\,\tfrac{3}{2};\,z^2] = \frac{1}{2z}\log\frac{(1+z)}{(1-z)}. \tag{1.5.3}$$

Also

$$_2F_1[1,b;\,1;\,z/b] = 1 + \sum_{n=1}^{\infty} 1(1+1/b)(1+2/b)\ldots(1+(n-1)/b)\frac{z^n}{n!}. \tag{1.5.4}$$

Hence
$$\lim_{b\to 0}\{_2F_1[1,b;\,1;\,z/b]\} = \sum_{n=0}^{\infty}\frac{z^n}{n!} = e^z. \tag{1.5.5}$$

From this result, expressions for the various trigonometric and hyperbolic functions can be deduced. Thus

$$_2F_1[\tfrac{1}{2}+\tfrac{1}{2}a, \tfrac{1}{2}-\tfrac{1}{2}a;\, \tfrac{3}{2};\, (\sin z)^2] = \sin az, \tag{1.5.6}$$

$$_2F_1[\tfrac{1}{2}a, -\tfrac{1}{2}a;\, \tfrac{1}{2};\, (\sin z)^2] = \cos az, \tag{1.5.7}$$

$$_2F_1[\tfrac{1}{2}+\tfrac{1}{2}a, \tfrac{1}{2}-\tfrac{1}{2}a;\, \tfrac{3}{2};\, (\sinh z)^2] = \sinh az, \tag{1.5.8}$$

$$_2F_1[\tfrac{1}{2}a, -\tfrac{1}{2}a;\, \tfrac{1}{2};\, (\sinh z)^2] = \cosh az, \tag{1.5.9}$$

$$_2F_1[\tfrac{1}{2}, \tfrac{1}{2};\, \tfrac{3}{2};\, z^2] = \frac{1}{z}\sin^{-1}z, \tag{1.5.10}$$

$$_2F_1[\tfrac{1}{2}, 1;\, \tfrac{3}{2};\, -z^2] = \frac{1}{z}\tan^{-1}z. \tag{1.5.11}$$

The Legendre polynomial $P_n(x)$ is defined as the coefficient of $z^n$ in the expansion, in ascending powers of $z$, of $(1-2xz+z^2)^{-\frac{1}{2}}$.

By direct expansion, we can prove that the coefficient is in fact

$$_2F_1[-n, 1+n;\, 1;\, \tfrac{1}{2}-\tfrac{1}{2}x] = P_n(x). \tag{1.5.12}$$

This result is known as Murphy's formula.† An alternative way of deducing this result is to start from Rodriguez's formula which states that

$$P_n(x) = \frac{1}{2^n n!} \frac{d^n}{dx^n} \{(x^2 - 1)^n\}. \tag{1.5.13}$$

Then we have

$$P_n(x) = \frac{1}{2^n n!} \frac{d^n}{dx^n} [(1-x)^n \{(1-x) - 2\}^n],$$

$$= \frac{(-1)^n}{2^n n!} \frac{d^n}{dx^n} \left\{ \sum_{m=0}^{n} 2^{n-m} (1-x)^{n+m} \frac{(-n)_m}{m!} (-1)^m \right\},$$

$$= \frac{(-1)^n}{2^n n!} \sum_{m=0}^{n} \frac{(-n)_m}{m!} (-1)^m 2^{n-m} (n+m-1)_m (1-x)^m (-1)^n,$$

$$= \sum_{m=0}^{n} \frac{(-n)_m (-1)^m}{m! \, n! \, 2^m} (n+m-1)_m (1-x)^m,$$

$$= {}_2F_1[-n, 1+n; 1; \tfrac{1}{2} - \tfrac{1}{2}x]. \tag{1.5.14}$$

The general Legendre function of the first kind is

$$P_n^m(z) = \frac{(z+1)^{\frac{1}{2}m} (z-1)^{-\frac{1}{2}m}}{\Gamma(1-m)} \, {}_2F_1[-n, 1+n; 1-m; \tfrac{1}{2} - \tfrac{1}{2}z], \tag{1.5.15}$$

$$= \frac{2^m}{\Gamma(1-m)} (z^2 - 1)^{-\frac{1}{2}m} \, {}_2F_1[-m-n, 1-m+n; 1-m; \tfrac{1}{2} - \tfrac{1}{2}z], \tag{1.5.16}$$

(by 1.3.15), for $|\arg(z \pm 1)| < \pi$, that is, for $z$ not on the real axis from $+1$ to $-\infty$.

The general Legendre function of the second kind is

$$Q_n^m(z) = e^{i\pi m} \sqrt{\pi} \, \frac{\Gamma(1+m+n)}{\Gamma(\frac{3}{2}+n)} \, 2^{-1-n} (z+1)^{-\frac{1}{2}m} (z-1)^{\frac{1}{2}m-n-1}$$

$$\times {}_2F_1[1+n, 1-m+n; 2+2n; 2/(1-z)]. \tag{1.5.17}$$

The Gegenbauer function is

$$C_n^\nu(t) = (t+1)^{\frac{1}{2}-\nu} \frac{(2\nu)_n}{n!} 2^{\nu-\frac{1}{2}} \, {}_2F_1[\tfrac{1}{2} - \nu - n, \tfrac{1}{2} + \nu + n; \tfrac{1}{2} + \nu; \tfrac{1}{2}(1-t)], \tag{1.5.18}$$

for $n$ an integer $\geqslant 0$.

† Whittaker & Watson (1947), § 15.22.

Other elementary forms are

$$_2F_1[a, a+\tfrac{1}{2};\ \tfrac{1}{2};\ z] = \tfrac{1}{2}(1+z^{\frac{1}{2}})^{-2a} + \tfrac{1}{2}(1-z^{\frac{1}{2}})^{-2a}, \qquad (1.5.19)$$

$$_2F_1[a-\tfrac{1}{2}, a;\ 2a;\ z] = \{\tfrac{1}{2} + \tfrac{1}{2}(1-z)^{\frac{1}{2}}\}^{1-2a} \qquad (1.5.20)$$

and $$\qquad _2F_1[2a, a+1;\ a;\ z] = (1+z)/(1-z)^{2a+1}. \qquad (1.5.21)$$

The Confluent Hypergeometric function or Kummer's function is

$$M(a, c, z) \equiv \sum_{n=0}^{\infty} \frac{(a)_n}{(c)_n}\frac{z^n}{n!}. \qquad (1.5.22)$$

This can also be deduced as a special case of the Gauss function. In fact

$$\lim_{b \to \infty} \{_2F_1[a, b;\ c;\ z/b]\} = M(a, c, z). \qquad (1.5.23)$$

Thus all those functions which have one variable and one or two parameters are special cases of Kummer's function and are also special cases of the more general Gauss function with its three parameters. These functions include all the Bessel functions, the Error functions, the Hermite polynomials, the Airy functions, the Coulomb Wave functions, the Weber functions, the Whittaker functions and in fact all the commonly used functions of Mathematical Physics.† Thus it should always be remembered that every theorem about general hypergeometric functions will have interesting special cases, giving theorems about these simpler functions.

## 1.6 Some integral representations

When $a = 0$ but $b \neq 0$, the hypergeometric equation reduces to

$$z(1-z)\frac{d^2y}{dz^2} + \{c - (b+1)z\}\frac{dz}{dy} = 0. \qquad (1.6.1)$$

This equation can be integrated directly to give

$$\frac{dy}{dz} = Az^c(1-z)^{c-b-1}, \qquad (1.6.2)$$

and so $$\qquad y = A\int_{\zeta_0}^{\zeta} z^c(1-z)^{c-b-1}\,dz. \qquad (1.6.3)$$

Similarly if $a \neq 0$ but $b = 0$, by the symmetry in $a$ and $b$ we have

$$y = A\int_{\zeta_0}^{\zeta} z^c(1-z)^{c-a-1}\,dz. \qquad (1.6.4)$$

---

† See Erdélyi (1953) or Slater (1960) for complete lists of these functions, the relationships between them and their interconnexions.

We shall now seek a similar integral representation for the general Gauss function. Let

$$I = \int_0^1 t^{b-1}(1-t)^{c-b-1}(1-zt)^{-a}\,dt, \qquad (1.6.5)$$

where $|z| < 1$. This integral exists and is convergent if $\mathrm{Rl}\,(b) > 0$, and $\mathrm{Rl}\,(c-b) > 0$. Now

$$(1-zt)^{-a} = \sum_{n=0}^{\infty} \frac{(a)_n t^n z^n}{n!}.$$

Hence
$$I = \int_0^1 \sum_{n=0}^{\infty} \frac{(a)_n z^n}{n!} t^{b+n-1}(1-t)^{c-b-1}\,dt$$

$$= \sum_{n=0}^{\infty} \frac{(a)_n}{n!} z^n \int_0^1 t^{b+n-1}(1-t)^{c-b-1}\,dt$$

$$= \sum_{n=0}^{\infty} \frac{(a)_n}{n!} z^n \frac{\Gamma(b+n)\,\Gamma(c-b)}{\Gamma(c+n)}$$

$$= \frac{\Gamma(c-b)\,\Gamma(b)}{\Gamma(c)}\, {}_2F_1[a,b;\,c;\,z].$$

Hence we can say that

$${}_2F_1[a,b;\,c;\,z] = \frac{\Gamma(c)}{\Gamma(b)\,\Gamma(c-b)} \int_0^1 t^{b-1}(1-t)^{c-b-1}(1-zt)^{-a}\,dt,$$
$$(1.6.6)$$

provided that $|z| < 1$, $\mathrm{Rl}\,(c-b) > 0$, and $\mathrm{Rl}\,(b) > 0$. This integral is known as Pochammer's integral. It can be transformed in many ways, to give other elementary integrals for the Gauss function. For example, let us put $\qquad -s = t/(t-1)$
for $t$, then we find that

$${}_2F_1[a,b;\,c;\,1-z] = \frac{\Gamma(c)}{\Gamma(b)\,\Gamma(c-b)} \int_0^{\infty} s^{b-1}(1+s)^{a-c}(1+sz)^{-a}\,ds,$$
$$(1.6.7)$$

for $\mathrm{Rl}\,(c) > \mathrm{Rl}\,(b) > 0$, $|\arg z| < \pi$.

Similarly, if we write $1/s$ for $t$, we find that

$${}_2F_1[a,b;\,c;\,1/z] = \frac{\Gamma(c)}{\Gamma(b)\,\Gamma(c-b)} \int_1^{\infty} (s-1)^{c-b-1} s^{a-c}(s-1/z)^{-a}\,ds,$$
$$(1.6.8)$$

for $1+\mathrm{Rl}\,(a) > \mathrm{Rl}\,(c) > \mathrm{Rl}\,(b)$, and $|\arg(z-1)| < \pi$. If we substitute $e^{-t}$ for $t$, we find that

$${}_2F_1[a,b;\,c;\,z] = \frac{\Gamma(c)}{\Gamma(b)\,\Gamma(c-b)} \int_0^{\infty} e^{-bt}(1-e^{-t})^{c-b-1}(1-z\,e^{-t})^{-a}\,dt,$$
$$(1.6.9)$$

for $\mathrm{Rl}\,(c) > \mathrm{Rl}\,(b) > 0$.

If we substitute $\sin^2 t$ for $t$,

$$_2F_1[a,b;\,c;\,z] = \frac{2\Gamma(c)}{\Gamma(b)\,\Gamma(c-b)}\int_0^{\frac{1}{2}\pi} \frac{(\sin t)^{2b-1}\,(\cos t)^{2c-2b-1}}{(1-z\sin^2 t)^a}\,dt,$$

$$(1.6.10)$$

again under the condition that $\mathrm{Rl}\,(c) > \mathrm{Rl}\,(b) > 0$. Other integrals in the same group are

$$_2F_1[a,b;\,c;\,z] = \frac{2^{1-c}\Gamma(c)}{\Gamma(b)\,\Gamma(c-b)}\int_0^{\pi} \frac{(\sin t)^{2b-1}\,(1+\cos t)^{c-2b}}{(1-\frac{1}{2}z+\frac{1}{2}z\cos t)^a}\,dt,$$

$$(1.6.11)$$

if we put $(\frac{1}{2}\cos t - \frac{1}{2})$ for $t$,

$$_2F_1[a,b;\,c;\,z] = \frac{2\Gamma(c)}{\Gamma(b)\,\Gamma(c-b)}\int_0^{\infty} \frac{(\cosh t)^{2a-2c+1}\,(\sinh t)^{2c-2b-1}}{\{(\cosh t)^2 - z\}^a}\,dt,$$

$$(1.6.12)$$

if we put $1/(\cosh^2 t)$ for $t$,

$$_2F_1[a,b;\,c;\,z] = \frac{2^{b-a}\Gamma(c)}{\Gamma(b)\,\Gamma(c-b)}\int_0^{\infty} \frac{(\sinh t)^{2a-2c+1}\,(\cosh t - 1)^{2c-a-b-1}}{(\frac{1}{2}-z+\frac{1}{2}\cosh t)^a}\,dt,$$

$$(1.6.13)$$

if we put $1/(\frac{1}{2}+\frac{1}{2}\cosh t)$ for $t$, and

$$_2F_1[a,b;\,c;\,z] = \frac{2\Gamma(c)}{\Gamma(b)\,\Gamma(c-b)}\int_0^{\infty} \frac{(\sinh t)^{2b-1}\,(\cosh t)^{2a-2c+1}}{\{(\cosh t)^2 - z(\sinh t)^2\}^a}\,dt,$$

$$(1.6.14)$$

if we put $\tanh^2 t$ for $t$. These last four integrals all hold under the convergence conditions $\mathrm{Rl}\,(c) > \mathrm{Rl}\,(b) > 0$.

The convergence conditions for the Pochammer integral (1.6.6) can be relaxed if we take the integral round more complicated contours. Thus, let us consider the integral

$$I_1 \equiv \int_0^{(1+)} t^{b-1}(1-t)^{c-b-1}(1-tz)^{-a}\,dt, \qquad (1.6.15)$$

round a contour which encircles the point $+1$, once in an anticlockwise direction. This contour can be deformed into the real axis $(0, 1-\epsilon)$, the circle $C$ centre $+1$, and radius $\epsilon$, and the real axis again $(1-\epsilon, 0)$.

Then

$$I_1 = [1-\exp\{2(c-b)\,i\pi\}]\int_0^{1-\epsilon} t^{b-1}(1-t)^{c-b-1}(1-tz)^{-a}\,dt$$

$$+ \int_C t^{b-1}(1-t)^{c-b-1}(1-tz)^{-a}\,dt.$$

But $\qquad \int_{1-\epsilon}^{1} t^{b-1}(1-t)^{c-b-1}(1-tz)^{-a}\,dt \to 0 \quad\text{as}\quad \epsilon \to 0,$

and $\qquad \int_{C} t^{b-1}(1-t)^{c-b-1}(1-tz)^{-a}\,dt \to 0 \quad\text{as}\quad \epsilon \to 0.$

Hence

$$I_1 = [1-\exp\{2(c-b)\,i\pi\}]\int_0^1 t^{b-1}(1-t)^{c-b-1}(1-tz)^{-a}\,dt,$$

that is

$${}_2F_1[a,b;c;z] = \frac{i\Gamma(c)\exp\{i\pi(b-c)\}}{2\Gamma(b)\,\Gamma(c-b)\sin\{\pi(c-b)\}}$$

$$\times \int_0^{(1+)} t^{b-1}(1-t)^{c-b-1}(1-tz)^{-a}\,dt, \quad (1.6.16)$$

for $c-b \neq 1,2,3,\dots$, $|\arg(1-z)| < \pi$, $\mathrm{Rl}(b) > 0$, only.

By considering the integral round a similar contour taken from $+1$, round $O$, on the real axis, we can show that

$${}_2F_1[a,b;c;z] = \frac{-i\Gamma(c)\exp(-i\pi b)}{2\Gamma(b)\,\Gamma(c-b)\sin(\pi b)}\int_1^{(0+)} t^{b-1}(1-t)^{c-b-1}(1-tz)^{-a}\,dt,$$

$$(1.6.17)$$

for $b \neq 1,2,3,\dots$, $|\arg(-z)| < \pi$, and $\mathrm{Rl}(c-b) > 0$, only.

Finally if we consider the integral taken round the double looped Pochammer contour of Fig 1.1, we can dispense with both convergence conditions and show that

$${}_2F_1[a,b;c;z] = \frac{-\Gamma(c)\exp(-i\pi c)}{4\Gamma(b)\,\Gamma(c-b)\sin(\pi b)\sin\{\pi(c-b)\}}$$

$$\times \int^{(1+,0+,1-,0-)} t^{b-1}(1-t)^{c-b-1}(1-tz)^{-a}\,dt, \quad (1.6.18)$$

for all $\arg z$ and for $b$, and $c-b \neq 1,2,3,\dots$ only.

### 1.6.1 The Barnes-type integral.
Let us consider the contour integral

$$I_C = \int_C \frac{\Gamma(a+s)\,\Gamma(b+s)\,\Gamma(-s)}{\Gamma(c+s)}(-z)^s\,ds. \quad (1.6.1.1)$$

We can see that the integral has sequences of poles at

$$s = -a-n, \quad s = -b-n, \quad\text{and}\quad s = n, \quad\text{for}\quad n = 0,1,2,\dots,$$

so we shall take the integral round the contour $C$, consisting of the imaginary axis indented to the right to exclude all the poles of $\Gamma(a+s)$ and $\Gamma(b+s)$, and to the left to include the first $N$ poles of $\Gamma(-s)$, only.

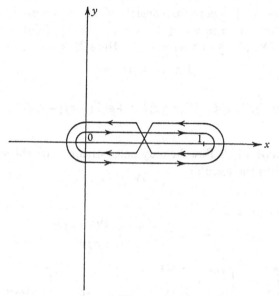

Fig. 1.1

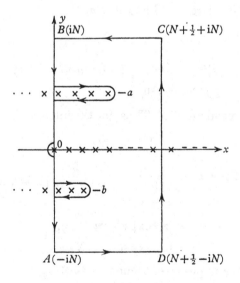

Fig. 1.2

We take that section of the axis which lies between the points $\pm iN$ only. The rest of the contour consists of the rectangle formed from the three straight lines $y = \pm N$, and $x = N + \frac{1}{2}$, joining the points $iN$, $N + \frac{1}{2} + iN$, $N + \frac{1}{2} - iN$ and $-iN$. Here $N$ is any integer greater than
$$|\text{Im}\,(a)| + |\text{Im}\,(b)|.$$
Now
$$I = 2\pi i \sum_{n=0}^{N} \text{Residue}\,\{\Gamma(a+s)\,\Gamma(b+s)\,\Gamma(-s)\,(-z)^s/\Gamma(c+s)\},$$
$$(1.6.1.2)$$

within the contour $C$. But the only poles within $C$ are those of $\Gamma(-s)$, and $\Gamma(-s)$ has the residue
$$\frac{(-1)^{n-1}}{n!}$$
at its pole $s = n$. Hence
$$I_C = -2\pi i \sum_{n=0}^{N} \frac{\Gamma(a+n)\,\Gamma(b+n)\,z^n}{\Gamma(c+n)\,n!}. \qquad (1.6.1.3)$$
But
$$I_C = \int_C = -\int_{-iN}^{+iN} + \int_{-iN}^{N+\frac{1}{2}-iN} + \int_{N+\frac{1}{2}-iN}^{N+\frac{1}{2}+iN} + \int_{N+\frac{1}{2}+iN}^{iN},$$
$$= -I_N + J_1 + J_2 + J_3, \qquad (1.6.1.4)$$
say.

Now, when $|s|$ is large, and $|\arg s| < \pi$,
$$\left| \frac{\Gamma(a+s)\,\Gamma(b+s)}{\Gamma(c+s)\,\Gamma(1+s)} \right|$$
$$= |s|^{\text{Rl}(a+b-c-1)} \exp\{-\text{Im}\,(a+b-c-1)\arg s\}\{1+o(1)\},$$
$$< A N^{\text{Rl}(a+b-c-1)},$$

where $A$ is independent of $N$. Thus, for the integral $J_1$, we have
$$s = x - iN,$$
and so
$$|\Gamma(-s)\,\Gamma(1+s)| = \frac{2\pi}{\exp\{\pi(N+ix)\} - \exp\{-\pi(N+ix)\}},$$
$$< 4\pi\,e^{-\pi N}.$$
Also
$$|(-z)^s| = |z|^x \exp\{N\arg(-z)\},$$
$$< |z|^x \exp\{(\pi-\epsilon)N\},$$
where $\epsilon$ is small and positive. Hence
$$|J_1| < \int_0^{N+\frac{1}{2}} A N^{\text{Rl}(a+b-c-1)}\,4\pi\,e^{-\pi N}\,|z|^x\,e^{(\pi-\epsilon)N}\,dx,$$
$$< 4\pi\,e^{-\epsilon N} A N^{\text{Rl}(a+b-c-1)}\,(N+\tfrac{1}{2}),$$

since $|z|^x \leqslant 1$, if $|z| < 1$. Now $e^{-\varepsilon N}$ tends to zero faster than

$$N^{\mathrm{Rl}(a+b-c-1)} \to \infty,$$

as $N \to \infty$. Hence $J_1 \to 0$, as $N \to \infty$. Similarly $J_3 \to 0$, as $N \to \infty$. For $J_2$, we have

$$s = N + \tfrac{1}{2} + iy,$$

where $-N \leqslant y \leqslant N$. Hence

$$|\Gamma(-s)\,\Gamma(1+s)| = \frac{\pi}{|\cosh{(\pi y)}|} < 2\pi\,e^{-\pi\,|y|}.$$

Also $$|(-z)^s| = |z|^{N+\frac{1}{2}} \exp\{-y\arg{(-z)}\}.$$
Hence

$$|J_2| < \int_{-N}^{N} A\,N^{\mathrm{Rl}(a+b-c-1)}\,2\pi\,e^{-\pi\,|y|}\,|z|^{N+\frac{1}{2}}\exp\{-y\arg{(-z)}\}\,dy,$$

$$< 2\pi A\,|z|^{N+\frac{1}{2}}\,N^{\mathrm{Rl}(a+b-c-1)}\int_{-N}^{N} e^{-\varepsilon|y|}\,dy,$$

$$< 4\pi A\,|z|^{N+\frac{1}{2}}\,N^{\mathrm{Rl}(a+b-c-1)}\,2N\,e^{-\varepsilon N}.$$

Now $|z|^{N+\frac{1}{2}}$ tends to zero faster than $N^{\mathrm{Rl}(a+b-c-1)} \to \infty$, since $|z| < 1$. Hence $J_2 \to 0$, as $N \to \infty$. Further

$$I_N \to I = \int_{-i\infty}^{i\infty} \frac{\Gamma(a+s)\,\Gamma(b+s)\,\Gamma(-s)}{\Gamma(c+s)} (-z)^s\,ds \quad \text{as} \quad N \to \infty,$$

so that from (1.6.1.3) and (1.6.1.4) we have finally

$$I = -2\pi i \sum_{n=0}^{\infty} \frac{\Gamma(a+n)\,\Gamma(b+n)\,z^n}{\Gamma(c+n)\,n!}, \qquad (1.6.1.5)$$

that is

$$_2F_1[a,b;\,c;\,z] = \frac{\Gamma(c)}{2\pi i\,\Gamma(a)\,\Gamma(b)} \int_{-i\infty}^{i\infty} \frac{\Gamma(a+s)\,\Gamma(b+s)\,\Gamma(-s)}{\Gamma(c+s)} (-z)^s\,ds,$$

$$(1.6.1.6)$$

provided that $|z| < 1$, and that $|\arg{(-z)}| < \pi$.

This integral is truly convergent, and not a Cauchy value, as the integrand is continuous near the origin. From this we can see that we can think of $_2F_1[a,b;\,c;\,z]$, when represented by this integral, as an analytic function, regular in the entire $z$-plane cut along the real $x$ axis from $O$ to $\infty$. It is the analytic continuation of the series $_2F_1[a,b;\,c;\,z]$, in any closed domain $D$ of this cut $z$-plane, since the integral converges uniformly in $z$ and in $s$ within $D$, and so it is regular in the cut $z$-plane. Sometimes, as an alternative, the semi-circular contour of Fig. 4.1 below, with centre $O$, and radius $R$ lying to the right of the imaginary $y$ axis is taken instead of the rectangular contour used here. The proof only differs in details.†

### 1.6.2 The Borel integral. We shall show that if

$$F(z) = \sum_{n=0}^{\infty} a_n z^n, \qquad (1.6.2.1)$$

and this series is convergent for $|z| < R$, and

$$\Phi(z) = \sum_{n=0}^{\infty} a_n \frac{z^n}{n!}, \qquad (1.6.2.2)$$

then
$$I \equiv \int_0^{\infty} e^{-t} \Phi(zt)\, dt, \qquad (1.6.2.3)$$

exists and
$$I = F(z), \qquad (1.6.2.4)$$

for $|z| < R$. For, since $F(z)$ is a convergent series, we must have

$$\lim_{n \to \infty} |(a_n)^{1/n}| = 1/R,$$

so that
$$|a_n| < (M/R)^n,$$

where $M$ is a positive constant. From this we deduce, using Stirling's approximation, that

$$\lim_{n \to \infty} \left| \left( \frac{a_n}{n!} \right)^{1/n} \right| \leqslant \frac{M}{R} \lim_{n \to \infty} (n!)^{-1/n},$$

$$\leqslant \frac{M}{R} \lim_{n \to \infty} (\sqrt{2\pi n}\, n^{n+\frac{1}{2}}\, e^{-n})^{-1/n},$$

and this tends to zero as $n$ tends to infinity. Hence the series $\Phi(z)$ is convergent for $|z| < R$, and the integral

$$\int_0^{\infty} e^{-t} |\Phi(zt)|\, dt$$

exists. Now
$$\int_0^{\infty} e^{-t} |\Phi(zt)|\, dt \leqslant \int_0^{\infty} e^{-t} \sum_{n=0}^{\infty} \frac{|a_n|\,|zt|^n}{n!}\, dt.$$

But
$$\sum_{n=0}^{\infty} \int_0^{\infty} \frac{|a_n|\,|z|^n}{n!}\, e^{-t} |t|^n\, dt = \sum_{n=0}^{\infty} |a_n|\,|z|^n,$$

† See Bailey's Tract (1935), § 1.6.

which is convergent for $|z| < R$. Hence the order of summation and integration can be reversed, and we have

$$\int_0^\infty e^{-t}\, \Phi(zt)\, dt = \sum_{n=0}^\infty a_n z^n.$$ (1.6.2.5)

If in every direction $(O, z)$ from the origin, we find the first singular point of $F(z)$ and through this point draw the line $LL'$ perpendicular to $Oz$, then the region contained within the set of lines $LL'$ is called the Borel polygon of summability, and any series of the type $F(z)$, for which such a Borel polygon exists, is said to be summable in the Borel sense or summable-$B$. The Borel integral is equal to $F(z)$ at all points within its polygon, since it is convergent and analytic every-where within this region, and we have seen that it is equal to $F(z)$ at some point within this region.†

In particular, since the Gauss function is convergent for $|z| < 1$, then the corresponding integral exists and is in fact

$$_2F_1[a, b;\ c;\ z] = \int_0^\infty e^{-t} \sum_{n=0}^\infty \frac{(a)_n\, (b)_n\, (zt)^n}{(c)_n\, n!\, n!}\, dt,$$ (1.6.2.6)

within the region $|z| < 1$, since the only singularities of $F(z)$ lie on the line $|z| = 1$.

## 1.7 The Gauss summation theorem

We have already noted that a linear relation exists between any three Gauss functions of the type

$$_2F_1[a+l, b+m;\ c+n; z],$$

where $l$, $m$ and $n$ are integers, and in § 1.4, we proved one such result which is of particular importance. This is

$$(c-a)\,(c-b)\,_2F_1[a, b;\ c+1;\ z] = c(c-a-b)\,_2F_1[a, b;\ c;\ z]$$
$$+ ab(1-z)\,_2F_1[a+1, b+1;\ c+1;\ z],$$ (1.7.1)

where $|z| < 1$. In this result let $z \to 1$ 'from below', that is through real values of $z$ less than one. Then, provided that $\mathrm{Rl}\,(c-a-b) > 0$, and that $a$, $b$, $c$ are not zero nor negative integers, all three series exist and have finite values. Hence in the limit, as $z \to 1$, we find that

$$_2F_1[a, b;\ c;\ 1] = \frac{(c-a)\,(c-b)}{c(c-a-b)}\,_2F_1[a, b;\ c+1;\ 1].$$ (1.7.2)

† See Dienes, *The Taylor Series* (1931), pp. 302–311, for further notes on the Borel polygon.

Let us apply this formula $n$ times. Then we have

$$_2F_1[a,b;\ c;\ 1] = \frac{(c-a)_n\,(c-b)_n}{(c)_n\,(c-a-b)_n}\,{}_2F_1[a,b;\ c+n;\ 1]. \qquad (1.7.3)$$

This result is Gauss's reduction formula. Now

$$(c)_n \equiv \frac{\Gamma(c+n)}{\Gamma(c)}.$$

Hence

$$\frac{(c-a)_n\,(c-b)_n}{(c)_n\,(c-a-b)_n} = \frac{\Gamma(c)\,\Gamma(c-a-b)\,\Gamma(c-a+n)\,\Gamma(c-b+n)}{\Gamma(c-a)\,\Gamma(c-b)\,\Gamma(c+n)\,\Gamma(c-a-b+n)},$$

$$\rightarrow \frac{\Gamma(c)\,\Gamma(c-a-b)}{\Gamma(c-a)\,\Gamma(c-b)},$$

as $n \to \infty$. Also

$$\left|{}_2F_1\left[a,b;\ c+n;\ 1\right]\right| \leqslant 1 + \sum_{m=1}^{\infty} \frac{(|a|)_m\,(|b|)_m}{m!\,(n-|c|)_m},$$

$$\leqslant 1 + \frac{|a|\,|b|}{n-|c|}\,{}_2F_1[|a|+1,\,|b|+1;\ n-|c|+1;\ 1],$$

$$\leqslant 1 + \frac{|a|\,|b|}{n-|c|}\,M, \text{ for } n > |c|, \qquad (1.7.4)$$

where $M$ is a constant, and this tends to one as $n$ tends to infinity. Hence

$$\frac{(c-a)_n\,(c-b)_n}{(c)_n\,(c-a-b)_n}\,{}_2F_1[a,b;\ c+n;\ 1] \rightarrow \frac{\Gamma(c)\,\Gamma(c-a-b)}{\Gamma(c-a)\,\Gamma(c-b)}, \qquad (1.7.5)$$

as $n \to \infty$, so that, by (1.7.3),

$$_2F_1[a,b;\ c;\ 1] = \frac{\Gamma(c)\,\Gamma(c-a-b)}{\Gamma(c-a)\,\Gamma(c-b)}, \qquad (1.7.6)$$

provided that $\mathrm{Rl}\,(c-a-b) > 0$.

This important result is known as Gauss's theorem. As a special case, when $b$ is a negative integer $-m$, we have the result known as Vandermonde's theorem

$$_2F_1[a,\ -m;\ c;\ 1] = (c-a)_m/(c)_m. \qquad (1.7.7)$$

This holds true for all values of $a$ and $c$, since the series terminates, except negative integer values of $c$ less than $a$, or $m$, when the series is not defined. Thus

$$_2F_1[-4,\ -2;\ -3;\ 1] = (-3+4)_2/(-3)_2, = 1.2.3/\{(-3)\,(-2)\},$$

$$= 1,$$

but

$$_2F_1[-4,\ -3;\ -2;\ 1] = (-2+4)_3/(-2)_3 = 2.3.4/\{(-2)\,(-1)\,0\}.$$

An alternative method of proof of Gauss's theorem is to put $z = 1$ in the Pochammer integral (1.6.6). This integral then reduces to a simple Beta function,

$$\int_0^1 t^{b-1}(1-t)^{c-b-1}(1-t)^{-a}\,dt,$$

and this can be evaluated immediately in terms of Gamma functions.

We shall next show that

$$\lim_{z \to 1} [{}_2F_1[a,b;\,a+b;\,z]/\log\{1/(1-z)\}] = \frac{\Gamma(a+b)}{\Gamma(a)\,\Gamma(b)}, \qquad (1.7.8)$$

and that

$$\lim_{z \to 1} [{}_2F_1[a,b;\,c;\,z]/\{(1-z)^{c-a-b}\}] = \frac{\Gamma(c)\,\Gamma(a+b-c)}{\Gamma(a)\,\Gamma(b)}, \qquad (1.7.9)$$

for $\mathrm{Rl}\,(c) > \mathrm{Rl}\,(a+b)$.

Both these results depend on the following lemma: if

(i) $\displaystyle\sum_{n=1}^{\infty} a_n z^n$ and $\displaystyle\sum_{n=1}^{\infty} b_n z^n$ are two series convergent for $|z| < 1$,

(ii) $\displaystyle\sum_{n=1}^{\infty} |a_n|$ is a divergent series,

(iii) $\displaystyle\sum_{n=1}^{\infty} |a_n|\,z^n < K \left| \sum_{n=1}^{\infty} a_n z^n \right|$ when $0 < |z| < 1$, and $K$ is independent of $z$,

and (iv) $\lambda \equiv \displaystyle\lim_{n \to \infty} (b_n/a_n)$ exists, then shall

$$\lambda = \lim_{z \to 1} \left\{ \sum_{n=1}^{\infty} b_n z^n \middle/ \left( \sum_{n=1}^{\infty} a_n z^n \right) \right\}.$$

Now $\log\{1/(1-z)\} = z + z^2/2 + \ldots + z^n/n + \ldots$ . Let $a_n = 1/n$. This is positive for all positive values of $n$, and $\displaystyle\sum_{n=1}^{\infty} a_n$ is divergent. Also

$$\sum_{n=1}^{\infty} |a_n|\,z^n < 2 \left| \sum_{n=1}^{\infty} a_n z^n \right| = 2\,|\log\{1/(1-z)\}|,$$

since

$$\sum_{n=1}^{\infty} z^n/n = \log\{1/(1-z)\},$$

and the conditions of the lemma are satisfied for $a_n$.

Let

$${}_2F_1[a,b;\,c;\,z] = \sum_{n=0}^{\infty} b_n z^n.$$

This is a convergent series for $|z| < 1$, and

$$b_n = \frac{\Gamma(a+b)\,\Gamma(a+n)\,\Gamma(b+n)}{\Gamma(a)\,\Gamma(b)\,\Gamma(a+b+n)\,n!} \sim \frac{\Gamma(a+b)}{\Gamma(a)\,\Gamma(b)} \frac{1}{n},$$

for $n$ large enough. Hence

$$\lim_{n \to \infty} (b_n/a_n) = \frac{\Gamma(a+b)}{\Gamma(a)\,\Gamma(b)},$$

which satisfies condition (iv) above, and our first result (1.7.8) follows.

For the second result (1.7.9) we have

$$(1-z)^{c-a-b} = \sum_{n=0}^{\infty} a_n z^n,$$

where    $a_n = (a)_n/n! = \Gamma(a+b-c+n)/\{\Gamma(a+b-c)\,n!\}.$

Hence    $a_n \sim n^{a+b-c+n-\frac{1}{2}} n^{-n-\frac{1}{2}}/\Gamma(a+b-c),$

$$\sim n^{a+b-c-1}/\Gamma(a+b-c),$$

if $\mathrm{Rl}(c) > \mathrm{Rl}(a+b).$

Hence $\Sigma\,|a_n|$ is divergent, but $\Sigma a_n z^n$ is convergent to the value $(1-z)^{c-a-b}$ for $|z| < 1$, and

$$\Sigma\,|a_n|\,z^n < 2|(1-z)^{c-a-b}|.$$

Also    $_2F_1[a,b;\,c;\,z] \equiv \Sigma b_n z^n.$

Hence    $b_n = \dfrac{\Gamma(c)\,\Gamma(a+n)\,\Gamma(b+n)}{\Gamma(a)\,\Gamma(b)\,\Gamma(c+n)\,n!} \sim \dfrac{\Gamma(c)}{\Gamma(a)\,\Gamma(b)}\,n^{a+b-c-1}$

and    $\lim\limits_{n \to \infty} (b_n/a_n) = \dfrac{\Gamma(c)\,\Gamma(a+b-c)}{\Gamma(a)\,\Gamma(b)},$

from which our second result follows.

Alternatively, without using the lemma, we could say that, by Euler's identity, (1.3.15),

$$\lim_{z \to 1} \{(1-z)^{a+b-c}\,_2F_1[a,b;\,c;\,z]\} = \lim_{z \to 1} \{_2F_1[c-a,c-b;\,c;\,z]\}$$

$$= \frac{\Gamma(c)\,\Gamma(a+b-c)}{\Gamma(a)\,\Gamma(b)}, \quad \text{as } z \to 1.$$

The result (1.7.8) also can be proved without the lemma. Here† the limit is

$$\lim_{z \to 1} \frac{\partial}{\partial z}\{_2F_1[a,b;\,a+b;\,z]\}\Big/\frac{\partial}{\partial z}\{\log 1/(1-z)\},$$

$$= \lim_{z \to 1} \left\{\frac{ab}{a+b}\,(1-z)\,_2F_1[a+1,b+1;\,a+b+1;\,z]\right\},$$

$$= \lim_{z \to 1} \left\{\frac{ab}{a+b}\,_2F_1[a,b;\,a+b+1;\,z]\right\}, \quad \text{by (1.3.15)}$$

$$= \frac{ab}{a+b}\,\frac{\Gamma(a+b+1)}{\Gamma(a+1)\,\Gamma(b+1)}, \quad \text{by (1.7.6)}$$

† By de l'Hospital's theorem.

where $a+b+1 > a+b$,

$$= \frac{\Gamma(a+b)}{\Gamma(a)\,\Gamma(b)}.$$

These alternative proofs for (1.7.8) and (1.7.9) are due to T. Chaundy.

### 1.7.1 Another special summation theorem.

Within the circle $|z| < \frac{1}{2}$, we have the expansion

$$(1-z)^{-a}\,_2F_1[a, b; c; -z/(1-z)] = \sum_{r=0}^{\infty} \frac{(a)_r\,(b)_r}{(c)_r\,r!}(-1)^r z^r (1-z)^{-a-r}$$

$$= \sum_{r=0}^{\infty} \sum_{s=0}^{\infty} \frac{(a)_r\,(b)_r\,(a+r)_s}{(c)_r\,r!\,s!}(-1)^r z^{r+s}.$$

(1.7.1.1)

The coefficient of $z^n$ in this double series is

$$\sum_{r=0}^{n} \frac{(a)_r\,(b)_r\,(a+r)_{n-r}\,(-1)^r}{(c)_r\,(n-r)!\,r!} = \frac{(a)_n}{n!} \sum_{r=0}^{n} \frac{(b)_r\,(-n)_r}{(c)_r\,r!}$$

$$= \frac{(a)_n}{n!}\,_2F_1[-n, b; c; 1]$$

$$= \frac{(a)_n\,(c-b)_n}{(c)_n\,n!},$$

(1.7.1.2)

by Vandermonde's theorem (1.7.7), since

$$(a+r)_{n-r} = \frac{(a)_n}{(a)_r} \quad \text{and} \quad \frac{1}{(n-r)!} = \frac{(-n)_r\,(-1)^r}{n!}.$$

Many such useful formulae connecting coefficients of this type will be found in Appendix I.

Thus we have proved that

$$(1-z)^{-a}\,_2F_1[a, b; c; -z/(1-z)] = \,_2F_1[a, c-b; c; z], \quad (1.7.1.3)$$

for $|z| < \frac{1}{2}$. By analytic continuation, this result can be extended to all the values of $z$ in the region $|z| < 1$, $\mathrm{Rl}\,(z) < \frac{1}{2}$. In terms of the solutions $y_1$ to $y_{24}$ of Gauss's equation (1.2.1), we have now proved that

$$y_1 = y_{21}.$$

If we let $z \to -1$, and apply Abel's theorem we shall find that when $\mathrm{Rl}\,(b-a) > -1$,

$$_2F_1[a, c-b; c; -1] = 2^{-a}\,_2F_1[a, b; c; \tfrac{1}{2}]. \quad (1.7.1.4)$$

If, in Pochammer's integral (1.6.6), we put

$$z = -1 \quad \text{and} \quad a = 1+b-c,$$

we find that

$$\frac{\Gamma(b)\,\Gamma(c-b)}{\Gamma(c)}\,{}_2F_1[b,1+b-c;\ c;\ -1] = \int_0^1 t^{b-1}(1-t^2)^{c-b-1}\,\mathrm{d}t.$$

$$(1.7.1.5)$$

This integral is a Beta function, and so it can be evaluated in terms of Gamma functions. From this we find the summation theorem

$${}_2F_1[a,b;\ 1+a-b;\ -1] = 2^{-a}\frac{\Gamma(1+a-b)\,\Gamma(1+\tfrac{1}{2}a)}{\Gamma(1+a)\,\Gamma(1+\tfrac{1}{2}a-b)}.$$

$$(1.7.1.6)$$

This is Kummer's theorem.

Similarly, if we put $b = 1-a$ and $z = \tfrac{1}{2}$, we are led to

$$\frac{2^a\,\Gamma(1-a)\,\Gamma(c+a-1)}{\Gamma(c)}\,{}_2F_1[a,1-a;\ c;\ \tfrac{1}{2}] = \int_0^1 (2t-t^2)^{-a}\,(1-t)^{c+a-2}\,\mathrm{d}t.$$

$$(1.7.1.7)$$

If, in this integral, we take $(1-t)^2$ as the new variable, it also becomes a Beta function, whence we can deduce Bailey's summation theorem for a ${}_2F_1[\tfrac{1}{2}]$ series. This is

$${}_2F_1[a,1-a;\ c;\ \tfrac{1}{2}] = \frac{\Gamma(\tfrac{1}{2}c)\,\Gamma(\tfrac{1}{2}c+\tfrac{1}{2})}{\Gamma(\tfrac{1}{2}c+\tfrac{1}{2}a)\,\Gamma(\tfrac{1}{2}+\tfrac{1}{2}c-\tfrac{1}{2}a)}.$$

$$(1.7.1.8)$$

Alternatively, we can apply Kummer's theorem to the left side of (1.7.1.4) above and deduce this same result.

Kummer's theorem (1.7.1.6) can also be applied to (1.7.1.4) in another way. When $c = \tfrac{1}{2}(a+b+1)$, this leads us to the summation theorem

$${}_2F_1[a,b;\ \tfrac{1}{2}(a+b+1);\ \tfrac{1}{2}] = \frac{\Gamma(\tfrac{1}{2})\,\Gamma(\tfrac{1}{2}+\tfrac{1}{2}a+\tfrac{1}{2}b)}{\Gamma(\tfrac{1}{2}+\tfrac{1}{2}a)\,\Gamma(\tfrac{1}{2}+\tfrac{1}{2}b)}.$$

$$(1.7.1.9)$$

This is the result usually called Gauss's second summation theorem.

## 1.8 Analytic continuation formulae

As we saw above, the Gauss equation (1.2.1) can have only two independent solutions in any one domain. Hence there exist linear relationships connecting any three of the twenty-four solutions. In particular, by Euler's identity (1.3.15), we find that the twenty-four solutions can be reduced to six equivalent groups, two solutions of each group being valid in each of the three regions of convergence. Thus

$$y_1 = y_3 = y_{21} = y_{22}$$

$$(1.8.1)$$

for $|z| < 1$ and $\mathrm{Rl}\,(z) < \frac{1}{2}$,

$$y_2 = y_4 = y_{23} = y_{24} \qquad (1.8.2)$$

for $|z| < 1$ and $\mathrm{Rl}\,(z) < \frac{1}{2}$,

$$y_5 = y_7 = y_{17} = y_{18} \qquad (1.8.3)$$

for $|z-1| < 1$ and $\mathrm{Rl}\,(z) > \frac{1}{2}$,

$$y_6 = y_8 = y_{19} = y_{20} \qquad (1.8.4)$$

for $|z-1| < 1$ and $\mathrm{Rl}\,(z) > \frac{1}{2}$,

$$y_9 = y_{11} = y_{13} = y_{15} \qquad (1.8.5)$$

for $|z| > 1$ and $|z-1| > 1$,

$$y_{10} = y_{12} = y_{14} = y_{16} \qquad (1.8.6)$$

for $|z| > 1$ and $|z-1| > 1$.

Thus we can express any one of these solutions in terms of the two solutions valid in the same domain. We can express $y_5$ valid for $|z-1| < 1$, in terms of $y_1$ and $y_2$ valid for $|z| < 1$, and we can use these results to connect a solution in any one of the three zones with two solutions in any other of the three zones. There are several different methods of proving such relations. The first one we give depends on the Gauss summation theorem.

Suppose that we require a relation which expresses $y_5$ valid for $|z-1| < 1$ in terms of $y_1$ and $y_2$ valid for $|z| < 1$. We know that such a relation exists and that it must have the form

$$y_5 = Ay_1 + By_2,$$

that is

$$
{}_2F_1[a, b;\, 1+a+b-c;\, 1-z]
$$
$$
= A\,{}_2F_1[a, b;\, c;\, z] + Bx^{1-c}\,{}_2F_1[1+a-c,\, 1+b-c;\, 2-c;\, z], \qquad (1.8.7)
$$

where we have to find values for $A$ and $B$. Now let $z \to 0$. Then we find that

$$
{}_2F_1[a, b;\, 1+a+b-c;\, 1] = A,
$$

provided that $\mathrm{Rl}\,(1+a+b-c) > \mathrm{Rl}\,(a+b)$, that is provided that $1 > \mathrm{Rl}\,(c)$. Hence, by Gauss's theorem, (1.7.6),

$$
A = \frac{\Gamma(1+a+b-c)\,\Gamma(1-c)}{\Gamma(1+a-c)\,\Gamma(1+b-c)}. \qquad (1.8.8)
$$

Next let $z \to 1$. Then, if $\mathrm{Rl}\,(c) > \mathrm{Rl}\,(a+b)$, we have

$$
1 = A\,{}_2F_1[a, b;\, c;\, 1] + B_2F_1[1+a-c,\, 1+b-c;\, 2-c;\, 1],
$$

from which we can deduce that

$$B = \frac{\Gamma(1+a+b-c)\,\Gamma(c-1)}{\Gamma(a)\,\Gamma(b)}, \qquad (1.8.9)$$

so that our complete result is

$$_2F_1[a,b;\ 1+a+b-c;\ 1-z] = \frac{\Gamma(1+a+b-c)\,\Gamma(1-c)}{\Gamma(1+a-c)\,\Gamma(1+b-c)}\ _2F_1[a,b;\ c;\ z]$$

$$+\frac{\Gamma(1+a+b-c)\,\Gamma(c-1)}{\Gamma(a)\,\Gamma(b)}\,z^{1-c}\ _2F_1[1+a-c,1+b-c;\ 2-c;\ z],$$

$$(1.8.10)$$

provided that $1 > \mathrm{Rl}\,(c) > \mathrm{Rl}\,(a+b)$.

This last result can be extended to provide the analytic continuation of $_2F_1[a,b;\ c;\ z]$ over the whole of the complex $z$-plane, excluding only the negative real axis, and for all values of $a$, $b$ and $c$ real or complex, excluding zero and those integer values which cause one of the functions to be completely indeterminate.

For $0 \leqslant |z| \leqslant 1$, we have

$$A = \frac{\Gamma(c)\,\Gamma(c-a-b)}{\Gamma(c-a)\,\Gamma(c-b)}, \qquad (1.8.11)$$

if $\mathrm{Rl}\,(c) > \mathrm{Rl}\,(a+b)$ as $z \to 1$. But on the other hand, if

$$\mathrm{Rl}\,(c) < \mathrm{Rl}\,(a+b) \quad \text{as } z \to 1,$$

then $\qquad (1-z)^{c-a-b}\dfrac{\Gamma(c)\,\Gamma(a+b-c)}{\Gamma(a)\,\Gamma(b)} \sim A + B(1-z)^{c-a-b},$

so that $\qquad\qquad B = \dfrac{\Gamma(c)\,\Gamma(a+b-c)}{\Gamma(a)\,\Gamma(b)}. \qquad (1.8.12)$

Again, when $\mathrm{Rl}\,(c) < 1$, as $z \to 0$, an equation is obtained of the form

$$1 = AH + BK,$$

that is

$$1 = A\,_2F_1[a,b;\ 1+a+b-c;\ 1] + B\,_2F_1[c-a,c-b;\ 1+c-a-b;\ 1]. \qquad (1.8.13)$$

But, when $\mathrm{Rl}\,(c) > 1$, as $z^{1-c} \to \infty$,

$$1 \sim Az^{1-c}\frac{\Gamma(1+a+b-c)\,\Gamma(c-1)}{\Gamma(a)\,\Gamma(b)} + Bz^{1-c}\frac{\Gamma(1+c-a-b)\,\Gamma(c-1)}{\Gamma(c-a)\,\Gamma(c-b)}, \qquad (1.8.14)$$

which gives an equation of the form $AL + BM = 0$. Hence in any case, there are always two equations to be solved for the two constants needed, for all values of $a$, $b$ and $c$, except zero and those integral values noted above. These equations are usually summed up by the following scheme,

|  | $\mathrm{Rl}(c) > \mathrm{Rl}(a+b)$ | $\mathrm{Rl}(c) < \mathrm{Rl}(a+b)$ |
|---|---|---|
| $\mathrm{Rl}(c) > 1$ | $A = $ constant | $B = $ constant |
|  | $AL + BM = 0$ | $AL + BM = 0$ |
| $\mathrm{Rl}(c) < 1$ | $A = $ constant | $B = $ constant |
|  | $AH + BK = 1$ | $AH + BK = 1$ |

## 1.8.1 Analytic continuation using Barnes's integrals.

Another method of proof of equation (1.8.10) relies on the Barnes contour integral. We shall show that

$$\frac{\Gamma(a)\,\Gamma(b)}{\Gamma(c)}\,y_1 = \frac{\Gamma(a)\,\Gamma(b-a)}{\Gamma(c-a)}\,y_9 + \frac{\Gamma(b)\,\Gamma(a-b)}{\Gamma(c-b)}\,y_{10}. \quad (1.8.1.1)$$

Now, if $m$ is any positive integer, we have

$$I_1 \equiv \frac{1}{2\pi i}\int_{-i\infty}^{i\infty} \frac{\Gamma(a+s)\,\Gamma(b+s)\,\Gamma(-s)}{\Gamma(c+s)}(-z)^s\,ds,$$

$$= \frac{1}{2\pi i}\int_{-m-i\infty}^{-m+i\infty} \frac{\Gamma(a+s)\,\Gamma(b+s)\,\Gamma(-s)}{\Gamma(c+s)}(-z)^s\,ds$$

$$+ \sum_{n=0}^{p} (\text{Residues of the integrand at } s = -a-n)$$

$$+ \sum_{n=0}^{q} (\text{Residues of the integrand at } s = -b-n), \quad (1.8.1.2)$$

that is $$I_1 = I_2 + \Sigma_p + \Sigma_q, \quad (1.8.1.3)$$

where $p$ and $q$ are integers $\leqslant m$, which tend to infinity with $m$. The path of integration of the first integral is indented, as in Fig. 1.2 above, to avoid any poles, and the path of integration of the second integral is indented to avoid the pole at $s = -m$ which would otherwise lie upon it, but will now lie to its left.

Now the residue at $s = -a-n$ is

$$\frac{(-1)^n\,\Gamma(a+n)\,\Gamma(b-a-n)}{\Gamma(1+n)\,\Gamma(c-a-n)}(-z)^{-a-n}.$$

Hence

$$\Sigma_p \to \frac{\Gamma(a)\,\Gamma(b-a)}{\Gamma(c-a)}(-z)^{-a}\,{}_2F_1[a, 1+a-c;\ 1+a-b;\ 1/z],$$

$$(1.8.1.4)$$

as $m \to \infty$. Similarly,

$$\Sigma_q \to \frac{\Gamma(b)\,\Gamma(a-b)}{\Gamma(c-b)}\,(-z)^{-b}\,{}_2F_1[b, 1+b-c; 1+b-a; 1/z],$$

$$(1.8.1.5)$$

as $m \to \infty$, and $\qquad I_1 = \dfrac{\Gamma(a)\,\Gamma(b)}{\Gamma(c)}\,{}_2F_1[a, b; c; z].$ $\qquad(1.8.1.6)$

Now

$$I_2 = -\frac{1}{2\pi i}z^{-m}\int_{-i\infty}^{i\infty}\frac{\Gamma(a-m+s)\,\Gamma(b-m+s)}{\Gamma(c-m+s)\,\Gamma(1-m+s)}\frac{\pi}{\sin(\pi s)}(-z)^s\,ds,$$

$$(1.8.1.7)$$

and, when $|\arg(-z)| \leqslant \pi-\epsilon$, for $\epsilon > 0$, then

$$|(-z)^s| = |\exp[\{\log|z| + i\arg(-z)\}s]|,$$

$$\leqslant |z|^{\,\mathrm{Rl}(s)}\,e^{(\pi-\epsilon)\,\mathrm{Im}(s)},$$

$$< |z|\,e^{(\pi-\epsilon)\,\mathrm{Im}(s)}. \qquad (1.8.1.8)$$

Hence

$$|I_2| \leqslant \left|-\frac{z^{-m}}{2\pi i}\right|\left|\int_{-i\infty}^{i\infty}|\mathrm{Im}(s)|^{\,\mathrm{Rl}(a+b-c-1)}\,2\pi\,e^{-\pi\,\mathrm{Im}(s)}|z|\,e^{(\pi-\epsilon)\,\mathrm{Im}(s)}\,ds,$$

$$(1.8.1.9)$$

$$< \frac{|z|^{1-m}}{2\pi}\int_{-i\infty}^{i\infty}2\pi\,e^{-\frac{1}{2}|\mathrm{Im}(s)|\,\epsilon}\,ds, \qquad (1.8.1.10)$$

for $|\mathrm{Im}(s)|$ large. Now this integral is bounded for all $z$ and $m$ when $|\arg(-z)| \leqslant \pi-\epsilon$. Hence $\qquad I_2 \to 0,$

as $m \to \infty$, and we have the final result

$$\frac{\Gamma(a)\,\Gamma(b)}{\Gamma(c)}\,{}_2F_1[a, b; c; z] = \frac{\Gamma(a)\,\Gamma(b-a)}{\Gamma(c-a)}\,(-z)^{-a}\,{}_2F_1[a, 1+a-c; 1+a-b; 1/z]$$

$$+ \frac{\Gamma(b)\,\Gamma(a-b)}{\Gamma(c-b)}\,(-z)^{-b}\,{}_2F_1[b, 1+b-c; 1+b-a; 1/z],$$

$$(1.8.1.11)$$

if $|\arg(-z)| < \pi$.

There are, in all, twelve such results connecting any one of the three pairs of solutions $(y_1, y_2)$, $(y_5, y_6)$, and $(y_9, y_{10})$, with any one of the remaining four solutions. These relations express any one of these six solutions in terms of a pair of solutions in another domain.

In a similar way, we can prove the following eight relations which hold between these six solutions. They connect any three solutions which are not defined in the same domain. If $\mathrm{Im}(z) > 0$, the upper signs should be taken in the exponentials, throughout, and if Im $(z) < 0$, the lower signs should be taken.

$$\exp\left(\pm i\pi b\right)\frac{\Gamma(b)\,\Gamma(1+a-c)}{\Gamma(1+a+b-c)}\,y_5$$

$$= \frac{\Gamma(b)\,\Gamma(c-b)}{\Gamma(c)}\,y_1 + \exp\{\pm i\pi(1+b-c)\}\frac{\Gamma(1+a-c)\,\Gamma(c-b)}{\Gamma(1+a-b)}\,y_9,$$

$$(1.8.1.12)$$

$$\exp\left(\pm i\pi a\right)\frac{\Gamma(a)\Gamma(1+b-c)}{\Gamma(1+a+b-c)}\,y_5$$

$$= \frac{\Gamma(a)\,\Gamma(c-a)}{\Gamma(c)}\,y_1 + \exp\{\pm i\pi(1+a-c)\}\frac{\Gamma(1+b-c)\,\Gamma(c-a)}{\Gamma(1+b-a)}\,y_{10},$$

$$(1.8.1.13)$$

$$\exp\{\pm i\pi(1+b-c)\}\frac{\Gamma(1+b-c)\,\Gamma(a)}{\Gamma(1+a+b-c)}\,y_5$$

$$= \frac{\Gamma(1+b-c)\,\Gamma(1-b)}{\Gamma(2-c)}\,y_2 + \exp\{\pm i\pi(1+b-c)\}\frac{\Gamma(a)\,\Gamma(1-b)}{\Gamma(1+a-b)}\,y_9,$$

$$(1.8.1.14)$$

$$\exp\{\pm i\pi(1+a-c)\}\frac{\Gamma(1+a-c)\,\Gamma(b)}{\Gamma(1+a+b-c)}\,y_5$$

$$= \frac{\Gamma(1+a-c)\,\Gamma(1-a)}{\Gamma(2-c)}\,y_2 + \exp\{\pm i\pi(1+a-c)\}\frac{\Gamma(b)\,\Gamma(1-a)}{\Gamma(1+b-a)}\,y_{10},$$

$$(1.8.1.15)$$

$$\exp\{\pm i\pi(c-b)\}\frac{\Gamma(c-b)\,\Gamma(1-a)}{\Gamma(1+c-a-b)}\,y_6$$

$$= \frac{\Gamma(b)\,\Gamma(c-b)}{\Gamma(c)}\,y_1 + \exp\{\pm i\pi(1-b)\}\frac{\Gamma(1-a)\,\Gamma(b)}{\Gamma(1+b-a)}\,y_{10}, \qquad (1.8.1.16)$$

$$\exp\{\pm i\pi(c-a)\}\frac{\Gamma(c-a)\,\Gamma(1-b)}{\Gamma(1+c-a-b)}\,y_6$$

$$= \frac{\Gamma(a)\,\Gamma(c-a)}{\Gamma(c)}\,y_1 + \exp\{\pm i\pi(1-a)\}\frac{\Gamma(1-b)\,\Gamma(a)}{\Gamma(1+a-b)}\,y_9, \qquad (1.8.1.17)$$

$$\exp\{\pm i\pi(1-b)\}\frac{\Gamma(1-b)\,\Gamma(c-a)}{\Gamma(1+c-a-b)}\,y_6$$

$$= \frac{\Gamma(1-b)\,\Gamma(1+b-c)}{\Gamma(2-c)}\,y_2 + \exp\{\pm i\pi(1-b)\}\frac{\Gamma(c-a)\,\Gamma(1+b-c)}{\Gamma(1+b-a)}\,y_{10},$$

$$(1.8.1.18)$$

$$\exp\{\pm i\pi(1-a)\}\frac{\Gamma(1-a)\,\Gamma(c-b)}{\Gamma(1+c-a-b)}\,y_6$$

$$=\frac{\Gamma(1-a)\,\Gamma(1+a-c)}{\Gamma(2-c)}\,y_2+\exp\{\pm i\pi(1-a)\}\frac{\Gamma(c-b)\,\Gamma(1+a-c)}{\Gamma(1+a-b)}\,y_9.$$

$$(1.8.1.19)$$

## 1.9 Numerical evaluation of the Gauss function

The Gauss function has one variable $z$ and three parameters $a$, $b$ and $c$. Since all of these may be complex numbers, the actual production of any tables of the function's numerical values would seem to serve little useful purpose, as complete tabulation with provision for interpolation, even over strictly limited ranges of real values, would be too bulky to be handled easily. However this fact does not debar us from considering the question 'How do we set about the evaluation of the function for any given short range of values?'

The use of an electronic computer enables us to produce the numerical value of the function at any desired point, provided that we can find a method of evaluation which leads to numerical answers which have real meaning, that is, answers in which the accumulated errors, due to the length of the calculation, have not totally overwhelmed the significant figures in the answers produced. We give a rather extreme example to illustrate this point. Suppose that our computer, using floating decimal point facilities, can only carry three significant figures and that at each step in the calculation a possible error $\epsilon$ is left which is less than or equal to half a unit in the third figure retained, that is, if all our figures lie in the range $0.1 \leqslant x \leqslant 0.9$, then $\epsilon \leqslant \pm 0.0005$.

Under these conditions, a simple sum of two terms might produce an error big enough to obscure the third significant figure; for example, we can only say that $(0.1 \pm 0.0005) + (0.1 \pm 0.0005)$ lies between $0.199$ and $0.201$. If several hundreds of terms like these are summed the possible accumulated error grows considerably, even when all the terms to be summed are of the same approximate size. If the terms alternate in sign the situation is aggravated, and the loss of significant figures usually becomes serious very quickly indeed. Thus we must make quite sure that the process we are using has only random errors in the last figure retained, and no error of predictable sign occurs such as a rounding off error always in the same direction.

When $a$, $b$, $c$ and $x$ are all positive, a straight-forward summation of the series

$$_2F_1[a, b; c; x]$$

will produce the numerical value of the Gauss function for all finite positive values of $a$, $b$ and $c$, and for $0 \leqslant x \leqslant \frac{1}{2}$. When $\frac{1}{2} < x < 1$, the number of terms to be summed becomes too large, and the accumulated errors begin to obscure the significant figures, as noted above. So, in this range of values of $x$, it is better to evaluate the Gauss function of $1-x$, that is, the solution $y_5$ of the Gauss equation. The complete solution of the equation is then of the form $Ay_1 + By_5$ as usual.

When $x = 1$, we have the Gauss theorem (1.7.6) to provide an explicit sum. For example, using five figure tables,

$$1/\{_2F_1[0{\cdot}1, 0{\cdot}2; 0{\cdot}4; 1]\} = \frac{\Gamma(0{\cdot}3)\,\Gamma(0{\cdot}2)}{\Gamma(0{\cdot}4)\,\Gamma(0{\cdot}1)},$$

$$= \frac{0{\cdot}2\,\Gamma(1{\cdot}2)\,0{\cdot}3\,\Gamma(1{\cdot}3)}{0{\cdot}4\,\Gamma(1{\cdot}4)\,0{\cdot}1\,\Gamma(1{\cdot}1)},$$

$$= \frac{3 \times 0{\cdot}91817 \times 0{\cdot}89747}{2 \times 0{\cdot}88726 \times 0{\cdot}95135},$$

$$= \frac{2{\cdot}47209}{1{\cdot}68818},$$

$$= 1{\cdot}46435.$$

When $x > 1$, we can use the solutions $y_9$ and $y_{13}$ and the analytic continuation formulae of (1.8.1.12–19).

When values of the function have been found over appropriate ranges of positive values of $a$, $b$ and $c$, the tabulation can be extended in the usual way, to a wide range of negative values of $a$, $b$ and $c$, by the use of the recurrence relations (1.4.1–15), provided that we avoid always any integer or zero values of $a$, $b$ and $c$ for which the functions used are not defined. For example if we know that

$$_2F_1[0{\cdot}1, 0{\cdot}2; 0{\cdot}3; 0{\cdot}2] = 1{\cdot}015,$$

and that $\qquad _2F_1[1{\cdot}1, 0{\cdot}2; 0{\cdot}3; 0{\cdot}2] = 1{\cdot}545,$

we can deduce from the recurrence relation (1·4·1) that

$$_2F_1[-0{\cdot}9, 0{\cdot}2; 0{\cdot}3; 0{\cdot}2] = 1{\cdot}024.$$

The fact that the function is always symmetrical in $a$ and $b$ should never be forgotten. Thus we can also say that

$$_2F_1[0{\cdot}2, -0{\cdot}9; 0{\cdot}3; 0{\cdot}2] = 1{\cdot}024.$$

# 2

## THE GENERALIZED GAUSS FUNCTION

### 2.1 Historical notes

The idea of extending the number of parameters in the Gauss function seems to have occurred for the first time, in the work of Clausen (1828). He introduced a series with three numerator parameters and two denominator parameters. Over the next hundred years the well-known set of special summation theorems associated with the names of Saalschutz (1890), Dixon (1903) and Dougall (1907) were developed.

These are all for series in which $A = B+1$, and $z = 1$. It can be shown that Dougall's theorem, giving the sum of a $_7F_6(1)$ series, is the most general possible theorem of this kind. The whole theory as it existed then was analysed exhaustively and brought to perfection by W. N. Bailey, in a long series of papers during the decades of 1920–50. Indeed at this time L. J. Rogers is reported to have said 'Nothing remains to be done in the field of hypergeometric series'. The whole theory of the general function $_AF_B(z)$ was still untouched. The first attempts at a general transformation theory were already being made by Whipple (1934, 1937), and the concept of the asymptotic expansions for the function were already implicit in the work of Barnes (1907a). This side of the theory was developed by MacRobert (1938, 1939) before 1939, and later by Sears (1951a, b, c), Slater (1952c, 1955b, c, e) and Meijer (1941–56).

### 2.1.1 Definitions. The series

$$1 + \frac{a_1 a_2 \dots a_A}{b_1 b_2 \dots b_B} \frac{z}{1!} + \frac{a_1(a_1+1) a_2(a_2+1) \dots a_A(a_A+1)}{b_1(b_1+1) b_2(b_2+1) \dots b_B(b_B+1)} \frac{z^2}{2!} + \dots$$

$$\equiv \sum_{n=0}^{\infty} \frac{(a_1)_n (a_2)_n \dots (a_A)_n}{(b_1)_n (b_2)_n \dots (b_B)_n} \frac{z^n}{n!} \quad (2.1.1.1)$$

is called the generalized Gauss function, or generalized hypergeometric function. It has $A$ numerator parameters $a_1, a_2, a_3, \dots, a_A$, $B$ denominator parameters $b_1, b_2, b_3, \dots, b_B$, and one variable $z$. Any of these quantities may be real or complex but the $b$ parameters must not be negative integers, as in that case the series is not defined. The sum of this series, when it exists, is denoted by the symbol

$$_AF_B[a_1, a_2, a_3, ..., a_A; b_1, b_2, b_3, ..., b_B; z]. \qquad (2.1.1.2)$$

If any of the $a$ parameters is a negative integer, the function reduces to a polynomial. This notation was due to Barnes (1907$b$).

These notations can be shortened still further to

$$\sum_{n=0}^{\infty} \frac{((a)_A)_n z^n}{((b)_B)_n n!} = {}_AF_B[(a); (b); z]. \qquad (2.1.1.3)$$

Similarly, for a product of several Gamma functions, we can write

$$\frac{\Gamma(a_1)\,\Gamma(a_2)\,\Gamma(a_3) \dots \Gamma(a_A)}{\Gamma(b_1)\,\Gamma(b_2)\,\Gamma(b_3) \dots \Gamma(b_B)} = \Gamma\begin{bmatrix} a_1, a_2, ..., a_A \\ b_1, b_2, ..., b_B \end{bmatrix}$$

$$= \Gamma[(a); (b)], \qquad (2.1.1.4)$$

where it is understood that there are always $A$ of the $a$ parameters and $B$ of the $b$ parameters, if this is not shown explicitly. A dash will be used to denote the omission of a zero factor in such a sequence of parameters. Thus, $(a)' - a_m$ will indicate the sequence

$$a_1 - a_m, a_2 - a_m, ..., a_{m-1} - a_m, a_{m+1} - a_m, a_{m+2} - a_m, ..., a_A - a_m.$$

Such contracted notations will be found to be absolutely vital to the proper understanding of the more advanced parts of the theory.

There are several alternative notations for the general hypergeometric function. Thus, to avoid the difficulty of restricting the $b$ parameters to values which are not negative integers, several authors use

$$_A\mathfrak{F}_B[a_1, a_2, ..., a_A; b_1, b_2, ..., b_B; z] = {}_AF_B[(a); (b); z]/\Gamma[b_1, b_2, ..., b_B].$$

$$(2.1.1.5)$$

This form of the function is defined numerically at those points where $_AF_B(z)$ is not defined. An alternative symbol † is

$$_A\Phi_B[a_1, a_2, ..., a_A; b_1, b_2, ..., b_B; z] = {}_AF_B[(a); (b); z]/\Gamma[b_1, b_2, ..., b_B].$$

$$(2.1.1.6)$$

This symbol will not be used in the present work since it would be confused with the very similar symbol used for the corresponding basic general hypergeometric series, to be considered in Chapter 3.

† See Erdélyi (1939$a$).

Several special symbols have been used to represent the asymptotic forms of the function. One of the most widely used of these is

$$E(A; a_1, a_2, ..., a_A; B; b_1, b_2, ..., b_B; -1/z)$$
$$= \Gamma[(a); (b)] {}_AF_B[(a); (b); z]. \quad (2.1.1.7)$$

This is MacRobert's $E$-function.[†] Much of the later general theory has been developed in terms of Meijer's $G$-function,[‡]

$$G^1_{A\ B+1}\left(-z\ \middle|\ \begin{matrix} 1-a_1, 1-a_2, ..., 1-a_A \\ 0, 1-b_1, 1-b_2, ..., 1-b_B \end{matrix}\right)$$
$$= \Gamma[(a); (b)] {}_AF_B[(a); (b); z]. \quad (2.1.1.8)$$

In the general series ${}_AF_B(z)$, if the sum of the numerator parameters exceeds the sum of the denominator parameters by one, that is if

$$b_1 + b_2 + ... + b_B = a_1 + a_2 + ... + a_A + 1, \quad (2.1.1.9)$$

the series is said to be Saalschutzian. If $A = B+1$, and

$$1 + a_1 = b_1 + a_2 = b_2 + a_3 = ... = b_B + a_{B+1}, \quad (2.1.1.10)$$

the series is said to be well-poised, and if all pairs but one of the pairs of parameters satisfy these relations, the series is said to be nearly-poised. Since the order of the parameters can always be interchanged in the series without altering it, the pair of unequal parameters can always be brought to occupy either the first or the last places in the sequence. Such series are then said to be nearly-poised series of the first or second kind respectively. These three terms, Saalschutzian, well-poised and nearly-poised are all due to Whipple (1925, 1926a).

**2.1.2 Differential equations.** The series ${}_AF_B[(a); (b); z]$ satisfies the differential equation

$$\left\{ z\frac{d}{dz}\left(z\frac{d}{dz}+b_1-1\right)\left(z\frac{d}{dz}+b_2-1\right)...\left(z\frac{d}{dz}+b_B-1\right) \right.$$
$$\left. -z\left(z\frac{d}{dz}+a_1\right)\left(z\frac{d}{dz}+a_2\right)...\left(z\frac{d}{dz}+a_A\right)\right\}y = 0. \quad (2.1.2.1)$$

The rank of this equation is the greater of $A$ and $B+1$. When $A \leqslant B$, this becomes Poole's equation[§]

$$\sum_{\nu=1}^{B} z^{\nu-1}(a_\nu z - b_\nu)\frac{d^\nu y}{dz^\nu} + a_0 y + z^B\frac{d^{B+1}y}{dz^{B+1}} = 0. \quad (2.1.2.2)$$

---

† MacRobert (1938).　　　　　　　‡ Meijer (1934), p. 11; (1946a), p. 229.
§ Poole (1935).

This equation has a regular singularity at $z = 0$, and an irregular singularity at $z = \infty$.

When $A = B + 1$, the equation (2.1.2.1) becomes

$$\sum_{\nu=1}^{B} z^{\nu-1}(a_\nu z - b_\nu)\frac{d^\nu y}{dz^\nu} + a_0 y + z^B(1-z)\frac{d^{B+1}y}{dz^{B+1}} = 0. \quad (2.1.2.3)$$

This is a Fuchsian equation with regular singularities at $z = 0$, $z = 1$, and $z = \infty$. We shall see later that it has $B + 1$ linearly independent solutions for $|z| < 1$, that is the function

$$_{B+1}F_B[(a); (b); z]$$

and the $B$ similar functions

$$z^{1-b_\nu}\;_{B+1}F_B\left[\begin{matrix}1+a_1-b_\nu, 1+a_2-b_\nu, \ldots, 1+a_{B+1}-b_\nu;\\ 1+b_1-b_\nu, 1+b_2-b_\nu, \ldots * \ldots, 1+b_B-b_\nu, 2-b_\nu;\end{matrix}\; z\right]$$

$$\equiv z^{1-b_\nu}\;_{B+1}F_B[1+(a)-b_\nu; 1+(b)'-b_\nu, 2-b_\nu; z] \quad (2.1.2.4)$$

for $\nu = 1, 2, 3, \ldots, B$. Here the $*$ indicates in an alternative way, that the expression $1+b_\nu-b_\nu$ is omitted from the sequence in the denominator. In fact, it has become the factorial term in the series. Again it is assumed that no two of the $b$-parameters differ by an integer, or some of the solutions would become infinite.

If $\Delta \equiv z\,d/dz$, in terms of such operators, the generalized hypergeometric equation (2.1.2.1) can be written

$$\Delta(\Delta+b_1-1)(\Delta+b_2-1)\ldots(\Delta+b_B-1)y$$

$$= z(\Delta+a_1)(\Delta+a_2)\ldots(\Delta+a_A)y. \quad (2.1.2.5)$$

Let us suppose that $\quad y = z^\rho \sum_{n=0}^{\infty} u_n(\rho)z^n$,

is any solution of the equation (2.1.2.1), and let us substitute this series and its derivatives with respect to $z$ into the above equation (2.1.2.5). Then we shall find that the 'indicial equation' (that is the sum of the coefficients of $z^\rho$ equated to zero) is

$$\rho(\rho+b_1-1)(\rho+b_2-1)\ldots(\rho+b_B-1) = 0. \quad (2.1.2.6)$$

Hence solutions are given by

$$\rho = 0, \quad 1-b_1, \quad 1-b_2, \ldots, 1-b_B.$$

The other coefficients must be connected by the relation

$$(\rho+n)(\rho+n+b_1-1)\ldots(\rho+n+b_B-1)u_n(\rho)$$

$$= (\rho+n+a_1-1)(\rho+n+a_2-1)\ldots(\rho+n+a_A-1)u_{n-1}(\rho). \quad (2.1.2.7)$$

Hence $\qquad u_n(\rho) = (\rho + (a) - 1)_n / \{(\rho + (b) - 1)_n \, n!\},$ $\qquad$ (2.1.2.8)

and the complete set of solutions is given in (2.1.2.4) above.

The radius of convergence of the resulting series is 0, 1 or $\infty$, according as $A - 1 > B$, $A - 1 = B$ or $A - 1 < B$. If $A - 1 > B$, the formal solution in series breaks down, but if $A - 1 \leqslant B$, there are $B + 1$ solutions obtainable, provided that $u_n(0)$ and $u_n(1 - B_\nu)$ do not become infinite for any values of $n$ or $\nu$. When $A - 1 < B$, each solution is a power series in $z$, which is convergent everywhere in the $z$-plane, multiplied by the factor $z^0$ or $z^{1-b_\nu}$. When $A - 1 = B$, each solution is a power series multiplied by $z^0$ or $z^{1-b_\nu}$ as before, but in this case, the solutions are convergent for $|z| < 1$, only. However, they can all be extended outside this circle of convergence, by analytic continuation, and the use of integrals of the Borel type.

### 2.1.3. Integration of the generalized function.
The integration of the generalized hypergeometric function $_AF_B(z)$ with respect to $z$ is quite straightforward. We have

$$\int_0^Z {}_AF_B[(a); (b); z]\,\mathrm{d}z$$

$$= \frac{(b_1 - 1)(b_2 - 1) \dots (b_B - 1)}{(a_1 - 1)(a_2 - 1) \dots (a_A - 1)} \{ {}_AF_B[(a) - 1; (b) - 1; z] - 1\},$$

$$\qquad (2.1.3.1)$$

and in general

$$\int_0^Z \int_0^Z \dots \int_0^Z {}_AF_B[(a); (b); z]\,\mathrm{d}^n z$$

$$= (-1)^{(A+B)n} \frac{(1 - (b))_n}{(1 - (a))_n} {}_AF_B[(a) - n; (b) - n; Z]$$

$$- \sum_{r=1}^n \frac{(1 - (b))_r (-1)^{(A+B)r} Z^{n-r}}{(1 - (a))_r (n - r)!}. \qquad (2.1.3.2)$$

Integration with respect to a parameter is not so simple, since even

$$\int_0^A (a)_n \,\mathrm{d}a$$

involves all the symmetric functions up to order $n$ and the formal result for an integral of the type

$$\int_0^S {}_AF_B[(a) + s; (b); z]\,\mathrm{d}s$$

would be of the same degree of complexity as the coefficient of $z^n$ in the expansion

$$\prod_{s=1}^{\infty} (1 - z^s)^m = \sum_{n=0}^{\infty} A_n z^n.$$

## 2.2 The convergence of the general series

The series ${}_AF_B(z)$ converges for all values of $z$, real or complex when $A \leqslant B$, for, if $u_n z^n$ is the $n$'th term of our series,

$$\left| \frac{u_{n+1}}{u_n} \right| \leqslant \frac{|a_1 + n| \, |a_2 + n| \ldots |a_A + n| \, |z|}{|b_1 + n| \, |b_2 + n| \ldots |b_B + n| \, (1 + n)}$$

$$\leqslant \frac{|z| \, n^{A-B-1}(1 + |a_1|/n)(1 + |a_2|/n) \ldots (1 + |a_A|/n)}{(1 + 1/n)(1 + |b_1|/n)(1 + |b_2|/n) \ldots (1 + |b_B|/n)}, \quad (2.2.1)$$

and this expression tends to zero as $n \to \infty$, provided that $A \leqslant B$. Also, we can see that if $A = B + 1$, the series is convergent when $|z| < 1$. It also converges when $z = 1$, if

$$\mathrm{Rl} \left( \sum_{\nu=1}^{B} b_\nu - \sum_{\nu=1}^{A} a_\nu \right) > 0,$$

and when $z = -1$, if $\quad \mathrm{Rl} \left( \sum_{\nu=1}^{B} b_\nu - \sum_{\nu=1}^{A} a_\nu \right) > -1.$

If $A > B + 1$, the series never converges, except when $z = 0$, and the function is only defined when the series terminates, that is when one or more of the $a$ parameters is zero or a negative integer.

In particular, the series

$${}_3F_2[a, b, c; d, e; z]$$

is convergent if $|z| < 1$,

or if $\qquad z = 1 \quad$ and $\quad \mathrm{Rl}\,(d + e - a - b - c) > 0,$

or if $\qquad z = -1 \quad$ and $\quad \mathrm{Rl}\,(d + e - a - b - c) > -1.$

## 2.2.1 Contiguous hypergeometric functions. Any two hypergeometric functions

$${}_AF_B[(a); (b); z] \quad \text{and} \quad {}_AF_B[(a'); (b'); z].$$

are said to be contiguous, when all their parameters are equal except one pair, and this pair of parameters differs only by unity, for example,

$${}_0F_2[; b_1, b_2 + 1; z] \quad \text{is contiguous to} \quad {}_0F_2[; b_1, b_2; z].$$

Similarly, any two hypergeometric functions are said to be associated, when their parameters differ by integers only. Thus

$$_2F_3[a_1, a_2; b_1, b_2, b_3; z] \quad \text{and} \quad _2F_3[a_1 + m, a_2 + n; b_1 - p, b_2 - q, b_3 - r; z]$$

are associated functions, for all integer values of $m, n, p, q$ and $r$. When $A = B + 1$, a linear relationship can always be found between any $B + 2$ contiguous functions. Similar linear relationships also exist between any generalized Gauss function and its associated functions.

### 2.2.2 Special cases of generalized hypergeometric functions.

All the functions of mathematical physics can be expressed in terms of generalized Gauss functions. They form a table of increasing complexity as the number of parameters increases. Thus we have the exponential, trigonometric and hyperbolic functions which are all based on series of the type

$$_0F_0[; ; z] = 1 + z + \frac{z^2}{2!} + \frac{z^3}{3!} + \ldots + \frac{z^n}{n!} + \ldots = e^z, \quad (2.2.2.1)$$

in which there is one variable and no parameters. If we introduce one parameter, we have two general types, the Binomial functions based on

$$_1F_0[a; ; z] = (1 - z)^{-a}, \quad (2.2.2.2)$$

$$= 1 + az + \frac{a(a+1)z^2}{2!} + \ldots + \frac{(a)_n z^n}{n!} + \ldots,$$

and the Bessel functions based on

$$_0F_1[; b; z] = \Gamma(b)(iz)^{b-1} J_{b-1}(2iz^{\frac{1}{2}}). \quad (2.2.2.3)$$

With two parameters, we have the confluent hypergeometric and Whittaker functions based on Kummer's functions

$$_1F_1[a; b; z] = M(a, b, z), \quad (2.2.2.4)$$

their asymptotic representations based on

$$_2F_0[b - a, 1 - a; ; 1/z] \sim z^{b-a} e^{-z} \Gamma(a) M(a, b, z)/\Gamma(b), \quad (2.2.2.5)$$

and the function $\qquad _0F_2[; a, b; z]$

which is only defined if $a$ and $b$ are not negative integers.

With three parameters, we have the Gauss function

$$_2F_1[a, b; c; z]$$

and the three functions $\qquad _3F_0[a, b, c; ; z],$

which is either a divergent series or a polynomial,

$$_1F_2[a;\, b,c;\, z]$$

which is the product of two confluent hypergeometric functions†, and

$$_0F_3[\ ;\, a,b,c;\, z].$$

Other special cases arising out of this group, with three parameters, are the products of two Bessel functions, like

$$J_\nu(iz^{\frac12})\,J_{-\nu}(iz^{\frac12}) = \frac{\sin\pi\nu}{\pi\nu}\,_1F_2[\tfrac12;\,\nu+1,\,-\nu+1;\,z], \qquad (2.2.2.6)$$

$$J_\nu^2(iz^{\frac12}) = \frac{(-z)^\nu}{\{\Gamma(\nu+1)\}^2}\,2^{-2\nu}\,_1F_2[\nu+\tfrac12;\,\nu+1,2\nu+1;\,z] \qquad (2.2.2.7)$$

and $\qquad I_\nu^2(iz^{\frac12}) = \dfrac{(-z)^\nu}{\{\Gamma(\nu+1)\}^2}\,2^{-2\nu}\,_1F_2[\nu+\tfrac12;\,\nu+1,2\nu+1;\,-z],$

$$(2.2.2.8)$$

Lommel's function,‡

$$s_{\mu,\nu}(2iz) = \frac{(2iz)^{\mu+1}}{(\mu-\nu+1)(\mu+\nu+1)}\,_1F_2[1;\,\tfrac12\mu-\tfrac12\nu+\tfrac32,\tfrac12\mu+\tfrac12\nu+\tfrac32;\,z]$$

$$(2.2.2.9)$$

and Struve's functions

$$\mathbf{H}_\nu(2iz) = \frac{(iz)^{\nu+1}}{\Gamma(\tfrac32)\,\Gamma(\nu+\tfrac32)}\,_1F_2[1;\,\tfrac32,\nu+\tfrac32;\,z] \qquad (2.2.2.10)$$

and $\qquad \mathbf{L}_\nu(2iz) = \dfrac{(iz)^{\nu+1}}{\Gamma(\tfrac32)\,\Gamma(\nu+\tfrac32)}\,_1F_2[1;\,\tfrac32,\nu+\tfrac32;\,-z]. \qquad (2.2.2.11)$

Finally, as an example of a series with five parameters, we quote the general multiplication theorem for Bessel functions,

$$J_\mu(iz)\,J_\nu(iz) = \frac{(iz)^{\frac12\mu+\frac12\nu}}{2^{\mu+\nu}\,\Gamma(1+\mu)\,\Gamma(1+\nu)}\,_2F_3\!\left[\begin{matrix}\tfrac12\mu+\tfrac12\nu+\tfrac12,\tfrac12\mu+\tfrac12\nu+1;\\ 1+\mu,1+\nu,1+\mu+\nu;\end{matrix}\ z\right].$$

$$(2.2.2.12)$$

**2.2.3 Reversal of the series.** Since it is possible to reverse the order of any finite series, for the terminating general hypergeometric series, we find that if $(a)_r$ occurs in the $r$th term of the original series, then $(a)_{m-r}$ occurs in the corresponding term $r$ places from the end. Hence we can reverse the order of the terms from the end to the beginning of the series simply by writing

$$(a)_{m-r} = \frac{(-1)^r\,(a)_m}{(1-a-m)_r} \qquad (2.2.3.1)$$

† Whipple (1927 b).　　　　‡ Watson (1948 a), p. 346.

in place of $(a)_r$. In particular $r!$ becomes $(-1)^r (1)_r$ and the factorial $(-m)_r$ becomes $(1)_r$. In general, we can write

$$_{A+1}F_B[(a), -m; (b); z]$$

$$= \frac{((a))_m (-z)^m}{((b))_m} {}_{B+1}F_A[1-(b), 1-m; 1-(a); (-1)^{A+B}/z].   (2.2.3.2)$$

Any finite hypergeometric series can be reversed in this way. In particular, a well-poised series when so reversed remains well-poised, a nearly-poised series of the first kind becomes a nearly-poised series of the second kind and a nearly-poised series of the second kind becomes a nearly-poised series of the first kind.

It is also possible to split a finite series into two parts, and reverse the order of one of the parts only, thus

$$_{A+1}F_B[(a), -2m; (b); z] = {}_{A+1}F_B[(a), -2m; (b); z]_m$$

$$+(-1)^m z^{2m} \frac{((a))_{2m}}{((b))_{2m}} {}_{B+1}F_A\left[1-(b)-m, 1-2m; 1-(a)-m; \frac{(-1)^{A+B}}{z}\right]_m.$$

$$(2.2.3.3)$$

Here the suffix $m$ indicates that only the first $m$ terms of the $F$ series are to be included in the expansion.

## 2.3 Special summation theorems

In chapter one, we gave detailed proofs of the two fundamental summation theorems of Gauss and Vandermonde. In this section we shall give proofs of four further special summation theorems, those which carry the names of Saalschutz, Kummer, Dixon and Dougall. Then we shall discuss the underlying general transformation theorem, known as Bailey's theorem, from which there can be deduced proofs of these four special summation theorems, and of many other results.

**2.3.1 Saalschutz's theorem.** We have already proved Euler's identity (1.3.15). This can be rewritten as

$$_2F_1[c-a, c-b; c; z] = (1-z)^{a+b-c} {}_2F_1[a, b; c; z].   (2.3.1.1)$$

Now the coefficient of $z^n$ on the left-hand side of this identity is

$$\frac{(c-a)_n (c-b)_n}{(c)_n n!},$$

and this expression must be equal to the coefficient of $z^n$ on the right-hand side, that is to the coefficient of $z^n$ in the double series

$$\sum_{r=0}^{\infty} \frac{(a)_r (b)_r z^r}{(c)_r r!} \sum_{s=0}^{\infty} \frac{(c-a-b)_s z^s}{s!}$$

which is formed by the expansion of $(1-z)^{a+b-c}$ in powers of $z$.

If we put $s = n - r$, we find that this coefficient is in fact

$$\sum_{r=0}^{n} \frac{(a)_r (b)_r (c-a-b)_{n-r}}{(c)_r r! (n-r)!}$$

$$= \frac{(c-a-b)_n}{n!} \sum_{r=0}^{n} \frac{(a)_r (b)_r (-n)_r}{r!(c)_r (1-c+a+b-n)_r}$$

$$= \frac{(c-a-b)_n}{n!} {}_3F_2[a, b, -n; c, 1+a+b-c-n; 1]. \quad (2.3.1.2)$$

Hence

$$\frac{(c-a)_n (c-b)_n}{(c)_n (c-a-b)_n} = {}_3F_2[a, b, -n; c, 1+a+b-c-n; 1]. \quad (2.3.1.3)$$

This is Saalschutz's theorem.† It gives the sum of the series

$$ {}_3F_2[a, b, c; d, e; 1] = \Gamma \begin{bmatrix} d, 1+a-e, 1+b-e, 1+c-e \\ 1-e, d-a, d-b, d-c \end{bmatrix}, \quad (2.3.1.4)$$

provided that one of the numerator parameters is a negative integer $-n$, and that

$$d+e = 1+a+b+c, \quad (2.3.1.5)$$

that is to say, that the series is Saalschutzian. When $n \to \infty$, the result (2.3.1.3) reduces to Gauss's theorem (1.1.5), since

$$(-n)_r/(1+a+b-c-n)_r \to 1 \quad \text{as} \quad n \to \infty.$$

The difference between the sum of the denominator parameters and the sum of the numerator parameters is called the parametric excess of the series, and it is usually denoted by $s$. In this case,

$$s = d+e-a-b-c,$$

and in a Saalschutzian series $s = 1$, always.

## 2.3.2 Kummer's theorem.

First we shall prove Kummer's quadratic transform

$$ {}_2F_1[a, b; 1+a-b; z]$$

$$= (1-z)^{-a} {}_2F_1[\tfrac{1}{2}a, \tfrac{1}{2}+\tfrac{1}{2}a-b; 1+a-b; -4z/(1-z)^2] \quad (2.3.2.1)$$

† Saalschutz (1890).

where we must have $\qquad |4z| \leqslant |1-z|^2$

in order that both the series are convergent. Now, within the circle $|z| < 3 - 2\sqrt{2}$, both series can be expanded in increasing powers of $z$. The coefficient of $z^N$ on the right is

$$\sum_{n=0}^{N} \frac{(\tfrac{1}{2}a)_n (\tfrac{1}{2} + \tfrac{1}{2}a - b)_n (-4)^n (a+2n)_{N-n}}{(1+a-b)_n \, n! \, (N-n)!}$$

$$= \frac{(a)_N}{N!} \sum_{n=0}^{N} \frac{(\tfrac{1}{2} + \tfrac{1}{2}a - b)_n (a+N)_n (-N)_n}{n! \, (1+a-b)_n (\tfrac{1}{2} + \tfrac{1}{2}a)_n}, \qquad (2.3.2.2)$$

since $\qquad \dfrac{1}{(N-n)!} = (-1)^n (-N)_n \qquad (2.3.2.3)$

and $\qquad (-4)^n (a+2n)_{N-n} = \dfrac{(a)_{N+n} \, 4^n}{(a)_{2n}}$

$$= \frac{(a)_N (a+N)_n \, 4^n}{4^n (\tfrac{1}{2}a)_n (\tfrac{1}{2} + \tfrac{1}{2}a)_n}. \qquad (2.3.2.4)$$

The series on the right of (2.3.2.2) is summable by Saalschutz's theorem, and its sum is

$$\frac{(\tfrac{1}{2} + \tfrac{1}{2}a)_N (1 - b - N)_N}{(1+a-b)_N (\tfrac{1}{2} - \tfrac{1}{2}a - N)_N} = \frac{(b)_N}{(1+a-b)_N}, \qquad (2.3.2.5)$$

so that the coefficient of $z^N$ on the right-hand side of (2.3.2.1) is in fact

$$\frac{(a)_N (b)_N}{(1+a-b)_N \, N!}$$

which is the coefficient of $z^N$ on the left-hand side of (2.3.2.1) above. By analytic continuation, this result is true everywhere within that loop of the curve

$$|4z| = |1-z|^2$$

which surrounds the origin.

Now let $z \to -1$. This point lies on the above curve, so that, by Abel's theorem on continuity†, we have

$$_2F_1[a, b; \, 1+a-b; \, -1] = 2^{-a} \, _2F_1[\tfrac{1}{2}a, \tfrac{1}{2} + \tfrac{1}{2}a - b; \, 1+a-b; \, 1]. \qquad (2.3.2.6)$$

We can sum the series on the right-hand side by Gauss's theorem, (1.1.5), and this gives us

$$2^{-a} \Gamma \begin{bmatrix} 1+a-b, & \tfrac{1}{2} \\ 1 + \tfrac{1}{2}a - b, & \tfrac{1}{2} + \tfrac{1}{2}a \end{bmatrix} = \Gamma \begin{bmatrix} 1+a-b, & 1 + \tfrac{1}{2}a \\ 1 + \tfrac{1}{2}a - b, & 1 + a \end{bmatrix}, \qquad (2.3.2.7)$$

since $\qquad \Gamma(\tfrac{1}{2} + \tfrac{1}{2}a) \, \Gamma(\tfrac{1}{2}a) = 2^{1-a} \pi^{\frac{1}{2}} \Gamma(a). \qquad (2.3.2.8)$

---

† Whittaker & Watson (1947), § 3.71.

Both the hypergeometric series are convergent if

$$Rl\,(1-2b) > -1,$$

that is if $Rl\,(b) < 1$.

Hence finally we get Kummer's theorem[†]

$$_2F_1[a,b;\ 1+a-b;\ -1] = \Gamma\begin{bmatrix}1+a-b,\ 1+\tfrac12 a\\ 1+\tfrac12 a-b,\ 1+a\end{bmatrix}, \quad (2.3.2.9)$$

if $Rl\,(b) < 1$. The series which is summed by this theorem, is known as Kummer's series. It is the simplest type of well-poised series.

### 2.3.3 Dixon's theorem.

This theorem was first proved in 1903.[‡] It gives the sum, in terms of Gamma functions, of the series

$$_3F_2[a,b,c;\ 1+a-b,1+a-c;\ 1]$$

provided that this series is convergent, that is provided that

$$Rl\,(a-2b-2c) > -3.$$

A simple direct proof, due to Watson,[§] is based on interchanges in the order of summation of a double series, and makes use of the summation theorems of both Gauss and Kummer. That proof was given in detail by Bailey[‖] and will not be repeated here. The following alternative proof is due originally to Bailey.[¶] We shall show first that

$$ab(1-z)\,_3F_2[a+1,b+1,c;\ 1+a-b,2+a-c;\ z]$$
$$+ (a-c+1)\,(a-2b-2c+2)\,_3F_2[a,b,c;\ 1+a-b,1+a-c;\ z]$$
$$= (a-2c+2)\,(a-b-c+1)\,_3F_2[a,b,c-1;\ 1+a-b,2+a-c;\ z].$$
$$(2.3.3.1)$$

This is a relation between three Saalschutzian $_3F_2(z)$ series. It can be proved very simply, by comparing the coefficients of $z^n$ on both sides of the equation, and checking that they are in fact equal. Next let $z \to 1$, in the above relation. For the series to be convergent now, we must have $Rl\,(a-2b-2c) > 0$. Then we find that

$$F \equiv {}_3F_2[a,b,c;\ 1+a-b,1+a-c;\ 1]$$

$$= \frac{(1+\tfrac12 a-c)\,(1+a-b-c)}{(1+a-c)\,(1+\tfrac12 a-b-c)}\,_3F_2[a,b,c-1;\ 1+a-b,2+a-c;\ 1].$$
$$(2.3.3.2)$$

---

[†] Kummer (1836), p. 53.          [‡] Dixon (1903).
[§] Watson (1924a).                [‖] Bailey (1935), § 3.1.
[¶] Bailey (1937a).

Now let us write $c-1$ for $c$, and repeat this process $n$ times. Then we have

$$F = \frac{(1+\frac{1}{2}a-c)_n\,(1+a-b-c)_n}{(1+a-c)_n\,(1+\frac{1}{2}a-b-c)_n}\,{}_3F_2[a,b,c-n;\,1+a-b,1+a-c+n;\,1].$$

$$(2.3.3.3)$$

When $n \to \infty$,    $\dfrac{(c-n)_r}{(1+a-c+n)_r} \to (-1)^r,$

so that, if $\mathrm{Rl}\,(b) < 1$,

$$F = \Gamma\begin{bmatrix}1+a-c,\ 1+\frac{1}{2}a-b-c\\1+\frac{1}{2}a-c,\,1+a-b-c\end{bmatrix}{}_2F_1[a,b;\,1+a-b;\,-1].$$

$$(2.3.3.4)$$

This ${}_2F_1(-1)$ series can be summed by Kummer's theorem $(2.3.2.9)$ so that finally we have

$${}_3F_2[a,b,c;\,1+a-b,1+a-c;\,1]$$

$$= \Gamma\begin{bmatrix}1+\frac{1}{2}a,1+\frac{1}{2}a-b-c,1+a-b,\ 1+a-c\\1+a,\ 1+a-b-c,\ 1+\frac{1}{2}a-b,1+\frac{1}{2}a-c\end{bmatrix}.\quad (2.3.3.5)$$

By analytic continuation, we can dispense with the condition $\mathrm{Rl}\,(b) < 1$, so that the result holds under the single condition for convergence that $\mathrm{Rl}\,(\frac{1}{2}a-b-c) > -1$.

This result is Dixon's theorem.† It gives the sum of a well-poised ${}_3F_2(1)$ series, and it includes as a special case the sum of the cubes of the binomial coefficients.

In particular, if $c = -n$, the result reduces to

$${}_3F_2[a,b,-n;\,1+a-b,1+a+n;\,1] = \frac{(1+a)_n\,(1+\frac{1}{2}a-b)_n}{(1+\frac{1}{2}a)_n\,(1+a-b)_n}.$$

$$(2.3.3.6)$$

Finally, when $c \to -\infty$, Dixon's theorem reduces to Kummer's theorem.

A generalization of Dixon's theorem is

$${}_3F_2[a,b,c;\,e,f;\,1]$$

$$= \Gamma\begin{bmatrix}e,f,s\\a,b+s,c+s\end{bmatrix}{}_3F_2[e-a,f-a,s;\,s+b,s+c;\,1],\quad (2.3.3.7)$$

where $s \equiv e+f-a-b-c$, and we must have $\mathrm{Rl}\,(s) > 0$, and $\mathrm{Rl}\,(a) > 0$, in order that the two series are convergent. These two series have the

† Dixon (1903).

property that if one of them is well-poised then so is the other one. Again we shall make use of Gauss's theorem, and then the proof follows the lines of Watson's proof of Dixon's theorem, referred to above. Let

$$F \equiv \Gamma \begin{bmatrix} a,b,c \\ e,f \end{bmatrix} {}_3F_2[a,b,c;\ e,f;\ 1], \qquad (2.3.3.8)$$

$$= \sum_{n=0}^{\infty} \Gamma \begin{bmatrix} a+n,b+n,c+n \\ e+n,f+n,1+n \end{bmatrix}$$

$$= \sum_{n=0}^{\infty} \Gamma \begin{bmatrix} b+n,c+n \\ e+f-a+n,1+n \end{bmatrix} {}_2F_1[e-a,f-a;\ e+f-a+n;\ 1],$$

by Gauss's theorem. This ${}_2F_1(1)$ series is convergent when the original ${}_3F_2(1)$ series is convergent. Hence we have

$$F = \sum_{n=0}^{\infty} \sum_{m=0}^{\infty} \Gamma \begin{bmatrix} b+n,c+n,e-a+m,f-a+m \\ e-a,f-a,e+f-a+n+m,m+1,n+1 \end{bmatrix}.$$

$$(2.3.3.9)$$

For large enough values of $m$ and $n$, all these factors have ultimately the same sign, so that the double series is absolutely convergent, the order of summation can be interchanged and Gauss's theorem can be applied again.

This time we find that

$$F = \sum_{m=0}^{\infty} \Gamma \begin{bmatrix} b,c,e-a+m,f-a+m \\ e+f-a+m,e-a,f-a,1+m \end{bmatrix} {}_2F_1[b,c;\ e+f-a+m;\ 1]$$

$$= \Gamma \begin{bmatrix} b,c \\ e-a,f-a \end{bmatrix} \sum_{m=0}^{\infty} \Gamma \begin{bmatrix} e-a+m,f-a+m,e+f-a+m, \\ e+f-a+m,e+f-a-b+m, \\ \qquad\qquad e+f-a-b-c+m \\ \qquad\qquad e+f-a-c+m,1+m \end{bmatrix},$$

$$(2.3.3.10)$$

so that

$$F = \Gamma \begin{bmatrix} e+f-a-b-c,f-a,e-a,b,c \\ e+f-a-b,e+f-a-c,e-a,f-a \end{bmatrix}$$

$$\times {}_3F_2 \begin{bmatrix} e-a,f-a,e+f-a-b-c;\ \\ e+f-a-b,e+f-a-c; \end{bmatrix} 1 \Bigg], \qquad (2.3.3.11)$$

which gives the required result.

From this we can deduce Watson's summation theorem. We have, putting $e = \frac{1}{2}(a+b+1)$ and $f = 2c$ in (2.3.3.7),

$$_3F_2[a, b, c; \tfrac{1}{2}(a+b+1), 2c; 1]$$

$$= \Gamma \begin{bmatrix} \frac{1}{2}a + \frac{1}{2}b + \frac{1}{2}, \, 2c, \, \frac{1}{2} - \frac{1}{2}a - \frac{1}{2}b + c \\ a, \frac{1}{2} - \frac{1}{2}a + \frac{1}{2}b + c, \frac{1}{2} + \frac{1}{2}a - \frac{1}{2}b + c \end{bmatrix}$$

$$\times {}_3F_2 \begin{bmatrix} 2c - a, \frac{1}{2} + \frac{1}{2}b - \frac{1}{2}a, \frac{1}{2} - \frac{1}{2}a - \frac{1}{2}b + c; \\ \frac{1}{2} - \frac{1}{2}a - \frac{1}{2}b + 2c, \frac{1}{2} - \frac{1}{2}a + \frac{1}{2}b + c; \end{bmatrix} 1 \right]. \quad (2.3.3.12)$$

This last series is well-poised so that it can be summed by Dixon's theorem (2.3.3.5) above, to give us Watson's theorem,

$$_3F_2[a, b, c; \tfrac{1}{2} + \tfrac{1}{2}a + \tfrac{1}{2}b, 2c; 1]$$

$$= \Gamma \begin{bmatrix} \frac{1}{2} + c - \frac{1}{2}a - \frac{1}{2}b, \frac{1}{2} + \frac{1}{2}a + \frac{1}{2}b, 2c, \frac{1}{2}a, \\ \frac{1}{2} - \frac{1}{2}a + \frac{1}{2}b + c, 1 + c - \frac{1}{2}a, \frac{1}{2} + 2c - \frac{1}{2}a - \frac{1}{2}b \\ \frac{1}{2} - \frac{1}{2}a + \frac{1}{2}b + c, c, a, \frac{1}{2} + \frac{1}{2}a - \frac{1}{2}b + c, \\ \frac{1}{2} - \frac{1}{2}b + c, 1 + 2c - a, \frac{1}{2}b - \frac{1}{2} \end{bmatrix}$$

$$= \Gamma \begin{bmatrix} \frac{1}{2}, \frac{1}{2} + c, \frac{1}{2} + \frac{1}{2}a + \frac{1}{2}b, \frac{1}{2} - \frac{1}{2}a - \frac{1}{2}b + c \\ \frac{1}{2} + \frac{1}{2}a, \frac{1}{2} + \frac{1}{2}b, \frac{1}{2} - \frac{1}{2}a + c, \frac{1}{2} - \frac{1}{2}b + c \end{bmatrix}, \quad (2.3.3.13)$$

provided that the series is convergent, that is that

$$\mathrm{Rl}\left(\tfrac{1}{2} - \tfrac{1}{2}a - \tfrac{1}{2}b + c\right) > 0.$$

When $c \to \infty$, this result reduces to Gauss's second summation theorem (1.7.1.9). This result was first proved by Watson[†] for terminating series. The more general result for non-terminating series is actually due to Whipple.[‡]

In a similar way, we can deduce Whipple's summation theorem, from (2.3.3.11), for the second series in (2.3.3.11) is summable by Watson's theorem if

$$c = s = e + f - a - b - c, \quad 1 + 2c - a - b = 2c,$$

and

$$1 + c - a = \tfrac{1}{2}(e - a) + \tfrac{1}{2}(f - a) + \tfrac{1}{2},$$

that is when $a + b = 1$, and $e + f = 1 + 2c$. Under these conditions we find that

$$_3F_2[a, 1 - a, c; e, 1 + 2c - e; 1]$$

$$= 2^{1-2c}\pi\Gamma \begin{bmatrix} e, 1 + 2c - e \\ \frac{1}{2}e + \frac{1}{2}a, \frac{1}{2} + \frac{1}{2}e - \frac{1}{2}a, \frac{1}{2} + c - \frac{1}{2}e + \frac{1}{2}a, 1 + c - \frac{1}{2}e - \frac{1}{2}a \end{bmatrix}$$

$$(2.3.3.14)$$

provided that $\mathrm{Rl}\,(e + f - a - b - c) > 0$, that is provided that $\mathrm{Rl}\,(c) > 0$.

---

† Watson (1925).        ‡ Whipple (1925), p. 113.

When $c \to \infty$, this result reduces to Bailey's summation theorem (1.7.1.8).

### 2.3.4 Dougall's theorem.

The last result in this group of special summation theorems, is Dougall's theorem†. This theorem states that

$$
{}_7F_6 \left[ \begin{array}{cccccc} a, 1+\tfrac{1}{2}a, & b, & c, & d, & e, & f; \\ & \tfrac{1}{2}a, 1+a-b, 1+a-c, 1+a-d, 1+a-e, 1+a-f; \end{array} 1 \right]
$$

$$
= \Gamma \left[ \begin{array}{l} 1+a-b, 1+a-c, 1+a-d, 1+a-f, 1+a-b-c-d, \\ \quad 1+a-b-c-f, 1+a-b-d-f, 1+a-c-d-f, \\ 1+a, 1+a-b-c, 1+a-b-d, 1+a-c-d, \\ \quad 1+a-b-f, 1+a-c-f, 1+a-d-f, 1+a-b-c-d-f \end{array} \right],
$$

$$
\tag{2.3.4.1}
$$

provided that the series terminates and that

$$
1 + 2a = b + c + d + e + f. \tag{2.3.4.2}
$$

This theorem gives the sum of a well-poised $_7F_6(1)$ series in which the sum of the denominator parameters exceeds that of the numerator parameters by two. The most elementary proof, on the lines of Dougall's original proof, is by induction.

The result is obviously true when $f = 0$. Suppose that it is true when $f = -1, -2, \ldots, 1-m$. The Gamma functions on the right are symmetrical in $b, c, d$ and $f$. Hence, by this symmetry, the result is true when $c = 0, -1, -2, \ldots, 1-m$, and $f$ has any value, and also when $d = 0, -1, -2, \ldots, 1-m$, and $f$ has any value. But

$$
d = 1 + 2a - b - c - e - f.
$$

Hence the result is also true when $c$ has one of the $2m$ values

$$
0, -1, -2, \ldots, 1-m, 1+2a-b-d-e, 1+2a-b-d-e+1, \ldots,
$$
$$
1 + 2a - b - d - e + m - 1,
$$

and $f$ has any value. In particular, it is true when $f = -m$. In this case, (2.3.4.1) can be rewritten as

$$
(1+a-b)_m (1+a-c)_m (1+a-d)_m (1+a-b-c-d)_m \, {}_7F_6(1)
$$
$$
= (1+a)_m (1+a-b-c)_m (1+a-b-d)_m (1+a-c-d)_m. \tag{2.3.4.3}
$$

This equation expresses an equality between two polynomials in $c$, each of degree $2m$, when $c$ assumes any one of the above $2m$ values. Suppose now that $c = a + m$, in (2.3.4.3). This value is not one of our

† Dougall (1907).

$2m$ values, but it is a pole of the last term of the $_7F_6(1)$ series. From this fact, that it is a pole, by using the last term only, we can check that (2.3.4.3) holds for $c = a+m$. Hence (2.3.4.3) holds for $2m+1$ values of $c$ and since it is only of degree $2m$ in $c$, it must hold for all values of $c$, when $f = -m$, if it holds when $f = 0, -1, -2, ..., 1-m$. But it holds for $f = 0$. Hence it holds for $f = -1$, and so finally it holds for $f = -m$, by induction, and for all values of $c$.

If we put $f = -m$, and substitute $1+2a-b-c-d+m$ for $e$, we can write the result as

$$_7F_6\left[\begin{matrix} a, 1+\tfrac{1}{2}a, & b, & c, & d, 1+2a-b-c-d+m, -m; \\ \tfrac{1}{2}a, 1+a-b, 1+a-c, 1+a-d, b+c+d-a-m, 1+a+m; \end{matrix} 1\right]$$

$$= \frac{(1+a)_m (1+a-b-c)_m (1+a-b-d)_m (1+a-c-d)_m}{(1+a-b)_m (1+a-c)_m (1+a-d)_m (1+a-b-c-d)_m}.$$

$$(2.3.4.4)$$

In this expression, we can let $m \to \infty$. We find then that we have

$$_5F_4\left[\begin{matrix} a, 1+\tfrac{1}{2}a, & b, & c, & d; \\ \tfrac{1}{2}a, 1+a-b, 1+a-c, 1+a-d; \end{matrix} 1\right]$$

$$= \Gamma\left[\begin{matrix} 1+a-b, 1+a-c, 1+a-d, 1+a-b-c-d \\ 1+a, 1+a-b-c, 1+a-b-d, 1+a-c-d \end{matrix}\right],$$

$$(2.3.4.5)$$

provided that the series is convergent, that is provided that

$$Rl\,(b+c+d-a) < 1.$$

This result gives the sum of a well-poised infinite $_5F_4(1)$ series. If the series terminates, that is, if $d = -m$, the theorem becomes

$$_5F_4\left[\begin{matrix} a, 1+\tfrac{1}{2}a, & b, & c, & -m; \\ \tfrac{1}{2}a, 1+a-b, 1+a-c, 1+a+m; \end{matrix} 1\right]$$

$$= \frac{(1+a)_m (1+a-b-c)_m}{(1+a-b)_m (1+a-c)_m}. \quad (2.3.4.6)$$

If $d \to -\infty$, (2.3.4.6) reduces further to

$$_4F_3\left[\begin{matrix} a, 1+\tfrac{1}{2}a, & b, & c; \\ \tfrac{1}{2}a, 1+a-b, 1+a-c; \end{matrix} -1\right] = \Gamma\left[\begin{matrix} 1+a-b, 1+a-c \\ 1+a, 1+a-b-c \end{matrix}\right],$$

$$(2.3.4.7)$$

provided that $Rl\,(b+c-\tfrac{1}{2}a) < 1$.

If now $c = -m$, we find that

$$_4F_3\left[\begin{array}{cccc} a, 1+\tfrac{1}{2}a, & b, & -m; \\ \tfrac{1}{2}a, 1+a-b, & 1+a+m; \end{array} -1\right] = \frac{(1+a)_m}{(1+a-b)_m},$$

(2.3.4.8)

and when we let $c \to -\infty$, this result becomes

$$_3F_2\left[\begin{array}{cc} a, 1+\tfrac{1}{2}a, & b; \\ \tfrac{1}{2}a, 1+a-b; \end{array} 1\right] = \Gamma\left[\begin{array}{c} 1+a-b \\ 1+a \end{array}\right]$$

(2.3.4.9)

which is a special case of Dixon's theorem. In particular,

$$_3F_2\left[\begin{array}{cc} a, 1+\tfrac{1}{2}a, & -m; \\ \tfrac{1}{2}a, 1+a+m; \end{array} 1\right] = (1+a)_m.$$

(2.3.4.10)

If we put $d = \tfrac{1}{2}a$, in (2.3.4.5), we can deduce Dixon's theorem directly.

A similar method of induction can be used to prove the following more general result, connecting two well-poised $_9F_8(1)$ series†,

$$_9F_8\left[\begin{array}{ccccc} a, 1+\tfrac{1}{2}a, & b, & c, & d, \\ \tfrac{1}{2}a, 1+a-b, & 1+a-c, & 1+a-d, \\ & & e, & f, & g, & h; \\ & & 1+a-e, & 1+a-f, & 1+a-g, & 1+a-h; & 1 \end{array}\right]$$

$$= \Gamma\left[\begin{array}{c} 1+a-e, 1+a-f, 1+a-g, 1+a-h, 1+a-f-g-h, \\ 1+a-f-g, 1+a-f-h, 1+a-g-h, 1+a-e-f-g-h, \\ 1+a-e-g-h, 1+a-e-f-h, 1+a-e-f-g \\ 1+a, 1+a-e-g, 1+a-e-f, 1+a-e-h \end{array}\right]$$

$$\times {}_9F_8\left[\begin{array}{cccc} k, 1+\tfrac{1}{2}k, k+b-a, k+c-a, k+d-a, & e, \\ \tfrac{1}{2}k, 1+a-b, 1+a-c, 1+a-d, 1+k-e, \\ & f, & g, & h; \\ & 1+k-f, 1+k-g, 1+k-h; & 1 \end{array}\right],$$

(2.3.4.11)

where

$$k = 1+2a-b-c-d,$$ (2.3.4.12)

$$b+c+d+e+f+g+h = 2+3a,$$ (2.3.4.13)

and $h$ is a negative integer.

Since this relation will appear as a deduction from the more general theory later (see § 2.4.4 and § 4.5.2), a detailed proof will not be given here.

† Bailey (1929a, b).

## 2.4 Bailey's transform

The theorem to be considered now, was first stated explicity by W. N. Bailey in 1944,[†] though the germ of the idea can be found in the work of Abel more than one hundred years earlier. If we have a doubly infinite array of objects $f(r, s)$ which can be represented as points in a plane, $(r, s)$, we can form a doubly infinite sum

$$\sum_{r=0}^{\infty} \sum_{s=0}^{\infty} f(r, s)$$

by summing the terms parallel to the $x$-axis and parallel to the $y$-axis, but we can also form a single infinite sum of finite quantities

$$\sum_{r=0}^{\infty} \sum_{s=0}^{r} f(r, s)$$

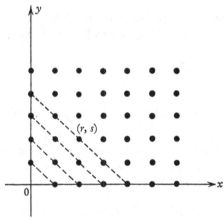

Fig. 2.1

by summing along the diagonal directions first, and then outwards, perpendicular to the diagonals, (see Fig. 2.1). This method of interchanging the order of summation to reduce two infinite sums to one infinite sum and one finite sum is fundamental in the study of general transformations of hypergeometric series.

Formally we have,

*if*
$$\beta_n = \sum_{r=0}^{n} \alpha_r u_{n-r} v_{n+r} \tag{2.4.1}$$

*and*
$$\gamma_n = \sum_{r=n}^{\infty} \delta_r u_{r-n} v_{r+n}, \tag{2.4.2}$$

† Bailey (1949), § 1.

*where $\alpha_r$, $\delta_r$, $u_r$ and $v_r$ are any functions of $r$ only, such that the series $\gamma_n$ exists, then*

$$\sum_{n=0}^{\infty} \alpha_n \gamma_n = \sum_{n=0}^{\infty} \beta_n \delta_n. \qquad (2.4.3)$$

For

$$\sum_{n=0}^{\infty} \alpha_n \gamma_n = \sum_{n=0}^{\infty} \sum_{r=n}^{\infty} \alpha_n \delta_r u_{r-n} v_{r+n}.$$

By hypothesis, this double series is convergent. Hence we can exchange the order of summation from rows to diagonals, and write

$$\sum_{n=0}^{\infty} \alpha_n \gamma_n = \sum_{r=0}^{\infty} \sum_{n=0}^{r} \alpha_n \delta_r u_{r-n} v_{r+n}$$

$$= \sum_{r=0}^{\infty} \beta_r \delta_r.$$

When the functions being used in this transformation are infinite series, the change in the order of summation has to be justified separately for each form of function, but when the functions used are finite series, as frequently happens, no such justification is needed.

In order to investigate transformations of general hypergeometric series, by the use of this theorem, let us suppose that

$$u_r = \frac{(u_1)_r (u_2)_r \cdots (u_U)_r u^r}{(e_1)_r (e_2)_r \cdots (e_E)_r r!} \qquad (2.4.4)$$

$$v_r = \frac{(v_1)_r (v_2)_r \cdots (v_V)_r v^r}{(f_1)_r (f_2)_r \cdots (f_F)_r} \qquad (2.4.5)$$

$$\delta_r = \frac{(d_1)_r (d_2)_r \cdots (d_D)_r x^r}{(g_1)_r (g_2)_r \cdots (g_G)_r} \qquad (2.4.6)$$

and

$$\alpha_r = \frac{(a_1)_r (a_2)_r \cdots (a_A)_r y^r}{(h_1)_r (h_2)_r \cdots (h_H)_r r!}. \qquad (2.4.7)$$

All these parameters and the variables $u$, $v$, $x$ and $y$ are independent of $r$. Then with these values of $u_r$, $v_r$ and $\delta_r$, since, if we write $s$ for $r+n$,

$$\gamma_n = \sum_{s=0}^{\infty} \delta_{s+n} u_s v_{s+2n},$$

we find that

$$\gamma_n = \frac{(d_1)_n (d_2)_n \cdots (d_D)_n (v_1)_{2n} (v_2)_{2n} \cdots (v_V)_{2n} x^n v^{2n}}{(g_1)_n (g_2)_n \cdots (g_G)_n (f_1)_{2n} (f_2)_{2n} \cdots (f_F)_{2n}}$$

$$\times {}_{U+D+V}F_{E+F+G}\left[\begin{matrix} u_1, u_2, \ldots, u_U, d_1+n, d_2+n, \ldots, \\ e_1, e_2, \ldots, e_E, g_1+n, g_2+n, \ldots, \\ d_D+n, v_1+2n, v_2+2n, \ldots, v_V+2n; \\ g_G+n, f_1+2n, f_2+2n, \ldots, f_F+2n; \end{matrix} uvx \right].$$

$$(2.4.8)$$

and

$$\beta_n = \frac{(u_1)_n (u_2)_n \cdots (u_U)_n (v_1)_n (v_2)_n \cdots (v_V)_n \, u^n v^n}{(e_1)_n (e_2)_n \cdots (e_E)_n (f_1)_n (f_2)_n \cdots (f_F)_n}$$

$$\times {}_{E+A+V+1}F_{U+H+F}\begin{bmatrix} 1-n-e_1, 1-n-e_2, \ldots, 1-n-e_E, a_1, a_2, \ldots, a_A, \\ 1-n-u_1, 1-n-u_2, \ldots, 1-n-u_U, h_1, h_2, \ldots, h_H, \\ v_1+n, v_2+n, \ldots, v_V+n, -n; \\ f_1+n, f_2+n, \ldots, f_F+n; \end{bmatrix} (-1)^{1+E-U} uvy \;\Bigg].$$

$$(2.4.9)$$

The complete theorem is, when stated in the contracted notation,

$$\sum_{n=0}^{\infty} \frac{((a))_n ((d))_n ((v))_{2n} \, x^n y^n v^{2n}}{((h))_n ((g))_n ((f))_{2n}}$$

$$\times {}_{U+D+V}F_{E+F+G}\begin{bmatrix} (u), (d)+n, (v)+2n; \\ (e), (g)+n, (f)+2n; \end{bmatrix} uvx \;\Bigg]$$

$$= \sum_{n=0}^{\infty} \frac{((d))_n ((u))_n ((v))_n \, u^n v^n x^n}{n!((e))_n ((f))_n ((g))_n}$$

$$\times {}_{A+E+V+1}F_{U+H+F}\begin{bmatrix} 1-n-(e), (a),(v)+n, -n; \\ 1-n-(u), (h), (f)+n; \end{bmatrix} (-1)^{1+E-U} uvy \;\Bigg].$$

$$(2.4.10)$$

This result contains as special cases very many relationships between generalized hypergeometric series. In the first place, whenever one of the inner series can be summed by one of the known summation theorems, the equation (2.4.10) will assume a simpler form. The two main summation theorems are Dougall's theorem and Saalschutz's theorem, since these two results contain most of the other summation theorems as special cases. They both involve terminating series, and we can apply them in turn to either side of (2.4.10).

### 2.4.1 Saalschutzian transformations.

Let us denote the series on the left of (2.4.10), ${}_{U+D+V}F_{E+F+G}(uvx)$, by $F_\gamma$ and the series on the right of (2.4.10), ${}_{A+E+V+1}F_{U+H+F}[(-1)^{1+E-U}uvy]$ by $F_\beta$, and let us suppose first that $F_\gamma$ can be summed by Saalschutz's theorem. There are then ten possible forms for $F_\gamma$, of which only four lead to series for $F_\beta$ which are also summable. These four cases are;

   (1) $F_\gamma = {}_3F_2[d_1+n, d_2+n, d_3+n; f+2n, g+n; 1]$

        $F_\beta = {}_{A+1}F_{H+1}[(a), -n; (h), f+n; -y]$

(2) $F_\gamma = {}_3F_2[v+2n, d+n, u; f+2n, g+n; 1]$

$F_\beta = {}_{A+2}F_{H+2}[(a), v+n, -n; (h), 1-u-n, f+n; y]$

(3) $F_\gamma = {}_3F_2[d_1+n, d_2+n, u; g_1+n, g_2+n; 1]$

$F_\beta = {}_{A+1}F_{H+1}[(a), -n; (h), 1-a-n; y]$

(4) $F_\gamma = {}_3F_2[d+n, u_1, u_2; g+n, e; 1]$

$F_\beta = {}_{A+2}F_{H+2}[(a), 1-e-n, -n; (h), 1-u_1-n, 1-u_2-n; y].$

In all these four series for $F_\gamma$, the parameters are subject to the usual restrictions for Saalschutz's theorem, that one of the numerator parameters is a negative integer, and that the sum of the numerator parameters exceeds that of the denominator parameters by one.

In case (1), when we sum $F_\gamma$ by Saalschutz's theorem, if we take

$$\delta_n = \frac{(d_1)_n (d_2)_n (-N)_n}{(g)_n}$$

then

$$\gamma_n = \frac{(g-d_1)_N (g-d_2)_N (d_1)_n (d_2)_n (-N)_n (-1)^n}{(g)_N (g-d_1-d_2)_N (1+d_1-g-N)_n (1+d_2-g-N)_n (1+d_1+d_2-g)_n}.$$

Let

$$\alpha_n = \frac{(a_1)_n (a_2)_n (\tfrac{1}{2}+\tfrac{1}{2}f)_n (f-1)_n (-1)^n}{n! (f-a_1)_n (f-a_2)_n (\tfrac{1}{2}f-\tfrac{1}{2})_n}.$$

Then we can sum $F_\beta$ as a well-poised ${}_5F_4(1)$ series provided that $f = 1+d_1+d_2-N-g$, to give

$$\beta_n = \frac{(f-a_1-a_2)_n}{n! (f-a_1)_n (f-a_2)_n},$$

and Bailey's transform gives us

$${}_4F_3\left[\begin{matrix} f-a_1-a_2, d_1, d_2, -N; \\ f-a_1, f-a_2, g; \end{matrix} 1\right] = \frac{(g-d_1)_N (g-d_2)_N}{(g)_N (g-d_1-d_2)_N}$$

$$\times {}_7F_6\left[\begin{matrix} f-1, \tfrac{1}{2}f+\tfrac{1}{2}, & a_1, & a_2, & d_1, & d_2, & -N; \\ \tfrac{1}{2}f-\tfrac{1}{2}, f-a_1, f-a_2, f-d_1, f-d_2, f+N; \end{matrix} 1\right],$$

$$\text{(2.4.1.1)}$$

where

$$g = 1+d_1+d_2-f-N.$$

This transforms a Saalschutian ${}_4F_3(1)$ series into a well-poised ${}_7F_6(1)$ series, with the special form of the second numerator parameter. It was given by Whipple.[†]

If further we assume that

$$2f = 1+a_1+a_2+d_1+d_2-N,$$

† Whipple (1926b), 7.7.

the ${}_4F_3(1)$ series reduces to a ${}_3F_2(1)$ series, which can then be summed by Saalschutz's theorem, and the sum of the ${}_7F_6(1)$ series is then found. This proves to be simply the deduction of Dougall's theorem from Saalschutz's theorem.

When $f = 1+a$, and $d_2 = \tfrac{1}{2}a$, (2.4.1.1) reduces to

$$
{}_5F_4\!\left[\begin{matrix} a, & b, & c, & d, & -N; \\ 1+a-b, & 1+a-c, & 1+a-d, & 1+a+N; \end{matrix}\ 1\right]
$$

$$
= \frac{(1+a)_N\,(1+\tfrac{1}{2}a-d)_N}{(1+\tfrac{1}{2}a)_N\,(1+a-d)_N}\ {}_4F_3\!\left[\begin{matrix} 1+a-b-c, & \tfrac{1}{2}a, & d, & -N; \\ 1+a-b, & 1+a-c, & d-\tfrac{1}{2}a-N; \end{matrix}\ 1\right].
$$

$$(2.4.1.2)$$

This transforms a well-poised ${}_5F_4(1)$ series into a Saalschutzian ${}_4F_3(1)$ series. When $b = 1+\tfrac{1}{2}a$, the ${}_4F_3(1)$ series reduces to a ${}_3F_2(1)$ series, which can be summed by Saalschutz's theorem, and we deduce again the sum of a well-poised ${}_5F_4(1)$ series.†

In the second case, we can sum both $F_\gamma$ and $F_\beta$ by Saalschutz's theorem. If we take

$$
\delta_n = \frac{(d)_n}{(1+v+d+u-f)_n},
$$

then
$$
\gamma_n = \Gamma\!\left[\begin{matrix} 1+v+d+u-f,\ 1+v-f,\ 1+d-f,\ 1+u-f \\ 1-f,\ 1+d+u-f,\ 1+v+u-f,\ 1+v+d-f \end{matrix}\right]
$$

$$
\times \frac{(d)_n\,(f-d-u)_n\,(\tfrac{1}{2}v)_n\,(\tfrac{1}{2}v+\tfrac{1}{2})_n}{(f-d)_n\,(1+v+d-f)_n\,(\tfrac{1}{2}f-\tfrac{1}{2}u)_n\,(\tfrac{1}{2}+\tfrac{1}{2}f-\tfrac{1}{2}u)_n}
$$

provided that neither $v$ nor $d$ is a negative integer.

Let
$$
\alpha_n = \frac{(f-v-u)_n}{n!}
$$

then
$$
\beta_n = \frac{(v)_n\,(\tfrac{1}{2}v+\tfrac{1}{2}u)_n\,(\tfrac{1}{2}+\tfrac{1}{2}v+\tfrac{1}{2}u)_n\,(f-v)_n}{(v+u)_n\,(\tfrac{1}{2}f)_n\,(\tfrac{1}{2}f+\tfrac{1}{2})_n\,n!}.
$$

Hence
$$
\Gamma\!\left[\begin{matrix} 1+v+d+u-f,\ 1+v-f,\ 1+d-f,\ 1+u-f \\ 1-f,\ 1+d+u-f,\ 1+d+v-f,\ 1+u+v-f \end{matrix}\right]
$$

$$
\times\ {}_5F_4\!\left[\begin{matrix} f-v-u, & d, f-d-u, & \tfrac{1}{2}v, \tfrac{1}{2}v+1; \\ f-d, & 1+v+d-f, & \tfrac{1}{2}+\tfrac{1}{2}f-\tfrac{1}{2}u, \tfrac{1}{2}f-\tfrac{1}{2}u; \end{matrix}\ 1\right]
$$

$$
=\ {}_5F_4\!\left[\begin{matrix} v, \tfrac{1}{2}u+\tfrac{1}{2}v, \tfrac{1}{2}+\tfrac{1}{2}u+\tfrac{1}{2}v, f-v, & d; \\ \tfrac{1}{2}+\tfrac{1}{2}f, \tfrac{1}{2}f, & f+u, 1+u+v+d-f; \end{matrix}\ 1\right].
$$

$$(2.4.1.3)$$

† Whipple (1926b), 5.2.

This transforms a nearly-poised terminating $_5F_4(1)$ series into a Saalschutzian $_5F_4(1)$ series.† Alternatively, we can sum $F_\beta$ by Dougall's theorem. If

$$\alpha_n = \frac{(v-u)_n\,(1+\tfrac{1}{2}v-\tfrac{1}{2}u)_n\,(a_1)_n\,(a_2)_n\,(a_3)_n}{(\tfrac{1}{2}v-\tfrac{1}{2}u)_n\,(1+v-u-a_1)_n\,(1+v-u-a_2)_n\,(1+v-u-a_3)_n\,n!}$$

then $\quad \beta_n = \dfrac{(v)_n\,(u+a_1)_n\,(u+a_2)_n\,(u+a_3)_n}{(1+v-u-a_1)_n\,(1+v-u-a_2)_n\,(1+v-u-a_3)_n\,n!}$

provided that $1+v-u = a_1+a_2+a_3+u$. Hence we find that

$$\Gamma\begin{bmatrix} u-v, d+2u-v, & 2u, & u+v \\ d+2u, & u, & d+u-v, 2u-v \end{bmatrix}$$

$$\times {}_5F_4\begin{bmatrix} v, u+a_1, u+a_2, u+a_3, & d; & 1 \\ 1+v-u-a_1, 1+v-u-a_2, 1+v-u-a_3, d+2u; \end{bmatrix}$$

$$= {}_9F_8\begin{bmatrix} v-u, 1+\tfrac{1}{2}v-\tfrac{1}{2}u, & a_1, & a_2, & a_3, \\ \tfrac{1}{2}v-\tfrac{1}{2}u, 1+v-u-a_1, 1+v-u-a_2, 1+v-u-a_3, \\ d, 1+v-2u-d, & \tfrac{1}{2}v, & \tfrac{1}{2}+\tfrac{1}{2}v; \\ 1+v-u-d, & u+d, 1+\tfrac{1}{2}v-u, \tfrac{1}{2}+\tfrac{1}{2}v-u; \end{bmatrix}1$$

$$(2.4.1.4)$$

provided that either $v$ or $d$ is a negative integer. This is a transformation of a well-poised $_9F_8(1)$ series into a nearly-poised $_5F_4(1)$ series, when both series terminate.‡

In particular, if we take $a_1 = -u$, the $_5F_4(1)$ series reduces to unity, and the $_9F_8(1)$ series becomes a $_7F_6(1)$ series. Hence we find that

$${}_7F_6\begin{bmatrix} v-u, 1+\tfrac{1}{2}v-\tfrac{1}{2}u, & -u, & d, 1+v-2u-d, \\ \tfrac{1}{2}v-\tfrac{1}{2}u, 1+v, & 1+v-u-d, & d+u, \\ & & \tfrac{1}{2}v, & \tfrac{1}{2}+\tfrac{1}{2}v & ; 1 \\ & & 1+\tfrac{1}{2}v-u, \tfrac{1}{2}+\tfrac{1}{2}v-u; \end{bmatrix}$$

$$= \Gamma\begin{bmatrix} u-v, d+2u-v, 2u, u+v \\ d+2u, u, d+u-v, 2u-v \end{bmatrix}, \quad (2.4.1.5)$$

where $d$ is a negative integer. This sum of a special $_7F_6(1)$ series is not a particular case of Dougall's theorem, unless $\tfrac{1}{2}v+\tfrac{1}{2} = u$.§ If we apply the transformation of a $_7F_6(1)$ series into a $_4F_3(1)$ series given above (2.4.1.1), we can deduce another special summation theorem‖

$${}_4F_3\begin{bmatrix} \tfrac{1}{2}v-\tfrac{1}{2}u, \tfrac{1}{2}+\tfrac{1}{2}v-\tfrac{1}{2}u, v+u, & -n; & 1 \\ \tfrac{1}{2}v, \tfrac{1}{2}+\tfrac{1}{2}v, 1+v-u; \end{bmatrix} = \frac{(u)_n}{(v)_n}. \quad (2.4.1.6)$$

† Bailey (1929b), 6.4.      ‡ Bailey (1929b), 8.1.
§ Bailey (1929b), 7.42.      ‖ Bailey (1929b), 8.2.

Various other special transformations can also be deduced from this result.

In the third case, we can sum both $F_\gamma$ and $F_\beta$ by Saalschutz's theorem. If we take

$$\delta_n = \frac{(d)_n (-N)_n}{(g)_n (1+u+d-g-N)_n}$$

then

$$\gamma_n = \frac{(g-d)_N (g-u)_N (d)_n (-N)_n}{(g)_N (g-d-u)_N (g-u)_n (1+d-g-N)_n}.$$

Let

$$\alpha_n = \frac{(a_1)_n (a_2)_n}{(a_1+a_2+u)_n n!}.$$

Then

$$\beta_n = \frac{(a_1+u)_n (a_2+u)_n}{(a_1+a_2+u)_n n!}.$$

Hence the transformation theorem leads us to the result

$$
{}_4F_3\left[\begin{array}{l} a_1+u, a_2+u, d, \qquad -N; \\ a_1+a_2+u, g, 1+u+d-g-N; \end{array} 1\right]
$$
$$
= \frac{(g-d)_N (g-u)_N}{(g)_N (g-d-u)_N} {}_4F_3\left[\begin{array}{l} a_1, \quad a_2, \qquad d, \qquad -N; \\ a_1+a_2+u, g-u, 1+d-g-N; \end{array} 1\right].
$$

$$(2.4.1.7)$$

This is a transformation between two terminating Saalschutzian ${}_4F_3(1)$ series. Again other summation theorems can be applied to $F_\beta$, to give other specialized transformations between ${}_4F_3(1)$ series and ${}_5F_4(1)$ series.

The fourth case stated above, leads only to a relation between two ${}_3F_2(1)$ series.

## 2.4.2 Vandermonde transformations.

Let us suppose next that the series for $F_\gamma$ can be summed by Vandermonde's theorem (1.7.7). Then, if

$$\delta_n = \frac{(d)_n}{(g)_n},$$

we find that

$$\gamma_n = \Gamma\left[\begin{array}{l} g, g-f-d \\ g-f, g-d \end{array}\right] \frac{(d)_n (1+f-d)_n (f)_{2n}(-1)^n}{(1+f+d-g)_n}.$$

If

$$\alpha_n = \frac{(-1)^n}{n! (a)_n},$$

the series for $F_\beta$ can also be summed by Vandermonde's theorem, to give

$$\beta_n = \frac{(f)_n (1+f-a)_n (-1)^n}{n! (a)_n}.$$

Hence we find that

$$
{}_3F_2\left[\begin{array}{c} f, 1+f-a, d; \\ a, g; \end{array} -1\right]
$$

$$
= \Gamma\left[\begin{array}{c} g, g-f-d \\ g-f, g-d \end{array}\right] {}_4F_3\left[\begin{array}{c} d, 1+f-g, \tfrac{1}{2}f, \tfrac{1}{2}f+\tfrac{1}{2}; \\ a, \tfrac{1}{2}+\tfrac{1}{2}f+\tfrac{1}{2}d-\tfrac{1}{2}g, 1+\tfrac{1}{2}f+\tfrac{1}{2}d-\tfrac{1}{2}g; \end{array} 1\right]
$$

$$(2.4.2.1)$$

provided that either $d$ or $f$ is a negative integer.

If further, $a = 1+f$, the ${}_3F_2(-1)$ series reduces to unity, and we have

$$
{}_4F_3\left[\begin{array}{c} d, 1+f-g, \qquad\qquad \tfrac{1}{2}f, \qquad\qquad \tfrac{1}{2}f+\tfrac{1}{2}; \\ 1+f, \tfrac{1}{2}+\tfrac{1}{2}f+\tfrac{1}{2}d-\tfrac{1}{2}g, 1+\tfrac{1}{2}f+\tfrac{1}{2}d-\tfrac{1}{2}g; \end{array} 1\right] = \Gamma\left[\begin{array}{c} g-f, g-d \\ g, g-f-d \end{array}\right],
$$

$$(2.4.2.2)$$

where either $d$ or $f$ is again a negative integer. This gives the sum of a finite special Saalschutzian ${}_4F_3(1)$ series.

Next let

$$
\alpha_n = \frac{(a)_n(-1)^n}{n!(h)_n(1+f+a-h)_n},
$$

then the series for $F_\beta$ is summable by Saalschutz's theorem, to give

$$
\beta_n = \frac{(f)_n(h-a)_n(1+f-h)_n}{n!(h)_n(1+f+a-h)_n}.
$$

Hence

$$
{}_4F_3\left[\begin{array}{c} f, 1+f-h, h-a, d; \\ h, 1+f+a-h, g; \end{array} 1\right] = \Gamma\left[\begin{array}{c} g, g-f-d \\ g-f, g-d \end{array}\right]
$$

$$
\times {}_5F_4\left[\begin{array}{c} a, d, \qquad 1+f-g, \qquad\qquad \tfrac{1}{2}f, \qquad\qquad \tfrac{1}{2}+\tfrac{1}{2}f; \\ h, 1+f+a-h, \tfrac{1}{2}+\tfrac{1}{2}f+\tfrac{1}{2}d-\tfrac{1}{2}g, 1+\tfrac{1}{2}f+\tfrac{1}{2}d-\tfrac{1}{2}g; \end{array} 1\right].
$$

$$(2.4.2.3)$$

This is a transformation between a nearly-poised ${}_4F_3(1)$ series and a Saalschutzian ${}_5F_4(1)$ series.[†] Either $f$ or $d$ must be a negative integer, $-N$. If we take $d = -N$, the other parameters $a, f, g$ and $h$ in the ${}_5F_4(1)$ series can be chosen in four ways, in order that this series can be reduced to a summable Saalschutzian ${}_3F_2(1)$ series. If we carry out these summations, writing $a$ for $f$ and $b$ for $g$, we find the four further summation theorems:

$$
{}_3F_2\left[\begin{array}{c} a, 1+\tfrac{1}{2}a, -N; \\ \tfrac{1}{2}a, \quad b; \end{array} 1\right] = \frac{(b-a-1-N)(b-a)_{N-1}}{(b)_N}. \quad (2.4.2.4)
$$

[†] It is due to Whipple (1926b), § 3.5.

This gives the sum of a nearly-poised $_3F_2(1)$ series, with the special form of the first parameter.

$$_3F_2\left[\begin{array}{ccc} a, & b, & -N; \\ 1+a-b, & 1+2b-N; \end{array} 1\right] = \frac{(a-2b)_N\,(1+\frac{1}{2}a-b)_N\,(-b)_N}{(1+a-b)_N\,(\frac{1}{2}a-b)_N\,(-2b)_N}.$$

$$(2.4.2.5)$$

This sums another type of nearly-poised $_3F_2(1)$ series.

$$_4F_3\left[\begin{array}{cccc} a, 1+\frac{1}{2}a, & b, & -N; \\ \frac{1}{2}a, 1+a-b, & 1+2b-N; \end{array} 1\right] = \frac{(a-2b)_N\,(-b)_N}{(1+a-b)_N\,(-2b)_N}.$$

$$(2.4.2.6)$$

This gives the sum of a nearly-poised $_4F_3(1)$ series with the special form of the first parameter.

$$_4F_3\left[\begin{array}{cccc} a, 1+\frac{1}{2}a, & b, & -N; \\ \frac{1}{2}a, 1+a-b, & 2+2b-N; \end{array} 1\right]$$

$$= \frac{(a-2b-1)_N\,(\frac{1}{2}+\frac{1}{2}a-b)_N\,(-b-1)_N}{(1+a-b)_N\,(\frac{1}{2}a-\frac{1}{2}-b)_N\,(-2b-1)_N}. \quad (2.4.2.7)$$

This gives the sum of another type of nearly-poised $_4F_3(1)$ series, also with the special form of the first parameter.

These four theorems can themselves be used to sum the series for $F_\beta$. Thus if we use the theorem (2.4.2.4) we find that if

$$\alpha_n = \frac{(f)_n\,(1+\frac{1}{2}f)_n\,(-1)^n}{n!\,(\frac{1}{2}f)_n\,(h)_n}$$

then
$$\beta_n = \frac{(f+2-h)_n\,(h-f-1)_n}{n!\,(h)_n\,(1+f-h)_n}.$$

Thus, if $\delta_n = (d)_n\,x^n$ so that

$$\gamma_n = (d)_n\,x^n/(1-x)^{d+n},$$

we find that

$$_3F_2\left[\begin{array}{ccc} f, 1+\frac{1}{2}f, d; & \frac{-x}{1-x} \\ \frac{1}{2}f, h; \end{array}\right] = (1-x)^d\,_3F_2\left[\begin{array}{cc} f+2-h, h-f-1, d; & x \\ 1+f-h, h; \end{array}\right].$$

$$(2.4.2.8)$$

If now we use (2.4.2.4) to sum the series for $F_\beta$, when

$$\alpha_n = \frac{(f)_n\,(1+\frac{1}{2}f)_n\,(-1)^n}{n!\,(\frac{1}{2}f)_n},$$

then we find that     $$\beta_n = \frac{(v-f-1)_{2n}}{n!\,(v)_{2n}\,(v-f)_n},$$

and if $\delta_n = (d_1)_n (d_2)_n$ the series for $F_\gamma$ can be summed by Gauss's theorem to give

$$\gamma_n = \Gamma \begin{bmatrix} v, v - d_1 - d_2 \\ v - d_1, v - d_2 \end{bmatrix} \frac{(d_1)_n (d_2)_n}{(v - d_1)_n (v - d_2)_n}.$$

Thus we find the transformation

$$\Gamma \begin{bmatrix} v, v - d_1 - d_2 \\ v - d_1, v - d_2 \end{bmatrix} {}_4F_3 \begin{bmatrix} f, 1 + \tfrac{1}{2}f, & d_1, & d_2; \\ & \tfrac{1}{2}f, v - d_1, v - d_2; \end{bmatrix} -1 \end{bmatrix}$$

$$= {}_4F_3 \begin{bmatrix} \tfrac{1}{2}v - \tfrac{1}{2}f - \tfrac{1}{2}, \tfrac{1}{2}v - \tfrac{1}{2}f, d_1, d_2; \\ \tfrac{1}{2}v + \tfrac{1}{2}, \tfrac{1}{2}v, v - f; \end{bmatrix} 1 \end{bmatrix}, \quad (2.4.2.9)$$

provided that $\mathrm{Rl}\,(v - d_1 - d_2) > 0$, and that $\mathrm{Rl}\,(f) < 0$.

Again, if we apply (2.4.2.5) to the summation of the series for $F_\beta$, and take

$$\alpha_n = \frac{(a)_n (-\tfrac{1}{2}e)_n}{n!\,(1 + a + \tfrac{1}{2}e)_n}$$

then we find that

$$\beta_n = \frac{(a + e)_n (1 + \tfrac{1}{2}a + \tfrac{1}{2}e)_n (\tfrac{1}{2}e)_n}{n!\,(1 + a + \tfrac{1}{2}e)_n (\tfrac{1}{2}a + \tfrac{1}{2}e)_n}.$$

Hence, if $\delta_n = x^n$, so that $\gamma_n = x^n / (1 - x)^e$, we deduce the result that

$${}_2F_1 \begin{bmatrix} a, & -\tfrac{1}{2}e; \\ & 1 + a + \tfrac{1}{2}e; \end{bmatrix} x \end{bmatrix} = (1 - x)^e \, {}_3F_2 \begin{bmatrix} a + e, 1 + \tfrac{1}{2}a + \tfrac{1}{2}e, & \tfrac{1}{2}e; \\ 1 + a + \tfrac{1}{2}e, & \tfrac{1}{2}a + \tfrac{1}{2}e; \end{bmatrix} x \end{bmatrix}.$$

$$(2.4.2.10)$$

This is a transformation of a particular well-poised ${}_3F_2(x)$ series into a well-poised ${}_2F_1(x)$ series. It is a combination of two results due to W. N. Bailey.†

## 2.4.3 Well-poised transformations.

We shall suppose next that $F_\gamma$ can be summed by Dougall's theorem, as a ${}_7F_6(1)$ series which is well-poised, in either a $d$ parameter, a $u$ parameter, or a $v$ parameter. Now $F_\gamma$ cannot be well-poised in a $d$ parameter, since in that case terms involving $\tfrac{1}{2}n$ would arise. Also when $F_\gamma$ is well-poised in a $u$ parameter, the corresponding series $F_\beta$ cannot be summed for any values of $\alpha_r$, which must be independent of $n$. Thus $F_\gamma$ can only be well-poised in a $v$ parameter, and in this case the parameters are always subject to the restriction

$$d_1 + n + d_2 + n + \ldots + d_D + n + u_1 + u_2 + \ldots + u_U + v_2 + 2n + v_3$$
$$+ 2n + \ldots + v_V + 2n = 1 + 2v_1 + 4n.$$

† Bailey (1929a), 4.07, 4.08.

There are thus only three possible cases in which $F_\gamma$ can be summed as a $_7F_6(1)$ series, which is well-poised in an $f$ parameter. Of these three cases, one results in a series for $F_\beta$ which cannot be summed, and the other two remaining cases are:

(1) $F_\gamma = {}_7F_6\left[\begin{array}{c} v_1 + 2n, 1 + \frac{1}{2}v_1 + n, v_2 + 2n, \qquad\qquad u_1, \\ \frac{1}{2}v_1 + n, 1 + v_1 - v_2, 1 + v_1 - u_1 + 2n, \\ u_2, \qquad d + n, \qquad -N + n; \\ 1 + v_1 - u_2 + 2n, 1 + v_1 - d + n, 1 + v_1 + N + n; \end{array} 1\right]$

$F_\beta = {}_{A+4}F_{H+4}\left[\begin{array}{c} (a), v_1 + n, v_2 + n, v_2 - v_1 - n, -n; \\ (h), 1 - u_1 - n, 1 - u_2 - n, 1 + v_1 - u_1 + n, \qquad y \\ 1 + v_1 - u_2 + n; \end{array}\right],$

where $\qquad 1 + 2v_1 = v_2 + u_1 + u_2 + d - N.$

(2) $F_\gamma = {}_7F_6\left[\begin{array}{c} v + 2n, 1 + \frac{1}{2}v + n, d_1 + n, d_2 + n, d_3 + n, \\ \frac{1}{2}v + n, 1 + v - d_1 + n, 1 + v - d_2 + n, \\ u, \qquad\qquad n - N; \\ 1 + v - d_3 + n, 1 + v - u + 2n, 1 + v + N + n; \end{array} 1\right]$

$F_\beta = {}_{A+2}F_{H+2}\left[\begin{array}{c} (a), v + n, \qquad -n; \\ (h), 1 - u - n, 1 + v - u + n; \end{array} y\right],$

where $\qquad 1 + 2v = u + d_1 + d_2 + d_3 - N.$

In the first case we sum both $F_\gamma$ and $F_\beta$ by Dougall's theorem. If

$$\delta_n = \frac{(1 + \frac{1}{2}v_1)_n (d)_n (-N)_n}{(\frac{1}{2}v_1)_n (1 + v_1 - d)_n (1 + v_1 + N)_n},$$

then

$$\gamma_n = \frac{(1 + v_1)_N (1 + v_1 - v_2 - d)_N (1 + v_1 - v_2 - u_1)_N (1 + v_1 - d - u_1)_N}{(1 + v_1 - v_2)_N (1 + v_1 - d)_N (1 + v_1 - u_1)_N (1 + v_1 - v_2 - d - u_1)_N}$$
$$\times \frac{(1 + v_1 - u_1 - u_2 - d)_n (1 + v_1 - u_1 - u_2 + N)_n}{(1 + v_1 - u_1 - u_2)_{2n} (1 + v_1 - d - u_2)_n (1 + v_1 + N - u_2)_n}$$
$$\times \frac{(u_2)_{2n} (d)_n (-N)_n}{(1 + v_1 + N - u_1)_n (1 + v_1 - d - u_1)_n},$$

where $\qquad u_1 + u_2 + d - N + v_2 = 1 + 2v_1.$

Let $v_2 - u_2 = v_1 - e_1$, and

$$\alpha_n = \frac{(v_1 - u_1)_n (1 + \frac{1}{2}v_1 - \frac{1}{2}e_1)_n (a)_n}{(\frac{1}{2}v_1 - \frac{1}{2}u_1)_n (1 + v_1 - u_1 - a)_n n!},$$

where $1 + 2v_1 - 2u_1 = 2u_2 + a$. Then $a = 1 - 2u_2$, and

$$\beta_n = \frac{(v_1)_n (v_2)_n (1 + u_1 - 2u_2)_n (1 - u_2)_n (\tfrac{1}{2}u_1 + \tfrac{1}{2}v_2)_n (\tfrac{1}{2} + \tfrac{1}{2}u_1 + \tfrac{1}{2}v_2)_n}{(1 + v_1 - v_2)_n (v_2 + u_2)_n (v_2 + u_1)_n (1 + \tfrac{1}{2}u_1 - \tfrac{1}{2}u_2)_n (\tfrac{1}{2} + \tfrac{1}{2}v_1 - \tfrac{1}{2}u_2)_n \, n!}$$

provided that $u_1, u_2$ and hence $v_2$ are not negative integers. Then†

$$\frac{(1 + v_1)_N (1 + v_1 - u_1 - d)_N (1 + v_1 - v_2 - u_1)_N (1 + v_1 - v_2 - d)_N}{(1 + v_1 - v_2)_N (1 + v_1 - d)_N (1 + v_1 - u_1)_N (1 + v_1 - v_2 - d - u_1)_N}$$

$$\times {}_9F_8 \left[ \begin{array}{c} v_1 - u_1, 1 + \tfrac{1}{2}v_1 - \tfrac{1}{2}u_1, 1 - 2u_2, v_2 - v_1 - N, v_2 + d - v_1, \qquad d, \\ \tfrac{1}{2}v_1 - \tfrac{1}{2}u_1, v_2 + u_2, 1 + v_1 + N - u_2, 1 + v_1 - d - u_2, 1 + v_1 - u_1 - d, \\ \tfrac{1}{2}v_2, \qquad \tfrac{1}{2}v_2 + \tfrac{1}{2}, \qquad\qquad -N; \\ 1 + \tfrac{1}{2}v_1 - \tfrac{1}{2}u_1 - \tfrac{1}{2}u_2, \tfrac{1}{2} + \tfrac{1}{2}v_1 - \tfrac{1}{2}u_1 - \tfrac{1}{2}u_2, 1 + v_1 - u_1 + N; \end{array} \; 1 \right]$$

$$= {}_9F_8 \left[ \begin{array}{c} v_1, 1 + \tfrac{1}{2}v_1, \qquad v_2, 1 + u_1 - 2u_2, 1 - u_2, \tfrac{1}{2}u_1 + \tfrac{1}{2}v_2, \\ \tfrac{1}{2}v_1, 1 + v_1 - v_2, v_2 + u_2, v_2 + u_1, 1 + \tfrac{1}{2}v_1 - \tfrac{1}{2}u_2, \\ \tfrac{1}{2} + \tfrac{1}{2}v_2 + \tfrac{1}{2}u_1, \qquad d, \qquad -N; \\ \tfrac{1}{2} + \tfrac{1}{2}v_1 - \tfrac{1}{2}u_2, 1 + v_1 - d, 1 + v_1 + N; \end{array} \; 1 \right],$$

$$(2.4.3.1)$$

where $v_1 - u_1 = v_2 - u_2$ and $1 + 2v_1 = u_1 + u_2 + v_2 + d - N$.

If we use Saalschutz's theorem (2.3.1.3) in place of Dougall's theorem, to sum one of the series, we find that

$${}_5F_4 \left[ \begin{array}{c} a, \qquad b, \qquad c, \qquad d, -m; \\ 1 + a - b, 1 + a - c, 1 + a - d, \quad w; \end{array} \; 1 \right] = \frac{(1 + 2k - a)_m (1 + k - a)_m}{(1 + k)_m (1 + 2k - 2a)_m}$$

$$\times {}_9F_8 \left[ \begin{array}{c} k, 1 + \tfrac{1}{2}k, \qquad \tfrac{1}{2}a, \qquad \tfrac{1}{2} + \tfrac{1}{2}a, \\ \tfrac{1}{2}k, 1 + k - \tfrac{1}{2}a, \tfrac{1}{2} + k - \tfrac{1}{2}a, \\ k + b - a, k + c - a, k + d - a, 1 + a - w, \qquad -m; \\ 1 + a - b, 1 + a - c, 1 + a - d, k + w - a, 1 + k + m; \end{array} \; 1 \right],$$

$$(2.4.3.2)$$

where $k = 1 + 2a - b - c - d$, and $w = 2a - 2k - m$. This is a transformation between a nearly-poised Saalschutzian ${}_5F_4(1)$ series and a special well-poised ${}_9F_8(1)$ series.

Similarly, if we use the summation theorem (2.4.2.6) to sum one of the series, we can deduce that

$${}_6F_5 \left[ \begin{array}{c} a, 1 + \tfrac{1}{2}a, \qquad b, \qquad c, \qquad d, -m; \\ \tfrac{1}{2}a, 1 + a - b, 1 + a - c, 1 + a - d, \quad w; \end{array} \; 1 \right] = \frac{(2k - a)_m (k - a)_m}{(1 + k)_m (2k - 2a)_m}$$

$$\times {}_9F_8 \left[ \begin{array}{c} k, 1 + \tfrac{1}{2}k, \qquad \tfrac{1}{2} + \tfrac{1}{2}a, 1 + \tfrac{1}{2}a, \\ \tfrac{1}{2}k, \tfrac{1}{2} + k - \tfrac{1}{2}a, k - \tfrac{1}{2}a, \\ k + b - a, k + c - a, k + d - a, 1 + a - w, \qquad -m; \\ 1 + a - b, 1 + a - c, 1 + a - d, k - a + w, 1 + k + m; \end{array} \; 1 \right],$$

$$(2.4.3.3)$$

† Bailey (1928), 7.41.

where again $k = 1 + 2a - b - c - d$ but $w = 1 + 2a - 2k - m$. This transforms a nearly-poised Saalschutzian $_6F_5(1)$ series into a special terminating well-poised $_9F_8(1)$ series.

If we use the theorem (2.4.2.5), we can deduce that

$$
_5F_4\left[\begin{array}{ccccc} a, & b, & c, & d, & -m; \\ & 1+a-b, & 1+a-c, & 1+a-d, & w; \end{array} 1\right]
$$

$$
= \frac{(k-a)_m\,(1+2k-a)_{m-1}\,(2k-a+2m)}{(1+k)_m\,(2k-2a)_m}
$$

$$
\times\,_9F_8\left[\begin{array}{c} k, 1+\tfrac{1}{2}k, \quad \tfrac{1}{2}+\tfrac{1}{2}a, \quad \tfrac{1}{2}a, \\ \tfrac{1}{2}k, \tfrac{1}{2}+k-\tfrac{1}{2}a, 1+k-\tfrac{1}{2}a, \\ k+b-a, k+c-a, k+d-a, 1+a-w, \qquad -m; \\ 1+a-b, 1+a-c, 1+a-d, k-a+w, 1+k+m; \end{array} 1\right],
$$

$$(2.4.3.4)$$

where $w = 1 + 2a - 2k - m$. This transforms a nearly-poised terminating $_5F_4(1)$ series into a special terminating well-poised $_9F_8(1)$ series. In this $_5F_4(1)$ series, it should be noted that the sum of the denominator parameters exceeds the sum of the numerator parameters by two.

Finally, if we apply the theorem (2.4.2.7), we can deduce that

$$
_6F_5\left[\begin{array}{ccccc} a, 1+\tfrac{1}{2}a, & b, & c, & d, & -m; \\ \tfrac{1}{2}a, & 1+a-b, & 1+a-c, & 1+a-d, & w; \end{array} 1\right]
$$

$$
= \frac{(k-a-1)_m\,(2k-a)_{m-1}\,(2k-a+2m-1)}{(1+k)_m\,(2k-2a-1)_m}
$$

$$
\times\,_9F_8\left[\begin{array}{c} k, 1+\tfrac{1}{2}k, \tfrac{1}{2}+\tfrac{1}{2}a, \quad 1+\tfrac{1}{2}a, \\ \tfrac{1}{2}k, \tfrac{1}{2}+k-\tfrac{1}{2}a, k-\tfrac{1}{2}a, \\ k+b-a, k+c-a, k+d-a, 1+a-w, \qquad -m; \\ 1+a-b, 1+a-c, 1+a-d, k-a+w, 1+k+m; \end{array} 1\right],
$$

$$(2.4.3.5)$$

where $w = 2 + 2a - 2k - m$. This expresses a nearly-poised $_6F_5(1)$ series in terms of a special terminating well-poised $_9F_8(1)$ series. Again, the sum of the denominator parameters in the $_6F_5(1)$ series exceeds the sum of the numerator parameters by two.

### 2.4.4 Dougall transforms.
The result (2.4.3.1) is a very general relation between two well-poised $_9F_8(1)$ series, but it is not the most

general one known. That is given by the second case of §2.4.3, when both $F_\gamma$ and $F_\beta$ are summed by Dougall's theorem. If we take

$$\delta_n = \frac{(1+\tfrac{1}{2}v)_n\,(d_1)_n\,(d_2)_n\,(d_3)_n\,(-N)_n}{(\tfrac{1}{2}v)_n\,(1+v-d_1)_n\,(1+v-d_2)_n\,(1+v-d_3)_n\,(1+v+N)_n},$$

then

$$\gamma_n = \frac{(1+v)_N\,(1+v-u-d_1)_N\,(1+v-u-d_2)_N\,(1+v-u-d_3)_N}{(1+v-u)_N\,(1+v-d_1)_N\,(1+v-d_2)_N\,(1+v-d_3)_N}$$

$$\times\,\frac{(d_1)_n\,(d_2)_n\,(d_3)_n\,(-N)_n}{(1+v-u-d_1)_n\,(1+v-u-d_2)_n\,(1+v-u-d_3)_n\,(1+v-u+N)_n}$$

provided that $u+d_1+d_2+d_3-N = 1+2v$.

If

$$\alpha_n = \frac{(v-u)_n\,(1+\tfrac{1}{2}v-\tfrac{1}{2}u)_n\,(a_1)_n\,(a_2)_n\,(a_3)_n}{n!\,(\tfrac{1}{2}v-\tfrac{1}{2}u)_n\,(1+v-u-a_1)_n\,(1+v-u-a_2)_n\,(1+v-u-a_3)_n},$$

then $\quad\beta_n = \dfrac{(v)_n\,(u+a_1)_n\,(u+a_2)_n\,(u+a_3)_n}{n!\,(1+v-u-a_1)_n\,(1+v-u-a_2)_n\,(1+v-u-a_3)_n}$

provided that $1+2v-2u = a_1+a_2+a_3+v$.

Hence we find that

$$\frac{(1+v)_N\,(1+v-u-d_1)_N\,(1+v-u-d_2)_N\,(1+v-u-d_3)_N}{(1+v-u)_N\,(1+v-d_1)_N\,(1+v-d_2)_N\,(1+v-d_3)_N}$$

$$\times\,{}_9F_8\!\left[\begin{array}{cccc} v-u,\,1+\tfrac{1}{2}v-\tfrac{1}{2}u, & a_1, & a_2, & a_3, \\ \tfrac{1}{2}v-\tfrac{1}{2}u,\,1+v-u-a_1, & 1+v-u-a_2, & 1+v-u-a_3, \\ d_1, & d_2, & d_3, & -N; \\ 1+v-u-d_1,\,1+v-u-d_2, & 1+v-u-d_3,\,1+v-u+N; \end{array}\,1\right]$$

$$=\,{}_9F_8\!\left[\begin{array}{cccc} v,\,1+\tfrac{1}{2}v, & u+a_1, & u+a_2, & u+a_3, \\ \tfrac{1}{2}v,\,1+v-u-a_1, & 1+v-u-a_2, & 1+v-u-a_3, \\ d_1, & d_2, & d_3, & -N; \\ 1+v-d_1,\,1+v-d_2, & 1+v-d_3,\,1+v+N; \end{array}\,1\right].$$

(2.4.4.1)

This is the relation between two general well-poised ${}_9F_8(1)$ series referred to above (2.3.4.11). This is one of the most general simple transformations of terminating well-poised series which has been found.

If we substitute $a_1+a_2+a_3+2N-1$ for $v$, and $1+2v-u+N-d_1-d_2$ for $d_3$, and then let $a_3 \to \infty$, we can deduce Whipple's result (2.4.1.1), which transforms a well-poised ${}_7F_6(1)$ into a Saalschutzian ${}_4F_3(1)$ series.

We shall now outline the extension of Whipple's result to nonterminating series. First, we substitute for $v$ and $a_3$ in terms of the other parameters and then rewrite (2.4.4.1) in the form

$$
{}_9F_8\left[\begin{matrix} a, 1+\tfrac{1}{2}a, & \beta+m, & c, & d, & e, \\ \tfrac{1}{2}a, 1+a-\beta-m, 1+a-c, 1+a-d, 1+a-e, \end{matrix}\right.
$$
$$
\left.\begin{matrix} f, & g, & -m; \\ 1+a-f, 1+a-g, 1+a+m; \end{matrix} 1\right]
$$

$$
= \frac{(1+a)_m\,(e-k)_m\,(f-k)_m\,(g-k)_m}{(-k)_m\,(1+a-e)_m\,(1+a-f)_m\,(1+a-g)_m}
$$

$$
\times {}_9F_8\left[\begin{matrix} k-m, 1+\tfrac{1}{2}k-\tfrac{1}{2}m, & k+\beta-a, k+c-m-a, k+d-m-a, & e, \\ \tfrac{1}{2}k-\tfrac{1}{2}m, 1+a-\beta-m, 1+a-c, 1+a-d, 1+k-m-e, \end{matrix}\right.
$$
$$
\left.\begin{matrix} f, & g, & -m; \\ 1+k-m-f, 1+k-m-g, 1+k; \end{matrix} 1\right],
$$

$$(2.4.4.2)$$

where $k = e+f+g-1-a$, $\beta = 2+3a-c-d-e-f-g$, and $m$ is a positive integer.

Suppose now that we try to let $m \to \infty$, through positive integer values. The first ${}_9F_8(1)$ series behaves correctly and tends to

$$
{}_7F_6\left[\begin{matrix} a, 1+\tfrac{1}{2}a, & c, & d, & e, & f, & g; \\ \tfrac{1}{2}a, 1+a-c, 1+a-d, 1+a-e, 1+a-f, 1+a-g; \end{matrix} 1\right]
$$

as $m \to \infty$. This series converges provided that

$$\mathrm{Rl}\,(2+2a-c-d-e-f-g) > 0.$$

Trouble arises however with the other ${}_9F_8(1)$ series, as now the ends of the finite series remain finite, while the middle terms become small. To overcome this difficulty, we split the ${}_9F_8(1)$ series into two parts, and reverse the order of the terms in the second half of this finite series, as in § 2.2.3. As $m \to \infty$, the first half of the series tends to

$$
{}_4F_3\left[\begin{matrix} k+\beta-a, & e, & f, & g; \\ 1+a-c, 1+a-d, 1+k; \end{matrix} 1\right],
$$

and the second half of the series, when reversed, tends to

$$
{}_4F_3\left[\begin{matrix} -a, & e-k, & f-a, & g-a; \\ 1+a-c-k, 1+a-d-k, 1-k; \end{matrix} 1\right]
$$

as $m \to \infty$.

The complete result is

$$
{}_7F_6\!\left[\begin{array}{cccccc} a, 1+\tfrac{1}{2}a, & c, & d, & e, & f, & g; \\ & \tfrac{1}{2}a, 1+a-c, 1+a-d, 1+a-e, 1+a-f, 1+a-g; \end{array} 1\right]
$$

$$
= \Gamma\!\left[\begin{array}{c} 1+a-e, 1+a-f, 1+a-g, 1+a-e-f-g \\ 1+a, 1+a-e-f, 1+a-f-g, 1+a-g-e \end{array}\right]
$$

$$
\times {}_4F_3\!\left[\begin{array}{cccc} 1+a-c-d, & e, & f, & g; \\ 1+a-c, 1+a-d, e+f+g-a; \end{array} 1\right]
$$

$$
+ \Gamma\!\left[\begin{array}{c} 1+a-c, 1+a-d, 1+a-e, 1+a-f, 1+a-g, \\ 1+a, 1+a-c-d, \qquad e, \qquad f, \qquad g, \\ e+f+g-1-a, 2+2a-c-d-e-f-g \\ 2+2a-c-e-f-g, 2+2a-d-e-f-g \end{array}\right]
$$

$$
\times {}_4F_3\!\left[\begin{array}{c} 2+2a-c-d-e-f-g, 1+a-f-g, \\ 2+a-e-f-g, \\ 1+a-g-e, 1+a-e-f; \\ 2+2a-c-e-f-g, 2+2a-d-e-f-g; \end{array} 1\right],
$$

(2.4.4.3)

provided that $\mathrm{Rl}\,(2+2a-c-d-e-f-g) > 0$.

This is the generalization of Whipple's result (2.4.1.1) when the series do not terminate. The ${}_7F_6(1)$ is well-poised and the two ${}_4F_3(1)$ series are both Saalschutzian and are always convergent. If

$$1+a-c-d, e, f \quad \text{or} \quad g$$

is a negative integer, the second part of the right-hand side of (2.4.4.3) vanishes and the result reduces to Whipple's transformation (2.4.1.1) of a well-poised ${}_7F_6(1)$ series into a Saalschutzian ${}_4F_3(1)$ series.

If $f+g = 1+a$ in (2.4.4.3), the first ${}_4F_3(1)$ series on the right vanishes because of the factor $\Gamma(1+a-f-g)$ in the denominator. The ${}_7F_6(1)$ series becomes a ${}_5F_4(1)$ series, and the result reduces to the summation theorem for a well-poised ${}_5F_4(1)$ series (2.3.4.5).

If $c = 0$, in (2.4.4.3), the ${}_7F_6(1)$ series reduces to 1, both the ${}_4F_3(1)$ series reduce to ${}_3F_2(1)$ series, and equation (2.4.4.3) becomes, after some reduction,

$$
{}_3F_2\!\left[\begin{array}{ccc} e, & f, & g; \\ 1+a, e+f+g-a; \end{array} 1\right]
$$

$$
= \Gamma\!\left[\begin{array}{c} 1+a, 1+a-f-g, 1+a-e-g, 1+a-e-f \\ 1+a-e, 1+a-f, 1+a-g, 1+a-e-f-g \end{array}\right]
$$

$$
- \Gamma\!\left[\begin{array}{c} 1+a, 1+a-e-f, 1+a-e-g, 1+a-f-g, e+f+g-a-1 \\ 1+a-e-f-g, 2+2a-e-f-g, e, f, g \end{array}\right]
$$

$$
\times {}_3F_2\!\left[\begin{array}{c} 1+a-f-g, 1+a-e-g, 1+a-e-f; \\ 2+a-e-f-g, 2+2a-e-f-g; \end{array} 1\right].
$$

(2.4.4.4)

This reduces to Saalschutz's theorem, when $e, f$ or $g$ is a negative integer. Thus (2.4.4.4) is a generalization of Saalschutz's theorem, when the restriction that the $_3F_2(1)$ series must be finite is removed. In Bailey (1935) p. 21, a second form of (2.4.4.4) is given which, however, is not symmetrical in $e, f$ and $g$.

### 2.4.5 Some possible extensions of Bailey's theorem.

There is no need, when using Bailey's theorem, to assume that the $\alpha_r$'s, $\beta_r$'s $\gamma_r$'s and $\delta_r$'s are all non-zero. It is quite possible to make use of the theorem when $\alpha_r$ is defined by

$$\alpha_r = \frac{((a))_r}{((b))_r\, r!} \quad \text{when } r \text{ is even,}$$
$$\alpha_r = 0 \qquad \text{when } r \text{ is odd,} \tag{2.4.5.1}$$

and

since

$$(a)_{2r} = (\tfrac{1}{2}a)_r\, (\tfrac{1}{2} + \tfrac{1}{2}a)_r\, 2^{2r}. \tag{2.4.5.2}$$

Similarly, we might define

$$\alpha_r = \frac{((a))_r}{((b))_r\, r!} \quad \text{when } r \text{ is a multiple of three,}$$
$$\alpha_r = 0 \qquad \text{when } r \text{ is not,} \tag{2.4.5.3}$$

and

since

$$(a)_{3r} = (\tfrac{1}{3}a)_r\, (\tfrac{1}{3} + \tfrac{1}{3}a)_r\, (\tfrac{2}{3} + \tfrac{1}{3}a)_r\, 3^{3r}. \tag{2.4.5.4}$$

In general, we could define

$$\alpha_r = \frac{((a))_r}{((b))_r\, r!} \quad \text{when } r \text{ is a multiple of } m,$$
$$\alpha_r = 0 \qquad \text{when } r \text{ is not,} \tag{2.4.5.5}$$

and

and use the fact that

$$(a)_{mr} = (a/m)_r\, ((1+a)/m)_r\, ((2+a)/m)_r \ldots ((m-1+a)/m)_r\, m^{mr}. \tag{2.4.5.6}$$

See Slater, M.A. *Thesis*, London (1949) for full details of such processes and the results they lead to.

Again we might study series involving

$$\alpha_r = \frac{((a))_{mr}}{((b))_{mr}\, (mr)!} \tag{2.4.5.7}$$

for all integer values of $r$, or for general real or complex values of $r$, since we can assume that

$$(a)_r = \frac{\Gamma(a+r)}{\Gamma(a)}. \tag{2.4.5.8}$$

A systematic study of such series has not yet been carried out, although isolated examples of such series occur in the literature.†

† See M. Jackson (1949b).

## 2.5 Products of hypergeometric series and Orr's theorem

Several relations connecting products of hypergeometric series, have been proved by Orr (1899). His most important results are these three theorems.

**Theorem I.** *If*

$$(1-z)^{a+b-c} \, _2F_1[2a, 2b; \, 2c; \, z] = \sum_{n=0}^{\infty} a_n z^n, \qquad (2.5.1)$$

*then*

$$_2F_1[a, b; \, c+\tfrac{1}{2}; \, z] \, _2F_1[c-a, c-b; \, c+\tfrac{1}{2}; \, z] = \sum_{n=0}^{\infty} \frac{(c)_n}{(c+\tfrac{1}{2})_n} a_n z^n. \qquad (2.5.2)$$

**Theorem II.** *If*

$$(1-z)^{a+b-c-\frac{1}{2}} \, _2F_1[2a, 2b; \, 2c; \, z] = \sum_{n=0}^{\infty} b_n z^n, \qquad (2.5.3)$$

*then*

$$_2F_1[a, b; \, c; \, z] \, _2F_1[c-a+\tfrac{1}{2}, c-b+\tfrac{1}{2}; \, c+1; \, z] = \sum_{n=0}^{\infty} \frac{(c+\tfrac{1}{2})_n}{(c+1)_n} b_n z^n. \qquad (2.5.4)$$

**Theorem III.** *If*

$$(1-z)^{a+b-c-\frac{1}{2}} \, _2F_1[2a-1, 2b; \, 2c-1; \, z] = \sum_{n=0}^{\infty} c_n z^n, \qquad (2.5.5)$$

*then*

$$_2F_1[a, b; \, c; \, z] \, _2F_1[c-a+\tfrac{1}{2}, c-b-\tfrac{1}{2}; \, c; \, z] = \sum_{n=0}^{\infty} \frac{(c-\tfrac{1}{2})_n}{(c)_n} c_n z^n. \qquad (2.5.6)$$

A special case when $c = a+b$, of the first of these theorems was published by Clausen as long ago as 1828.[†] It concerns the square of a Gauss series

$$_2F_1[a, b; \, a+b+\tfrac{1}{2}; \, z]^2 = \, _3F_2[2a, 2b, a+b; \, 2a+2b, a+b+\tfrac{1}{2}; \, z]. \quad (2.5.7)$$

The general case was discovered as an outcome of certain relations in planetary theory. Forty years later Orr published his proof based on a differential equation satisfied by the product of two hypergeometric series. Since then several alternative proofs have been given.[‡] The simplest proofs are those of Whipple, reproduced here.

---

† Cayley (1858).
‡ See Edwardes (1923); Watson (1924*b*); Whipple (1927*b*, 1929).

Let us compare the coefficients of $z^n$ in the three theorems. Then the identities which we have to prove are seen to be

$$
{}_3F_2\left[\begin{array}{c} 2a, 2b, -n; \\ 2c, 1+a+b-c-n; \end{array} 1\right]
$$
$$
= \frac{(c-a)_n (c-b)_n}{(c)_n (c-a-b)_n} {}_4F_3\left[\begin{array}{c} a, b, \tfrac{1}{2}-c-n, -n; \\ c+\tfrac{1}{2}, 1+a-c-n, 1+b-c-n; \end{array} 1\right],
$$
$$\tag{2.5.8}$$

$$
{}_3F_2\left[\begin{array}{c} 2a, 2b, -n; \\ 2c, \tfrac{1}{2}+a+b-c-n; \end{array} 1\right]
$$
$$
= \frac{(c-a+\tfrac{1}{2})_n (c-b+\tfrac{1}{2})_n}{(c+\tfrac{1}{2})_n (c+\tfrac{1}{2}-a-b)_n} {}_4F_3\left[\begin{array}{c} a, b, -c-n, -n; \\ c, \tfrac{1}{2}+a-c-n, \tfrac{1}{2}+b-c-n; \end{array} 1\right]
$$
$$\tag{2.5.9}$$

and

$$
{}_3F_2\left[\begin{array}{c} 2a-1, 2b, -n; \\ 2c-1, \tfrac{1}{2}+a+b-c-n; \end{array} 1\right]
$$
$$
= \frac{(c-a+\tfrac{1}{2})_n (c-b-\tfrac{1}{2})_n}{(c-\tfrac{1}{2})_n (c+\tfrac{1}{2}-a-b)_n} {}_4F_3\left[\begin{array}{c} a, b, 1-c-n, -n; \\ c, \tfrac{1}{2}+a-c-n, \tfrac{3}{2}+b-c-n; \end{array} 1\right].
$$
$$\tag{2.5.10}$$

First we can deduce (2.5.10) from (2.5.9), for the series on the right-hand side of (2.5.9) and (2.5.10) are both Saalschutzian. If we re-write (2.5.10) with $a+\tfrac{1}{2}$ and $c+\tfrac{1}{2}$ for $a$ and $c$ respectively, we can then see that the right-hand side of (2.5.10) can be transformed into the right-hand side of (2.5.9), by the application of (2.4.1.7), which is a relation between two terminating Saalschutzian ${}_4F_3(1)$ series, and so the third theorem can be deduced from the second theorem. Next we can deduce (2.5.8) from (2.5.9) and (2.5.10). Let us multiply (2.5.9) by $c/a$, and write $a+1$ and $c+1$ for $a$ and $c$ respectively in (2.5.10). Then we can subtract term by term, and we find that we have deduced (2.5.8) with $c+\tfrac{1}{2}$ written in place of $c$. Hence we can deduce the first theorem from the second one.

It remains only for us to prove the second theorem. We shall make use of three transformations, first, the transformation of a Saalschutzian ${}_4F_3(1)$ series, (2.4.1.7) used above, secondly, the transformation of a nearly-poised ${}_3F_2(1)$ series into a Saalschutzian ${}_4F_3(1)$ series, ((2.4.2.1) with $c = \tfrac{1}{2} + \tfrac{1}{2}a$), and thirdly the formula

$$
{}_3F_2[a, b, -n; e, f; 1]
$$
$$
= \frac{(e-a)_n (f-a)_n}{(e)_n (f)_n} {}_3F_2\left[\begin{array}{c} 1-s, a, -n; \\ 1+a-e-n, 1+a-f-n; \end{array} 1\right], \tag{2.5.11}
$$

where $s \equiv e+f-a-b+n$. This can be deduced from (2.4.1.7) by letting $d \to \infty$. Thus we find that

$$
{}_3F_2\left[\begin{array}{c} 2a, 2b, -n; \\ 2c, \tfrac{1}{2}+a+b-c-n; \end{array} 1\right]
$$

$$
= \frac{(\tfrac{1}{2}+a-b+c)_n\,(2c-2a)_n}{(2c)_n\,(\tfrac{1}{2}-a-b+c)_n}\, {}_3F_2\left[\begin{array}{c} 2a, \tfrac{1}{2}+a+b-c, -n; \\ \tfrac{1}{2}+a-b+c, 1+2a-2c-n; \end{array} 1\right]
$$

$$
= \frac{(\tfrac{1}{2}+a-b+c)_n}{(\tfrac{1}{2}-a-b+c)_n}\, {}_4F_3\left[\begin{array}{c} a, c-b, 2c+n, -n; \\ c, c+\tfrac{1}{2}, \tfrac{1}{2}+a-b+c; \end{array} 1\right]
$$

$$
= \frac{(\tfrac{1}{2}-a+c)_n\,(\tfrac{1}{2}-b+c)_n}{(\tfrac{1}{2}-a-b+c)_n\,(c+\tfrac{1}{2})_n}\, {}_4F_3\left[\begin{array}{c} a, b, -c-n, -n; \\ c, \tfrac{1}{2}+a-c-n, \tfrac{1}{2}+b-c-n; \end{array} 1\right].
$$

This proves the result (2.5.9), and so the second theorem is proved.

When $c = a+b$, corresponding to Clausen's theorem, deduced from the first theorem, we find from the second and third theorems, the two further results given by Orr,

$$
{}_2F_1[a, b;\ a+b-\tfrac{1}{2};\ z]\,{}_2F_1[a, b;\ a+b+\tfrac{1}{2};\ z]
$$
$$
= {}_3F_2[2a, 2b, a+b;\ 2a+2b-1, a+b+\tfrac{1}{2};\ z] \quad (2.5.12)
$$
and
$$
{}_2F_1[a, b;\ a+b-\tfrac{1}{2};\ z]\,{}_2F_1[a, b-1; a+b-\tfrac{1}{2};\ z]
$$
$$
= {}_3F_2[2a, 2b-1, a+b-1;\ 2a+2b-2, a+b-\tfrac{1}{2};\ z]. \quad (2.5.13)
$$

Five further theorems of a type similar to Orr's theorems have been proved by Bailey (1935a).

**Theorem IV.** *If*

$$
(1-z)^{a+b-c-\frac{1}{2}}\,{}_3F_2[2a, 2b, c;\ 2c, a+b+\tfrac{1}{2};\ z] = \sum_{n=0}^{\infty} a_n z^n, \quad (2.5.14)
$$

*then*

$$
{}_2F_1[a, b;\ a+b+\tfrac{1}{2};\ z]\,{}_2F_1[\tfrac{1}{2}+c-a, \tfrac{1}{2}+c-b;\ 2c-a-b+\tfrac{1}{2};\ z]
$$

$$
= \sum_{n=0}^{\infty} \frac{(c+\tfrac{1}{2})_n}{(2c-a-b+\tfrac{1}{2})_n}\, a_n z^n. \quad (2.5.15)
$$

By comparing the coefficients of $z^n$, we see that the identity to be proved is

$$
{}_4F_3[2a, 2b, c, -n;\ 2c, a+b+\tfrac{1}{2}, \tfrac{1}{2}-n+a+b-c;\ 1]
$$

$$
= \frac{(\tfrac{1}{2}+c-a)_n\,(\tfrac{1}{2}+c-b)_n}{(c+\tfrac{1}{2})_n\,(\tfrac{1}{2}-a-b+c)_n}\, {}_4F_3\left[\begin{array}{c} a, b, \tfrac{1}{2}+a+b-2c-n, -n; \\ \tfrac{1}{2}+a-c-n, \tfrac{1}{2}+b-c-n, a+b+\tfrac{1}{2}; \end{array} 1\right].
$$

$$
(2.5.16)
$$

This follows fairly easily from the application of (2.4.1.7), the transformation of two Saalschutzian $_4F_3(1)$ series, and (2.4.2.3), a transformation of a nearly-poised Saalschutzian $_4F_3(1)$ series.[†]

As special cases of theorem IV, we find[‡] that when $c = a+b-\frac{1}{2}$,

$$_2F_1[a,b;\ a+b+\tfrac{1}{2};\ z]\,_2F_1[a,b;\ a+b-\tfrac{1}{2};\ z]$$

$$= {}_3F_2[2a,2b,a+b;\ a+b+\tfrac{1}{2},2a+2b-1;\ z] \qquad (2.5.17)$$

and[§] if $c = a+b$,

$$_2F_1[a,b;\ a+b+\tfrac{1}{2};\ z]\,_2F_1[a+\tfrac{1}{2},b+\tfrac{1}{2};\ a+b+\tfrac{1}{2};\ z]$$

$$= (1-z)^{-\frac{1}{2}}\,_3F_2[2a,2b,a+b;\ a+b+\tfrac{1}{2},2a+2b;\ z], \quad (2.5.18)$$

which also follows as an immediate consequence of Clausen's theorem.

If $c = a-\frac{1}{2}$, theorem IV reduces to

$$_3F_2[2a,2b,a-\tfrac{1}{2};\ 2a-1,a+b+\tfrac{1}{2};z]$$

$$= (1-z)^{-b}\,_2F_1[b,a-b-\tfrac{1}{2};\ a+b+\tfrac{1}{2};\ z]. \quad (2.5.19)$$

**Theorem V.** *If*

$$(1-z)^{a+b-c-\frac{1}{2}}\,_3F_2[2a,2b,c;\ 2c,a+b+\tfrac{1}{2};\ z] = \sum_{n=0}^{\infty} a_n z^n, \quad (2.5.20)$$

*then*

$$_2F_1[b,c-b;\ c+\tfrac{1}{2};\ z]\,_2F_1[a+\tfrac{1}{2},c-a+\tfrac{1}{2};c+\tfrac{1}{2};z] = \sum_{n=0}^{\infty} \frac{(a+b+\tfrac{1}{2})_n}{(c+\tfrac{1}{2})_n}a_n z^n.$$

$$(2.5.21)$$

The identity implied here is

$$_4F_3[2a,2b,c,-n;\ 2c,a+b+\tfrac{1}{2},a+b+\tfrac{1}{2}-c-n;\ 1]$$

$$= \frac{(a+\tfrac{1}{2})_n(c+\tfrac{1}{2}-a)_n}{(a+b+\tfrac{1}{2})_n(\tfrac{1}{2}-a-b+c)_n}$$

$$\times {}_4F_3[b,c-b,\tfrac{1}{2}-c-n,-n;\ c+\tfrac{1}{2},\tfrac{1}{2}-a-n,\tfrac{1}{2}+a-c-n;\ 1].$$

$$(2.5.22)$$

This follows immediately from (2.5.15) of theorem IV and the application again of (2.4.1.7).

**Theorem VI.** *If*

$$(1-z)^{a+b-c}\,_3F_2[2a,2b,c+\tfrac{1}{2};\ 2c,a+b+\tfrac{1}{2};\ z] = \sum_{n=0}^{\infty} b_n z^n, \quad (2.5.23)$$

---

[†] Whipple (1926a), 10.11 and Whipple (1926b), 6.5.
[‡] Orr (1899), 59'.          [§] Orr (1899), 54'.

*then*

$$_2F_1[a, b; a+b+\tfrac{1}{2}; z] \,_2F_1[c-a, c-b; 2c-a-b+\tfrac{1}{2}; z]$$

$$= \sum_{n=0}^{\infty} \frac{(c)_n}{(2c-a-b+\tfrac{1}{2})_n} b_n z^n. \quad (2.5.24)$$

The identity implied here is

$$_4F_3[2a, 2b, c+\tfrac{1}{2}, -n; 2c, a+b+\tfrac{1}{2}, 1-n+a+b-c; 1]$$

$$= \frac{(c-a)_n (c-b)_n}{(c)_n (c-a-b)_n} \,_4F_3\left[\begin{array}{c} a, b, \tfrac{1}{2}+a+b-2c-n, -n; \\ a+b+\tfrac{1}{2}, 1+a-c-n, 1+b-c-n; \end{array} 1\right].$$

$$(2.5.25)$$

This theorem is analogous to Cayley's theorem, and the proof of this identity follows on similar lines.[†]

**Theorem VII.** *If*

$$_2F_1[a, b; c; z] \,_2F_1[a, b; d; z] = \sum_{n=0}^{\infty} c_n z^n, \quad (2.5.26)$$

*then*

$$_4F_3[a, b, \tfrac{1}{2}c+\tfrac{1}{2}d, \tfrac{1}{2}c+\tfrac{1}{2}b-\tfrac{1}{2}; a+b, c, d; 4z(1-z)] = \sum_{n=0}^{\infty} \frac{(c+d-1)_n}{(a+b)_n} c_n z^n.$$

$$(2.5.27)$$

**Theorem VIII.** *If*

$$_2F_1[a, b; c; z] \,_2F_1[a, d; c; z] = \sum_{n=0}^{\infty} d_n z^n, \quad (2.5.28)$$

*then*

$$(1-z)^{-a} \,_4F_3[a, b, d, c-a; \tfrac{1}{2}b+\tfrac{1}{2}d, \tfrac{1}{2}b+\tfrac{1}{2}d+\tfrac{1}{2}, c; -z^2/4(1-z)]$$

$$= \sum_{n=0}^{\infty} \frac{(c)_n}{(d+b)_n} d_n z^n. \quad (2.5.29)$$

The proofs of these two theorems follow on similar lines, but these are now based on the identity

$$_4F_3[-m, y, z, 1-m-u; 1-m-y, 1-m-z, w; 1]$$

$$= \Gamma\left[\begin{array}{c} w, w+u-1+2m \\ w+m, w-1+u+m \end{array}\right]$$

$$\times \,_5F_4\left[\begin{array}{c} 1-m-w, \tfrac{1}{2}-\tfrac{1}{2}m, -\tfrac{1}{2}m, 1-m-y-z, 1-m-u; \\ 1-m-y, 1-m-z, \tfrac{1}{2}z-\tfrac{1}{2}u-\tfrac{1}{2}w-m, 1-\tfrac{1}{2}u-\tfrac{1}{2}w-m; \end{array} 1\right],$$

$$(2.5.30)$$

which is the form assumed by (2.4.2.3) when $x \equiv f$ is a negative integer $-m$, and $n$ is not necessarily a positive integer.[‡]

† See Bailey (1932b).          ‡ Whipple (1926b), 6.6.

In particular, from theorem VII, if $d = a+b-c+1$, then

$$_2F_1[a, b; \ c; \ z]\,_2F_1[a, b; \ a+b-c+1; \ z]$$
$$= \,_4F_3[a, b, \tfrac{1}{2}a+\tfrac{1}{2}b, \tfrac{1}{2}a+\tfrac{1}{2}b+\tfrac{1}{2}; \ a+b, c, a+b-c+1; \ 4z(1-z)].$$
$$(2.5.31)$$

This relation holds inside that loop of the lemniscate

$$|4z(1-z)| = 1,$$

which surrounds the origin.

If $c = d = b$, we obtain Kummer's quadratic transform (2.3.2.1).

From theorem VIII, if $d = c-b$,

$$_2F_1[a, b; \ c; \ z]\,_2F_1[a, c-b; \ c; \ z]$$
$$= (1-z)^{-a}\,_4F_3[a, b, c-a, c-b; \ c, \tfrac{1}{2}c, \tfrac{1}{2}c+\tfrac{1}{2}; \ -z^2/\{4(1-z)\}] \quad (2.5.32)$$

and when $c = d = b$, we obtain†

$$(1-z)^{-a}\,_2F_1[a, b; \ a+b+\tfrac{1}{2}; \ -z^2/\{4(1-z)\}] = \,_2F_1[2a, a+b; \ 2a+2b; \ z].$$
$$(2.5.33)$$

Finally, if we generalize (2.5.29) by putting $z+n$ for $z$ in (2.5.28), we have

**Theorem IX.** *If* $c+c' = a+a'+b+b'$, *and*

$$_2F_1[a, b; \ c; \ z]\,_2F_1[a', b'; \ c'; \ z] = \sum_{n=0}^{\infty} e_n z^n,$$

*then*

$$(1-z)^{1-c-c'}\,_4F_3[a, c-b, \tfrac{1}{2}c+\tfrac{1}{2}c', \tfrac{1}{2}c+\tfrac{1}{2}c'-\tfrac{1}{2}; a+a', b+b', c; \ -4z/(1-z)^2]$$
$$= \sum_{n=0}^{\infty} \frac{(c')_n (c+c'-1)_n}{(a+a')_n (a+b')_n} e_n z^n. \quad (2.5.34)$$

When $a' = c-b, b' = 1-b$, and $c' = 1+a-b$, this gives us (2.5.32) again.

## 2.6 Partial sums of hypergeometric series

A number of results exist which express the sum of the first $n$ terms of a Gauss series with unit argument in terms of an infinite $_3F_2(1)$ series. The earliest of these results are due to Hill (1907, 1908) and Whipple (1930b). Ramanujan stated the identity

$$\frac{1}{n}+\left(\frac{1}{2}\right)^2 \frac{1}{n+1}+\left(\frac{1.3}{2.4}\right)^2 \frac{1}{n+2}+\ldots \text{ to infinity}$$

$$= \left\{\frac{\Gamma(n)}{\Gamma(n+\tfrac{1}{2})}\right\}^2 \left\{1+\left(\frac{1}{2}\right)^2+\left(\frac{1.3}{2.4}\right)^2+\ldots\right\} \text{ to } n \text{ terms.} \quad (2.6.1)$$

---

† Bailey (1928), 4.22.

Proofs of this theorem were given by Darling (1930) and Watson (1930$f$). Watson's proof which is reproduced here, is particularly simple. In the relation (2.3.3.1), connecting three $_3F_2(1)$ series, let us write $c = f+n-1$, where $n$ is a positive integer. Then we find that

$$_3F_2[a,b,f+n-1;\ e,f;\ 1] = \Gamma\begin{bmatrix}e,e-a-b\\e-a,e-b\end{bmatrix} _3F_2\begin{bmatrix}a,b,1-n;\\a+b-e+1,f;\end{bmatrix}1\ \end{bmatrix}.$$

(2.6.2)

Now let $e \to a+b+n$, and we find that

$_2F_1[a,b;f;\ 1]$ to $n$ terms

$$= \Gamma\begin{bmatrix}a+n,b+n\\n,a+b+n\end{bmatrix} _3F_2\begin{bmatrix}a,b,f+n-1;\\a+b+n,f;\end{bmatrix}1\ \end{bmatrix}. \quad (2.6.3)$$

Ramanujan's result (2.6.1) is the particular case of (2.6.3) in which $a = b = \frac{1}{2}$ and $f = 1$. The method of proof holds when $f \geqslant a+b$. Various generalizations of this result have been given by Whipple (1930$b$), Hodgkinson (1931) and Bailey (1931$b$). Bailey's generalization is the elegant result

$$\Gamma\begin{bmatrix}x+m,y+m\\m,x+y+m\end{bmatrix} _3F_2\begin{bmatrix}x,y,v+m-1;\\v,x+y+m;\end{bmatrix}1\ \end{bmatrix} \text{ to } n \text{ terms}$$

$$= \Gamma\begin{bmatrix}x+n,y+n\\n,x+y+n\end{bmatrix} _3F_2\begin{bmatrix}x,y,v+n-1;\\v,x+y+n;\end{bmatrix}1\ \end{bmatrix} \text{ to } m \text{ terms.} \quad (2.6.4)$$

Three alternative proofs are given by Bailey (1931$b$). The most straightforward of these is similar to that used in the usual proof of Dougall's theorem by induction.

We suppose, without loss of generality, that $n \geqslant m$. Then, in the terms of the series on the left, the factors $v+r$ in the denominator cancel with the factors in the numerator when $r \geqslant m-1$. Thus, if we multiply (2.6.4) straight across by $(v)_{m-1}$, we obtain two polynomials in $v$ of degree $m-1$. If we can now prove that these two polynomials are equal for $m$ values of $v$, we shall have established our result. But for each of the $m$ values $v = -n+1, -n, -n-1, ..., -n-m+2$, the partial series become complete hypergeometric series, which are summable by Saalschutz's theorem, and their equality can be seen immediately. Hence our result (2.6.4) is proved.

In particular, when $m \to \infty$ in (2.6.4), the theorem reduces to (2.6.3) above, which is now seen to be true for all values of the parameters. Many alternative forms of (2.6.2) can be deduced, simply by trans-

6

forming the series $_3F_2(1)$. Thus, if we use the relation (2.3.3.1) again, between three such series, we find that[†]

$_2F_1[a, b;\, c;\, 1]$ to $n$ terms

$$= \Gamma\begin{bmatrix} 1+a-c, 1+b-c \\ 1-c, 1+a+b-c \end{bmatrix}\left\{1 - \frac{(a)_n\,(b)_n}{(c-1)_n\,n!}\, {}_3F_2\begin{bmatrix} 1-a, 1-b, n; \\ 2-c, 1+n; \end{bmatrix}1\right\},$$

(2.6.5)

provided that $\mathrm{Rl}\,(1+a+b-c) > 0$.

Again if we make use of (2.3.3.7), a relation between two $_3F_2(1)$ series, we have[‡]

$_2F_1[a, b;\, c;\, 1]$ to $n$ terms

$$= \Gamma\begin{bmatrix} a+n, b+n \\ n, 1+a+b-c, c+n \end{bmatrix} {}_3F_2\begin{bmatrix} c-a, c-b, c+n-1; \\ c, c+n; \end{bmatrix}1\end{bmatrix}, \quad (2.6.6)$$

under the same restriction $\mathrm{Rl}\,(1+a+b-c) > 0$.

We can rewrite Whipple's transformation of a $_4F_3(1)$ series into a $_7F_6(1)$ series, (2.4.1.1), in the form

$_4F_3[t, x, y, z;\, u, v, w;\, 1]$

$$= \Gamma\begin{bmatrix} v+w-t, 1+x-u, 1+y-u, 1+z-u \\ 1+y+z-u, 1+z+x-u, 1+x+y-u, 1-u \end{bmatrix}$$

$$\times {}_7F_6\begin{bmatrix} a, 1+\tfrac{1}{2}a, w-t, v-t, & x, & y, & z; \\ \tfrac{1}{2}a, & w, & v, 1+y+z-u, 1+z+x-u, 1+x+y-u; \end{bmatrix}1\end{bmatrix},$$

(2.6.7)

where $a = x+y+z-u$, and $u+v+w-t-x-y-z = 1$, and one of the parameters $t, x, y$ or $z$ is a negative integer. Then, if we put $t = 1-n$, and let $u \to 1-n$, we find the result that

$$_3F_2[x, y, z;\, v, w;\, 1]_n = \Gamma\begin{bmatrix} v+w+n-1, x+n, y+n, z+n \\ n, y+z+n, z+x+n, x+y+n \end{bmatrix}$$

$$\times {}_7F_6\begin{bmatrix} a, 1+\tfrac{1}{2}a, w+n-1, v+n-1, & x, & y, & z; \\ \tfrac{1}{2}a, & v, & w, y+z+n, z+x+n, x+y+n; \end{bmatrix}1\end{bmatrix},$$

(2.6.8)

where $a = x+y+z+n-1$, and the $_3F_2(1)$ series on the left is Saalschutzian. Again we can deduce (2.6.2) from this result if we substitute for $w$ and let $z \to \infty$.

† Whipple (1930b).        ‡ Hodgkinson (1931).

Further, this well-poised $_7F_6(1)$ series can be transformed in various ways, into two Saalschutzian $_4F_3(1)$ series. In particular, using the formula (2.4.4.3) we find† that

$$_3F_2[x, y, z; v, w; 1]_n$$

$$= \Gamma\begin{bmatrix} x+n, y+n, z+n \\ n, v+n-1, w+n-1 \end{bmatrix} \left\{ \Gamma\begin{bmatrix} w, w-v \\ v+n-1, w-x, w-y, w-z \end{bmatrix} \right.$$

$$\times {}_4F_3\begin{bmatrix} v-x, v-y, v-z, v+n-1; \\ v, v+1-w, v+n; \end{bmatrix} 1 \end{bmatrix} + \Gamma\begin{bmatrix} v, v-w \\ w+n-1, v-x, v-y, v-z \end{bmatrix}$$

$$\times {}_4F_3\begin{bmatrix} w-x, w-y, w-z, w+n-1; \\ w, w+1-v, w+n; \end{bmatrix} 1 \end{bmatrix} \right\}. \quad (2.6.9)$$

In a similar way, using the two relations (2.4.3.1) and (2.4.4.1) connecting four well-poised $_9F_8(1)$ series, we can obtain two formulae each of which gives the sum to $n$ terms of a well-poised $_7F_6(1)$ series in terms of two infinite well-poised $_7F_6(1)$ series.

## 2.6.1 A partial summation theorem. We shall let

$$_{A+1}F_A[a_0, a_1, ..., a_A; 1+b_1, 1+b_2, ..., 1+b_A; 1]_N$$

$$\equiv \sum_{n=0}^{N} \frac{(a_0)_n (a_1)_n (a_2)_n...(a_A)_n}{n! (1+b_1)_n (1+b_2)_n...(1+b_A)_n}$$

$$\equiv \sum_{n=0}^{N} \alpha_n \qquad (2.6.1.1)$$

and $\qquad v_n \equiv (a_0+n)(a_1+n)(a_2+n)...(a_A+n)\alpha_n.$

Then $\qquad v_n - v_{n-1} = \alpha_n\{(a_0+n)(a_1+n)(a_2+n)...(a_A+n)$

$$-n(b_1+n)(b_2+n)...(b_A+n)\}, \quad (2.6.1.2)$$

since $\qquad \alpha_n = \dfrac{(a_0+n-1)(a_1+n-1)...(a_A+n-1)}{n(b_1+n)(b_2+n)...(b_A+n)}\alpha_{n-1}.$

Let us suppose now that the terms in powers of $n$ in (2.6.1.2) vanish. This implies that

$$\left. \begin{array}{c} a_0+a_1+a_2+...+a_A = b_1+b_2+...+b_A, \\ a_0a_1+a_1a_2+...+a_{A-1}a_A = b_1b_2+b_2b_3+...+b_{A-1}b_A, \\ \cdots\cdots\cdots\cdots\cdots\cdots\cdots\cdots\cdots\cdots\cdots\cdots\cdots\cdots\cdots\cdots\cdots\cdots\cdots \\ a_0a_1a_2...a_A = b_1b_2b_3...b_A, \end{array} \right\} \quad (2.6.1.3)$$

† Darling (1930), p. 335.

that is that $-a_0, -a_1, -a_2, \ldots, -a_A$ are the roots of the equation

$$x^{A+1} + C_1 x^A + C_2 x^{A-1} + \ldots + C_{A+1} = 0, \qquad (2.6.1.4)$$

and that $-b_1, -b_2, \ldots, -b_A$ are the roots of the equation

$$x^A + C_1 x^{A-1} + C_2 x^{A-2} + \ldots + C_A = 0. \qquad (2.6.1.5)$$

It follows that $\quad v_n - v_{n-1} = \alpha_n a_0 a_1 a_2 \ldots a_A,$

$$v_{n-1} - v_{n-2} = \alpha_{n-1} a_0 a_1 a_2 \ldots a_A,$$

$$\ldots \ldots \ldots \ldots \ldots \ldots \ldots \ldots \ldots$$

$$v_1 - v_0 = \alpha_1 a_0 a_1 a_2 \ldots a_A,$$

$$v_0 = \alpha_0 a_0 a_1 a_2 \ldots a_A.$$

So, by addition, $\quad v_n = \sum_{r=0}^{n} \alpha_r a_0 a_1 a_2 \ldots a_A, \qquad (2.6.1.6)$

that is

$$_{A+1}F_A[a_0, a_1, \ldots, a_A; 1+b_1, \ldots, 1+b_A; 1]_N$$

$$= \frac{(1+a_0)_N (1+a_1)_N (1+a_2)_N \ldots (1+a_A)_N}{N! \, (1+b_1)_N (1+b_2)_N (1+b_3)_N \ldots (1+b_A)_N} \qquad (2.6.1.7)$$

under the conditions (2.6.1.3).

Further, if $N \to \infty$, in (2.6.1.7), we have

$$_{A+1}F_A[a_0, a_1, a_2, \ldots, a_A; 1+b_1, \ldots, 1+b_A; 1]$$

$$= \Gamma \begin{bmatrix} 1+b_1, 1+b_2, \ldots, 1+b_A \\ 1+a_0, 1+a_1, 1+a_2, \ldots, 1+a_A \end{bmatrix}, \qquad (2.6.1.8)$$

under the same set of conditions. Since $a_0 + \Sigma a_\nu = \Sigma b_\nu$, this series is always convergent for all values of $a_0, a_1, \ldots, a_A$ and $b_1, b_2, \ldots, b_A$. In particular,

$$_2F_1[a, b; 1+a+b; 1]_n = \frac{(1+a)_n (1+b)_n}{n! \, (1+a+b)_n} \qquad (2.6.1.9)$$

and $\quad _3F_2[a, b, c; d, a+b+c-d; 1]_n = \frac{(1+a)_n (1+b)_n (1+c)_n}{n! \, (d)_n (a+b+c-d)_n}$

$$(2.6.1.10)$$

where $\quad bc + ca + ab = (d-1)(a+b+c-d-1).$

As $n \to \infty$, (2.6.1.9) becomes Gauss's theorem, and (2.6.1.10) becomes a disguised form of Dixon's theorem,

$$_3F_2[a, b, c; d, a+b+c-d; 1] = \Gamma \begin{bmatrix} d, a+b+c-d \\ 1+a, 1+b, 1+c \end{bmatrix}, \qquad (2.6.1.11)$$

where $\quad bc + ca + ab = (d-1)(a+b+c-d-1).$

# 3

# BASIC HYPERGEOMETRIC FUNCTIONS

## 3.1 Historical introduction

A different view of the problem of generalizing the Gauss function was taken by E. Heine, in his book.[†] In this work, he defined a basic number as

$$a_q \equiv \frac{1-q^a}{1-q},\qquad (3.1.1)$$

where $q$ and $a$ are real or complex numbers, so that as $q \to 1, \dfrac{1-q^a}{1-q} \to a$. This passage to the limit is of the type known as de l'Hospital's limit.

Using this concept, Heine then defined the basic analogue of the Gauss function as the infinite series

$$1 + \frac{(1-q^a)(1-q^b)z}{(1-q^c)(1-q)} + \frac{(1-q^a)(1-q^{a+1})(1-q^b)(1-q^{b+1})z^2}{(1-q^c)(1-q^{c+1})(1-q)(1-q^2)} + \dots,$$
$$(3.1.2)$$

where $|q| < 1$, so that as $q \to 1$, this series $\to {}_2F_1[a,b;\,c;\,z]$, the Gauss series.

A very early example of such a series is contained in Euler's identity[‡]

$$1 + \sum_{n=1}^{\infty} (-1)^n \{q^{\frac{1}{2}n(3n-1)} + q^{\frac{1}{2}n(3n+1)}\} = \prod_{n=1}^{\infty} (1-q^n), \qquad (3.1.3)$$

and several interesting results were given by Gauss,[§] for example

$$1 + \sum_{n=1}^{\infty} q^{\frac{1}{2}n(n+1)} = \prod_{n=1}^{\infty} \{(1-q^{2n})/(1-q^{2n-1})\}. \qquad (3.1.4)$$

The earliest example of an algebraic infinite product is the partition function

$$1 \Big/ \prod_{n=0}^{\infty} (1-aq^n)$$

discussed by Euler.[||] Gauss also studied these infinite products. For example he gave[¶]

$$(x + x^9 + x^{25} + \dots) - (x^3 + x^{27} + x^{75} + \dots)$$

$$= x(1-x^2)(1+x^{10})(1+x^{14})(1-x^{22})(1-x^{26})\dots$$

$$(1+x^{12})(1-x^{24})(1+x^{36})\dots(1+x^8)(1+x^{16})(1+x^{24})\dots \quad (3.1.5)$$

---

[†] Heine (1898). *Handbuch die Kugelfunctionen.*
[‡] Euler (1748).     [§] Gauss (1866).
[||] Euler (1748), p. 304.     [¶] Gauss (1866), p. 448, eq. 22; p. 454, eq. 57.

and

$$(1 - x^4)(1 - x^{10})(1 - x^{16})...(1 - x^2)(1 - x^8)(1 - x^{14})...$$
$$= (1 - x^2 - x^4 + x^{10} + x^{14} - x^{24} - ...)/\{(1 - x^6)(1 - x^{12})(1 - x^{18})...\}$$

(3.1.6)

that is, in the modern notation,

$$\prod_{n=1}^{\infty}\{(1 - x^{6n-2})(1 - x^{6n-4})(1 - x^{6n})\} = 1 + \sum_{n=1}^{\infty}(-1)^n x^{3n^2}(x^n + x^{-n}).$$

(3.1.7)

These results clearly foreshadowed the work of C. G. J. Jacobi (1804–1851), who, in his *Fundamenta Nova* (1829), defined the four theta functions

$$\vartheta_1(z, q) = 2\sum_{n=0}^{\infty}(-1)^n q^{(n+\frac{1}{2})^2}\sin\{(2n+1)z\}, \tag{3.1.8}$$

$$\vartheta_2(z, q) = 2\sum_{n=0}^{\infty} q^{(n+\frac{1}{2})^2}\cos\{(2n+1)z\}, \tag{3.1.9}$$

$$\vartheta_3(z, q) = 1 + 2\sum_{n=1}^{\infty} q^{n^2}\cos(2nz), \tag{3.1.10}$$

$$\vartheta_4(z, q) = 1 + 2\sum_{n=1}^{\infty}(-1)^n q^{n^2}\cos(2nz). \tag{3.1.11}$$

He deduced most of the known relations between these functions by purely algebraic methods. Jacobi also stated the theorem

$$1 + \sum_{n=1}^{\infty}(-1)^n x^{an^2}(z^{bn} + z^{-bn})$$
$$= \prod_{n=1}^{\infty}\{(1 - x^{a(2n-1)}z^b)(1 - x^{a(2n-1)}z^{-b})(1 - x^{2an})\}, \quad (3.1.12)$$

which is fundamental in most of the later work on infinite products. From this theorem it follows that

$$\vartheta_1(z, q) = -iq^{\frac{1}{4}}e^{iz}\prod_{n=1}^{\infty}\{(1 - q^{2n}e^{2iz})(1 - q^{2n-2}e^{-2iz})(1 - q^{2n})\},$$

(3.1.13)

$$\vartheta_2(z, q) = q^{\frac{1}{4}}e^{iz}\prod_{n=1}^{\infty}\{(1 + q^{2n}e^{2iz})(1 + q^{2n-2}e^{-2iz})(1 - q^{2n})\}, \quad (3.1.14)$$

$$\vartheta_3(z, q) = \prod_{n=1}^{\infty}\{(1 + q^{2n-1}e^{2iz})(1 + q^{2n-1}e^{-2iz})(1 - q^{2n})\}, \quad (3.1.15)$$

$$\vartheta_4(z, q) = \prod_{n=1}^{\infty}\{(1 - q^{2n-1}e^{2iz})(1 - q^{2n-1}e^{-2iz})(1 - q^{2n})\}. \quad (3.1.16)$$

The series in equations (3.1.8–12) can be expressed in the forms

$$\sum_{n=-\infty}^{\infty}(-1)^n x^{n(3n+1)} \quad \text{and} \quad \sum_{n=-\infty}^{\infty}(-1)^n x^{n(an+b)}.$$

They are thus very early examples of 'bilateral' series, that is to say, series which are infinite in both directions, though it is doubtful whether Gauss or Jacobi recognized this fact. Jacobi, in particular, gave very many examples of basic series, arising from the study of elliptic modular functions.

Most of the early examples were collected and investigated systematically, by Heine, and it became apparent in his work, that a theory almost exactly parallel and certainly as extensive as that for the Gauss functions could be developed for these basic hypergeometric functions. This development was carried out by F. H. Jackson (1870–1960), who throughout his long life, studied the concept of the basic number at length. He gave the basic analogues of most of the special summation theorems. In particular, he proved the basic analogue of Dougall's theorem as early as 1909, though the proof was not published until 1921.† He also developed the concepts of $q$-difference equations as basic analogues of the ordinary difference equations, and $q$-integration‡ as the analogue of integration.

Practically every branch of normal function theory has been extended to the basic number field, so that now we have basic exponential, trigonometric, and hyperbolic functions, basic analogues of Bessel, Weber and Airy functions, and basic Legendre, Laguerre, Hermite and Gegenbaur polynomials.§ Most of the actual applications of the $q$-concept have occurred in the field of pure mathematics, particularly in number theory, modular equations and elliptic integrals. The main obstacle to their application in applied mathematics was the difficulty of actual numerical evaluation in all but the simplest cases. With the advent of electronic computers, this difficulty has largely vanished, though much remains to be done in almost every part of $q$-function theory.

## 3.2 The convergence of Heine's series

In Heine's series (3.1.2), $a, b, c, q$ and $z$ may be real or complex numbers. If the $n$th term of the series is $u_n$, then we have

$$u_n = \frac{(1-q^a)(1-q^{a+1})\ldots(1-q^{a+n-1})(1-q^b)(1-q^{b+1})\ldots(1-q^{b+n-1})z^n}{(1-q)(1-q^2)\ldots(1-q^n)(1-q^c)(1-q^{c+1})\ldots(1-q^{c+n-1})}$$

$$(3.2.1)$$

---

† Jackson (1921 $a$).  ‡ F. H. Jackson (1951).
§ Jackson (1905 $a$, 1921 $b$, 1942); Hahn (1949 $a$, 1950, 1955).

so that
$$\frac{u_{n+1}}{u_n} = \frac{(1-q^{a+n})(1-q^{b+n})}{(1-q^{1+n})(1-q^{c+n})} z \tag{3.2.2}$$

and this expression $\to z$, as $n \to \infty$, provided that $|q| < 1$. Hence, when $|q| < 1$, Heine's series is absolutely convergent for $|z| < 1$.

The product
$$\prod_{n=0}^{\infty} (1-aq^n)$$

is absolutely convergent for all finite values of $a$, real or complex, when $|q| < 1$.

If $|q| > 1$, we can write $q = 1/p$, where $|p| < 1$. Then Heine's series becomes

$$1 + \frac{(1-p^{-a})(1-p^{-b})z}{(1-p^{-c})(1-p^{-1})} + \frac{(1-p^{-a})(1-p^{-a-1})(1-p^{-b})(1-p^{-b-1})z^2}{(1-p^{-c})(1-p^{-c-1})(1-p^{-1})(1-p^{-2})} + \dots$$

$$\equiv 1 + \frac{(1-p^a)(1-p^b)}{(1-p^c)(1-p)} zp^{1+c-a-b}$$

$$+ \frac{(1-p^a)(1-p^{a+1})(1-p^b)(1-p^{b+1})}{(1-p^c)(1-p^{c+1})(1-p)(1-p^2)} z^2 p^{2(1+c-a-b)} + \dots \tag{3.2.3}$$

This series is of exactly the same type as Heine's series, with $q$ replaced by $p$, and $z$ replaced by $zp^{1+c-a-b}$. This new series is convergent if $|p| < 1$, and $|zp^{1+c-a-b}| < 1$, that is $|z| < 1$. Thus it can always be supposed without any loss of generality, that $|q| < 1$, for, if it is not, the above process of putting $1/p$ for $q$ will always reduce the series to a similar series in which $|p| < 1$. This concept of inversion with respect to the base $q$ can be extended throughout the theory without much difficulty.

### 3.2.1 Notation. We shall write

$$(q^a; q)_n = (1-q^a)(1-q^{a+1})(1-q^{a+2})\dots(1-q^{a+n-1}). \tag{3.2.1.1}$$

In this notation, we have

$$\lim_{q \to 1} \left\{ \frac{(q^a; q)_n}{(q; q)_n} \right\} = (a)_n. \tag{3.2.1.2}$$

Thus, if $q = 0.9$, and $a = 0.1$, we have

$$(0.9^{0.1}; 0.9)_3 = (1 - 0.9^{0.1})(1 - 0.9^{1.1})(1 - 0.9^{2.1})$$

and
$$\lim_{q \to 1} \left\{ \frac{(q^{0.1}; q)_3}{(q; q)_3} \right\} = \lim_{q \to 1} \frac{(1-q^{0.1})(1-q^{1.1})(1-q^{2.1})}{(1-q)(1-q^2)(1-q^3)},$$

$$= 0.1 \times 1.1 \times 2.1,$$

$$= (0.1)_3.$$

Normally, we shall follow Watson's notation and write $a$ in place of $q^a$, for ease in printing, so that

$$(a; q)_n = (1 - a)(1 - aq)(1 - aq^2)...(1 - aq^{n-1})$$

$$= \prod_{m=0}^{\infty} \{(1 - aq^m)/(1 - aq^{m+n})\}, \qquad (3.2.1.3)$$

and in particular $\qquad (a; q)_0 = 1,$

for all values of $a$, real or complex.

Then Heine's series becomes the basic hypergeometric function which we shall write as

$$_2\Phi_1[a, b; c; q, z]$$

$$= 1 + \frac{(1-a)(1-b)}{(1-c)(1-q)} z + \frac{(1-a)(1-aq)(1-b)(1-bq)}{(1-c)(1-cq)(1-q)(1-q^2)} z^2 + ...$$

$$= \sum_{n=0}^{\infty} \frac{(a; q)_n (b; q)_n}{(c; q)_n (q; q)_n} z^n, \qquad (3.2.1.4)$$

where $|q| < 1$, and $|z| < 1$. Here $a$, $b$ and $c$ are the parameters, $z$ is the variable and $q$ is called the base of the series.

This symbol, due originally to Heine, and extended by F. H. Jackson, must not be confused with the symbol given in (2.1.1.6) above, which has been used by both Erdélyi and Meijer to represent an entirely different ordinary hypergeometric function.

Two separate passages to the limit are possible with the function $(a; q)_n$. The first of these is that in which $q \to 1$, considered above. The other passage to the limit is that in which $n \to \infty$. Here we find that

$$\lim_{n \to \infty} (a; q)_n = \prod_{n=0}^{\infty} (1 - aq^n), \qquad (3.2.1.5)$$

and if we make both passages to the limit together, we have

$$\lim_{q \to 1} \lim_{n \to \infty} \left\{ \frac{(a; q)_n}{(q; q)_n} \right\} = \lim_{q \to 1} \prod_{n=0}^{\infty} \frac{(1 - aq^n)}{(1 - q^n)}, \qquad (3.2.1.6)$$

$$= 1/\Gamma(\log a). \qquad (3.2.1.7)$$

Similarly, reversing the order of the two limiting processes, we have

$$\lim_{n \to \infty} \lim_{q \to 1} \left\{ \frac{(q^a; q)_n}{(q; q)_n} \right\} = \lim_{n \to \infty} (a)_n = 1/\Gamma(a). \qquad (3.2.1.8)$$

Alternative notations for the product $(a; q)_n$ have been used at various times. Thus we find

$$(a)_{q, n} \quad \text{(Bailey (1935), § 8.1)}$$

or $\qquad [a]_n \quad$ (Jackson (1941 a)).

The base $q$ can itself be a simple function, so that for example, we might write

$$(a; q^{\frac{1}{4}})_n = (1-a)(1-aq^{\frac{1}{4}})(1-aq)(1-aq^{\frac{3}{4}})..., \qquad (3.2.1.9)$$

and this must be distinguished carefully from

$$(aq^{\frac{1}{4}}; q)_n = (1-aq^{\frac{1}{4}})(1-aq^{\frac{5}{4}})(1-aq^{\frac{9}{4}}).... \qquad (3.2.1.10)$$

The disadvantages of the alternative notations can be seen clearly from the above example, for in such a case $(a)_{q^{\frac{1}{4}}, n}$ involves double suffixes which are very difficult to print, and $[a]_n$ does not state $q$ explicitly, so that no distinction can be made between the two products $(a; q^{\frac{1}{4}})_n$ and $(aq^{\frac{1}{4}}; q)_n$ in (3.2.1.9) and (3.2.1.10) above.

The general basic hypergeometric series is defined as

$$_A\Phi_B[a_1, a_2, ..., a_A; b_1, b_2, ... b_B; q, z]$$

$$\equiv \sum_{n=0}^{\infty} \frac{(a_1; q)_n (a_2; q)_n...(a_A; q)_n z^n}{(b_1; q)_n (b_2; q)_n...(b_B; q)_n (q; q)_n} \qquad (3.2.1.11)$$

in which there are always $A$ of the $a$ parameters, and $B$ of the $b$ parameters. In such a case, the product of products

$$(a_1; q)_n (a_2; q)_n...(a_A; q)_n$$

can be shortened still further to

$$((a; q))_n,$$

where it is understood that there are always $A$ of the $a$ parameters. So we may write (3.2.1.11) as

$$\sum_{n=0}^{\infty} \frac{((a; q))_n z^n}{((b; q))_n (q; q)_n} \equiv {}_A\Phi_B[(a);(b); q, z]. \qquad (3.2.1.12)$$

The general infinite products play a role in the present basic theory parallel to the role of the Gamma functions in the ordinary hypergeometric field. So, in a similar way, we can shorten the notation for them. We shall write

$$\prod_{n=0}^{\infty} \frac{(1-a_1 q^n)(1-a_2 q^n)...(1-a_A q^n)}{(1-b_1 q^n)(1-b_2 q^n)...(1-b_B q^n)}$$

$$\equiv \Pi \begin{bmatrix} a_1, a_2, ..., a_A; \\ b_1, b_2, ..., b_B; \end{bmatrix} q \end{bmatrix} \equiv \Pi[(a); (b); q], \qquad (3.2.1.13)$$

where again it is understood that there are always $A$ of the $a$ parameters, and $B$ of the $b$ parameters, as usual.

As an alternative to this notation, some writers have had

$$[a, q]_\infty = \prod_{n=0}^{\infty} (1 - aq^n).$$

We shall occasionally be forced to extend our notation even further, and write for example $_A\Phi_B[(a); (b); q, q^n z]$

to indicate a series in which $z^n$ is replaced by $z^n q^{n^2}$ in the $n$'th term. This can only be done if there is no possibility of confusion with $z^n q^{mn}$ occurring in the $n$'th term, so that the use of $n$ in such an index will have this special meaning and $n$ will be avoided in the normal index, $m$ being used in its place. In any doubtful case, the general term of the series should be written out in full.

**3.2.2 Some simple results.** In order to become a little more familiar with the basic hypergeometric notation, we shall discuss first a few simple and elegant results. As particular cases of the $_2\Phi_1(z)$ series, we have

$$\frac{z}{1-q}\, _2\Phi_1[q, q; q^2; q, z] = \frac{z}{1-q} + \frac{z^2}{1-q^2} + \ldots + \frac{z^n}{1-q^n} + \ldots$$

$$(3.2.2.1)$$

and

$$\frac{z}{1-q^{\frac{1}{2}}}\, _2\Phi_1[q, q^{\frac{1}{2}}; q; q, z] = \frac{z}{1-q^{\frac{1}{2}}} + \frac{z^2}{1-q^{\frac{3}{2}}} + \ldots + \frac{z^n}{1-q^{n+\frac{1}{2}}} + \ldots$$

$$(3.2.2.2)$$

If we divide (3.2.2.2) throughout by $z^{\frac{1}{2}}$ and replace $q$, $z$ by $q^2$, $q\,e^{2ix}$, where $x$ is real, the imaginary part of the series becomes

$$\frac{q^{\frac{1}{2}}\sin x}{1-q} + \frac{q^{\frac{3}{2}}\sin 3x}{1-q^3} + \ldots + \frac{q^{n+\frac{1}{2}}\sin(2n+1)x}{1-q^{2n+1}} + \ldots = \frac{Kk}{2\pi}\, \mathrm{sn}\frac{2Kx}{\pi}.$$

$$(3.2.2.3)$$

This is another illustration of the close connection between the theory of elliptic functions and that of basic hypergeometric functions, of which the elliptic functions are always special cases.

Again

$$_2\Phi_1[q, -1; -q; q, z] = 1 + \frac{2z}{1+q} + \frac{2z^2}{1+q^2} + \ldots + \frac{2z^n}{1+q^n} + \ldots$$

$$(3.2.2.4)$$

In a similar way, from this series (3.2.2.4) we can derive a series for $\mathrm{dn}\,(2Kx/\pi)$.

As the basic analogues for the binomial theorem, we have

$$_1\Phi_0[a; \; ; q, z] = 1 + \frac{1-a}{1-q}z + \frac{(1-a)(1-aq)}{(1-q)(1-q^2)}z^2 + \ldots \quad (3.2.2.5)$$

and    $$_1\Phi_0[a; \; ; q, qz] = 1 + \frac{1-a}{1-q}qz + \frac{(1-a)(1-aq)}{(1-q)(1-q^2)}q^2z^2 + \ldots.$$
$$(3.2.2.6)$$

By the subtraction of (3.2.2.6) from (3.2.2.5), we find that

$$_1\Phi_0[a; \; ; q, z] - {}_1\Phi_0[a; \; ; q, qz] = (1-a)z\,{}_1\Phi_0[aq; \; ; q, z]. \quad (3.2.2.7)$$

Similarly,

$$_1\Phi_0[a; \; ; q, z] - a\,{}_1\Phi_0[a; \; ; q, qz] = (1-a)\,{}_1\Phi_0[aq; \; ; q, z]. \quad (3.2.2.8)$$

If we eliminate the series on the right-hand side of (3.2.2.7) and (3.2.2.8), we have

$$(1-z)\,{}_1\Phi_0[a; \; ; q, z] = (1-az)\,{}_1\Phi_0[a; \; ; q, qz]. \quad (3.2.2.9)$$

Next let us apply the process (3.2.2.9), $n$ times in succession; then we get

$$_1\Phi_0[a; \; ; q, z] = \frac{(az; q)_n}{(z; q)_n}\,{}_1\Phi_0[a; \; ; q, q^n z], \quad (3.2.2.10)$$

so that, for $|q| < 1$, as $n \to \infty$, $q^n z \to 0$, and $_1\Phi_0[a; \; ; q, q^n z] \to 1$, and we have Heine's theorem

$$_1\Phi_0[a; \; ; q, z] = \prod_{r=0}^{\infty} \{(1-azq^r)/(1-zq^r)\}, \quad (3.2.2.11)$$

that is

$$1 + \frac{(1-a)}{(1-q)}z + \frac{(1-a)(1-aq)}{(1-q)(1-q^2)}z^2 + \ldots = \frac{(1-az)(1-azq)(1-azq^2)\ldots}{(1-z)(1-zq)(1-zq^2)\ldots}.$$
$$(3.2.2.12)$$

From this result, we can deduce immediately, that

$$_1\Phi_0[a; \; ; q, z]\,{}_1\Phi_0[b; \; ; q, az] = {}_1\Phi_0[ab; \; ; q, z]. \quad (3.2.2.13)$$

If $a = 0$, a special case of this result is

$$_0\Phi_0[; \; ; q, z] = 1 \bigg/ \prod_{r=0}^{\infty}(1-zq^r), \quad (3.2.2.14)$$

that is

$$1 + \frac{z}{1-q} + \frac{z^2}{(1-q)(1-q^2)} + \ldots = 1/\{(1-z)(1-zq)(1-zq^2)\ldots\}.$$
$$(3.2.2.15)$$

Further, when $z = q$,

$$_0\Phi_0[; \; ; q, q] = 1 \bigg/ \prod_{r=1}^{\infty}(1-q^r), \quad (3.2.2.16)$$

that is

$$1 + \frac{q}{1-q} + \frac{q^2}{(1-q)(1-q^2)} + \ldots = 1/\{(1-q)(1-q^2)(1-q^3)\ldots\}.$$

(3.2.2.17)

Alternatively, we can replace $z$ in (3.2.2.11) by $z/a$, and let $a \to \infty$. Then we find that

$$1 - \frac{z}{1-q} + \frac{qz^2}{(1-q)(1-q^2)} - \ldots + \frac{(-1)^n q^{\frac{1}{2}n(n-1)} z^n}{(q;q)_n} + \ldots$$

$$= (1-z)(1-qz)(1-q^2z)\ldots, \quad (3.2.2.18)$$

and finally, if $z = q$,

$$1 - \frac{q}{1-q} + \frac{q^3}{(1-q)(1-q^2)} - \ldots + \frac{(-1)^n q^{\frac{1}{2}n(n+1)}}{(q;q)_n} + \ldots$$

$$= (1-q)(1-q^2)(1-q^3)\ldots. \quad (3.2.2.19)$$

It has already been noted that there is a difficulty in the representation of series like those in (3.2.2.18) and (3.2.2.19) without ambiguity in our present notation, which is by no means perfect. Nevertheless, there are many quiet corners of the subject, like this one, which have given much pleasure and intellectual delight to many mathematicians during the past two centuries.

## 3.3 Special theorems on the summation of basic series

In the following sections we shall give the basic analogues, in so far as these exist, of all the special summation theorems of the ordinary hypergeometric series. In the first place, here are some necessary definitions, which are the analogues of the corresponding properties of ordinary series.

A series in which the product of each pair of numerator and denominator parameters is constant is called a well-poised basic series. For example,
$$_3\Phi_2[a, b, c; aq/b, aq/c; q, z]$$

is said to be a series well-poised in $a$. If this property holds except for one pair of parameters, this pair can be made to stand either first or last in the sequence of parameters, and such a series is then said to be a nearly-poised basic series of the first or second kind respectively. A series in which the product of the numerator parameters is $q$ times the product of the denominator parameters, is said to be a Saalschutzian basic series.

In general, if a transformation exists connecting several basic series, then, by a simple passage to the limit, as $q \to 1$, we can always deduce the corresponding transformation for ordinary hypergeometric series. The converse is not always true. While basic analogues exist of most of the main summation theorems and well-poised transformations, some of the transformations of nearly-poised and Saalschutzian series have no known basic analogues.

### 3.3.1 Jackson's theorem.

First we shall prove the basic analogue of Dougall's theorem, which is called Jackson's theorem.† From Jackson's theorem, we can deduce all the simpler basic summation theorems, just as, from Dougall's theorem, we could deduce all the simpler ordinary summation theorems.

Jackson's theorem states that

$$
{}_8\Phi_7 \left[ \begin{matrix} a, q\sqrt{a}, -q\sqrt{a}, & b, & c, & d, & e, & q^{-N}; \\ \sqrt{a}, & -\sqrt{a}, aq/b, aq/c, aq/d, aq/e, aq^{N+1}; \end{matrix} q, q \right]
$$

$$
= \frac{(aq; q)_N (aq/cd; q)_N (aq/bd; q)_N (aq/bc; q)_N}{(aq/b; q)_N (aq/c; q)_N (aq/d; q)_N (aq/bcd; q)_N}, \quad (3.3.1.1)
$$

where $N$ is a positive integer, and

$$
a^2 q^{N+1} = bcde. \quad (3.3.1.2)
$$

This condition corresponds to Dougall's condition,

$$
1 + 2a = b + c + d + e - N.
$$

The relation (3.3.1.2) states that the product of the denominator parameters must be $q^2$ times that of the numerator parameters. The series is terminating, since

$$
(q^{-N}; q)_N = (1 - q^{-N})(1 - q^{-N+1})\ldots(1 - q^{-N+N-1}).
$$

The series is well-poised since

$$
aq = (q\sqrt{a})\sqrt{a} = b(aq/b) = \ldots = q^{-N}(aq^{N+1}),
$$

and so we can say that Jackson's theorem gives the sum of a terminating well-poised ${}_8\Phi_7(q)$ series. The special forms of the second and third parameters produce in the $(r+1)$th terms of the series, the factor

$$
\frac{1 - aq^{2r}}{1 - a},
$$

† F. H. Jackson (1921 a).

just as in the $_7F_6(1)$ series, the parameters $1 + \frac{1}{2}a$ and $\frac{1}{2}a$, produce the factor $(a + 2r)/a$ in the $(r + 1)$th term of the series. Thus, if we write $q^a$ for $a$ and then let $q \to 1$, we have

$$\lim_{q \to 1} \left( \frac{1 - q^{a+2r}}{1 - q^a} \right) = \frac{a + 2r}{a}.$$

Thus, instead of a theorem on a $_5F_4(1)$ series, we shall expect to deduce a theorem on a $_6\Phi_5(q)$ series, and instead of a transformation of a $_9F_8(1)$ series, we shall hope to find a transformation between $_{10}\Phi_9(q)$ series.

The proof is by induction on the same lines as the standard proof of Dougall's theorem. Thus, the result is obviously true when $N = 0$. Suppose that the theorem is also true when $N = 1, 2, 3, ..., N_0 - 1$. By symmetry, the result is also true if $c$ or $d$ has one of the values $1, q^{-1}$, $q^{-2}, q^{-3}, ..., q^{-N_0+1}$, that is if $c$ or $a^2q/(bcef)$ has one of these values. In particular, it is true then, when $N = N_0$ and $c$ takes one of the $2N_0$ values above. But, when $N = N_0$, if we multiply the formula by

$$(aq/c; q)_{N_0} (aq/bcd; q)_{N_0},$$

we see that we have an equality between two polynomials in $c$, each of degree $2N_0$. Now $c = aq^{N_0}$ is a pole of the last term only of the series, and we can see that our equation holds for this value of $c$ also. Hence it is an identity in $c$, and the proof is completed by induction from $N = 0$, to $N = N_0$.

In (3.3.1.1) let us substitute

$$e = \frac{a^2q^{N+1}}{bcd},$$

and consider what happens as $N \to \infty$. In the $(r + 1)$th term, the factor

$$\frac{(q^{-N}; q)_r (a^2q^{N+1}/bcd; q)_r}{(bcdq^{-N}/a; q)_r (aq^{1+N}; q)_r}$$

occurs. Now as $N \to \infty$,

$$\frac{(a^2q^{N+1}/bcd; q)_r}{(aq^{1+N}; q)_r} \to 1.$$

Also

$$\frac{(q^{-N}; q)_r}{(bcdq^{-N}/a; q)_r} = \frac{(q^N - 1)(q^{N-1} - 1)...(q^{N-r+1} - 1)}{(q^N - bcd/a)(q^{N-1} - bcd/a)...(q^{N-r+1} - bcd/a)}$$

$$\to \left( \frac{a}{bcd} \right)^r \quad \text{as} \quad N \to \infty.$$

Thus the $_8\Phi_7(q)$ series becomes a $_6\Phi_5(aq/bcd)$ series, and Jackson's theorem becomes

$$_6\Phi_5\left[\begin{array}{cccccc} a, q\sqrt{a}, & -q\sqrt{a}, & b, & c, & d; \\ & \sqrt{a}, & -\sqrt{a}, & aq/b, aq/c, aq/d; \end{array} q, aq/bcd\right]$$

$$= \Pi\left[\begin{array}{c} aq, aq/cd, aq/bd, aq/bc; \\ aq/b, aq/c, aq/d, aq/bcd; \end{array} q\right]. \quad (3.3.1.3)$$

This gives the sum of a well-poised $_6\Phi_5(aq/bcd)$ series, and it is the basic analogue of the $_5F_4(1)$ summation theorem (2.3.4.5).

If $d = q^{-N}$, where $N$ is an integer, the series terminates, and the theorem reduces to

$$_6\Phi_5\left[\begin{array}{ccccc} a, q\sqrt{a}, & -q\sqrt{a}, & b, & c, & q^{-N}; \\ & \sqrt{a}, & -\sqrt{a}, aq/b, aq/c, aq^{N+1}; \end{array} q, aq^{1+N}/(bc)\right]$$

$$= \frac{(a; q)_N (a/bc; q)_N}{(a/b; q)_N (a/c; q)_N}. \quad (3.3.1.4)$$

This corresponds to (2.3.4.6), which is the sum of a well-poised terminating $_5F_4(1)$ series.

In (3.3.1.3) let $d = \sqrt{a}$. Then we shall find that

$$_4\Phi_3\left[\begin{array}{cccc} a, & -q\sqrt{a}, & b, & c; \\ & -\sqrt{a}, & aq/b, aq/c; \end{array} q, q\sqrt{a}/(bc)\right]$$

$$= \Pi\left[\begin{array}{c} aq, q\sqrt{a}/c, q\sqrt{a}/b, aq/bc; \\ aq/b, aq/c, q/a, q\sqrt{a}/bc; \end{array} q\right]. \quad (3.3.1.5)$$

If, in this series, we replace $a$, $b$ and $c$ by $q^a$, $q^b$ and $q^c$, respectively, and then let $q \to 1$, we obtain Dixon's theorem, so that (3.3.1.5) is in fact a basic analogue of Dixon's theorem (2.3.3.5). It should be noted that the sum of the series

$$_3\Phi_2[a, b, c; aq/b, aq/c; q, z]$$

which would provide an exact analogue of Dixon's series, cannot be found by this method.

### 3.3.2 The basic analogue of Saalschutz's theorem.

In Jackson's theorem (3.3.1.1), let us write $aq/d$ in place of $d$ and substitute $adq^n/bc$ for $e$. The theorem then becomes

$$_8\Phi_7\left[\begin{array}{ccccccc} a, q\sqrt{a}, & -q\sqrt{a}, & b, & c, adq^n/bc, & q^{-n}, & aq/d; \\ & \sqrt{a}, & -\sqrt{a}, aq/b, aq/c, bcq^{1-n}/d, aq^{n+1}, & d; \end{array} q, q\right]$$

$$= \frac{(aq; q)_n (d/c; q)_n (d/b; q)_n (aq/bc; q)_n}{(aq/b; q)_n (aq/c; q)_n (d; q)_n (d/bc; q)_n}. \quad (3.3.2.1)$$

Now let $a \to \infty$, in this finite series. We find that

$$
{}_3\Phi_2\left[\begin{matrix} b, c, q^{-n}; \\ d, bcq^{1-n}/d; \end{matrix} q, q\right] = \frac{(d/c;\ q)_n\,(d/b;\ q)_n}{(d;\ q)_n\,(d/bc;\ q)_n}. \tag{3.3.2.2}
$$

This is the basic analogue of Saalschutz's theorem (2.3.1.3).

We can now show that

$$
{}_2\Phi_1[a, b;\ c;\ q, z] = {}_2\Phi_1[c/a, c/b;\ c;\ q, abz/c]\,{}_1\Phi_0[ab/c;\ ;\ q, z]. \tag{3.3.2.3}
$$

This is the basic analogue of Euler's transform (2.3.1.1). If we compare the coefficients of $z^n$ on both sides of this equation, and use the above theorem (3.3.2.2), the proof follows immediately.

Again, in (3.3.2.2), let $n \to \infty$, through positive integral values. Then, for $n$ large, we have

$$
\frac{(q^{-n};\ q)_r\, q^r}{(bcq^{1-n}/d;\ q)_r} = \frac{(1-q^{-n})\,(1-q^{1-n})\dots(1-q^{r-n-1})\,q^r}{(1-bcq^{1-n}/d)\,(1-bcq^{2-n}/d)\dots(1-bcq^{r-n}/d)}
$$

$$
\sim \frac{(-1)^r\, q^{-nr}q^{1+2+\dots+(r-1)}\,q^r}{(-1)^r\,(bcq^{1-n}/d)^r\, q^{1+2+\dots+(r-1)}}
$$

$$
= \frac{q^{(1-n)r}}{(bc/d)^r\, q^{(1-n)r}},
$$

$$
\to (d/bc)^r \quad \text{as } n \to \infty, \tag{3.3.2.4}
$$

provided that $|d/bc| < 1$. Hence (3.3.2.2) leads us, as $n \to \infty$, to the result that

$$
{}_2\Phi_1[b, c;\ d;\ q, d/bc] = \Pi[d/c, d/b;\ d, d/bc;\ q]. \tag{3.3.2.5}
$$

This is the basic analogue of Gauss's theorem (1.1.5).

If now we put $c = q^{-N}$, where $N$ is an integer, in (3.3.2.5), or, alternatively, if we let $c \to \infty$, in (3.3.2.2), we get

$$
{}_2\Phi_1[b, q^{-N};\ d;\ q, dq^N/b] = \frac{(d/b;\ q)_N}{(d;\ q)_N}. \tag{3.3.2.6}
$$

This is a kind of analogue of Vandermonde's theorem, (1.7.7), but its usefulness is restricted by the presence of $q^N$ in the variable of the ${}_2\Phi_1(dq^N/b)$ function.

An alternative result can be obtained if we let $c \to 0$, in (3.3.2.2). This process leads us to the theorem

$$
{}_2\Phi_1[b, q^{-n};\ d;\ q, q] = \frac{(d/b;\ q)_n\, b^n}{(d;\ q)_n}, \tag{3.3.2.7}
$$

7

which is a very much more useful result, as the variable of the $_2\Phi_1(q)$ function is simply $q$, and does not depend on the number of terms in the series.

## 3.4   Applications of Bailey's transform to basic series

A direct basic analogue of the general transformation (2.4.10) will now be deduced from the application of Bailey's transform. Let us suppose that

$$u_r = \frac{(u_1; q)_r (u_2; q)_r \dots (u_U; q)_r u^r}{(e_1; q)_r (e_2; q)_r \dots (e_E; q)_r (q; q)_r}, \tag{3.4.1}$$

$$v_r = \frac{(v_1; q)_r (v_2; q)_r \dots (v_V; q)_r v^r}{(f_1; q)_r (f_2; q)_r \dots (f_F; q)_r}, \tag{3.4.2}$$

$$\delta_r = \frac{(d_1; q)_r (d_2; q)_r \dots (d_D; q)_r x^r}{(g_1; q)_r (g_2; q)_r \dots (g_G; q)_r} \tag{3.4.3}$$

and

$$\alpha_r = \frac{(a_1; q)_r (a_2; q)_r \dots (a_A; q)_r y^r}{(h_1; q)_r (h_2; q)_r \dots (h_H; q)_r (q; q)_r}. \tag{3.4.4}$$

Then

$$\gamma_n = \frac{(d_1; q)_n (d_2; q)_n \dots (d_D; q)_n (v_1; q)_{2n} (v_2; q)_{2n} \dots (v_V; q)_{2n}}{(g_1; q)_n (g_2; q)_n \dots (g_G; q)_n (f_1; q)_{2n} (f_2; q)_{2n} \dots (f_F; q)_{2n}}$$

$$\times x^n v^{2n} \; {}_{U+D+V}\Phi_{E+F+G} \left[ \begin{matrix} u_1, u_2, \dots, u_U, d_1 q^n, d_2 q^n, \dots, d_D q^n, \\ e_1, e_2, \dots, e_E, g_1 q^n, g_2 q^n, \dots, g_G q^n, \end{matrix} \right.$$

$$\left. \begin{matrix} v_1 q^{2n}, v_2 q^{2n}, \dots, v_V q^{2n}; \\ f_1 q^{2n}, f_2 q^{2n}, \dots, f_F q^{2n}; \end{matrix} \; q, uvx \right] \tag{3.4.5}$$

and

$$\beta_n = \frac{(u_1; q)_n (u_2; q)_n \dots (u_U; q)_n (v_1; q)_n (v_2; q)_n \dots (v_V; q)_n u^n v^n}{(q; q)_n (e_1; q)_n (e_2; q)_n \dots (e_E; q)_n (f_1; q)_n (f_2; q)_n \dots (f_F; q)_n}$$

$$\times {}_{E+A+V+1}\Phi_{U+H+F} \left[ \begin{matrix} q^{1-n}/e_1, q^{1-n}/e_2, \dots, q^{1-n}/e_E, a_1, a_2, \dots, a_A, v_1 q^n, \\ q^{1-n}/u_1, q^{1-n}/u_2, \dots, q^{1-n}/u_U, h_1, h_2, \dots, h_H, f_1 q^n, \end{matrix} \right.$$

$$\left. \begin{matrix} v_2 q^n, \dots, v_V q^n, q^{-n}; \\ f_2 q^n, \dots, f_F q^n; \end{matrix} \; q, uvy(-q)^{\frac{1}{2}(n-1)(1+E-U)} \right]. \tag{3.4.6}$$

The complete theorem, is then, when stated in the contracted notation,

$$\sum_{n=0}^{\infty} \frac{((a;q))_n\,((d;q))_n\,((v;q))_{2n}\,x^n y^n v^{2n}}{((h;q))_n\,((g;q))_n\,((f;q))_{2n}\,(q;q)_n}$$

$$\times {}_{U+V+D}\Phi_{E+F+G}\left[\begin{matrix}(u),(dq^n),(vq^{2n});\\(e),(gq^n),(fq^{2n});\end{matrix}\;q,\,uvx\right]$$

$$= \sum_{n=0}^{\infty} \frac{((d;q))_n\,((u;q))_n\,((v;q))_n\,u^n v^n x^n}{(q;q)_n\,((e;q))_n\,((f;q))_n\,((g;q))_n}$$

$$\times {}_{A+E+V+1}\Phi_{U+H+F}\left[\begin{matrix}(a),(q^{1-n}/e),(vq^n),q^{-n};\\(h),(q^{1-n}/u),(fq^n);\end{matrix}\;q,\,(-q)^{\frac{1}{2}(1+E-Q)(n-1)}uvy\right].$$

$$(3.4.7)$$

Here, and in (3.4.6) above, the factor $(n-1)$ has been put in the variable to show that the $n$'th term of the series contains the variable

$$(-q)^{(1+E-Q)\frac{1}{2}n(n-1)}\,u^n v^n y^n.$$

The result (3.4.7) is to be interpreted as having a meaning only when all the series involved are either convergent or terminating.

In particular, in Bailey's transform, let

$$u_n = 1/(q;q)_n,\quad v_n = 1/(x;q)_n \quad \text{and} \quad \delta_n = \frac{(y;q)_n\,(z;q)_n\,x^n}{y^n z^n}.$$

Then we can sum the series for $\gamma_n$ by the basic analogue of Gauss's theorem, (3.3.2.5), and we get

$$\gamma_n = \Pi\left[\begin{matrix}x/y,x/z;\\x,x/yz;\end{matrix}\;q\right]\frac{(y;q)_n\,(z;q)_n\,x^n}{(x/y;q)_n\,(x/z;q)_n\,y^n z^n}.\qquad(3.4.8)$$

Hence, from Bailey's transform, we have

$$\sum_{n=0}^{\infty} (y;q)_n\,(z;q)_n\,x^n y^{-n} z^{-n}\,\beta_n$$

$$= \Pi\left[\begin{matrix}x/y,x/z;\\x,x/yz;\end{matrix}\;q\right]\sum_{n=0}^{\infty}\frac{(y;q)_n\,(z;q)_n\,x^n \alpha_n}{(x/y;q)_n\,(x/z;q)_n\,y^n z^n},\qquad(3.4.9)$$

where $\quad \beta_n = \sum_{r=0}^{n}\dfrac{\alpha_r}{(x;q)_{n+r}\,(q;q)_{n-r}}\quad$ and $\quad \alpha_0 = 1.$

This relation will be used later (§7.3.1) in the deduction of some further identities of the Rogers–Ramanujan type.

### 3.4.1 Basic Saalschutzian transforms.

This basic analogue of Bailey's transform, (3.4.7), will now be applied to the deduction of transformations of basic series, using the basic analogue of Saalschutz's theorem (3.3.2.2). As before, we shall denote the series on the left of (3.4.7) $_{U+D+V}\Phi_{E+F+G}[uvx]$ by $\Phi_\beta$ and the series on the right-hand side of (3.4.7) $_{A+E+V+1}\Phi_{U+H+F}[-q^n uvy]$ by $\Phi_\gamma$. Let us suppose that $\Phi_\gamma$ is summable by (3.3.2.2). One possible form for $\Phi_\gamma$ is

$$\Phi_\gamma = {}_3\Phi_2\left[\begin{matrix} d_1 q^n, d_2 q^n, q^{n-N}; \\ fq^{2n+1}, d_1 d_2 q^{n-N}/f; \end{matrix} q, q \right], \qquad (3.4.1.1)$$

then we have

$$\Phi_\beta = {}_{A+1}\Phi_{H+1}[(a), q^{-N}; (h), fq^N; q, -y]. \qquad (3.4.1.2)$$

Let

$$\alpha_n = \frac{(a; q)_n (q\sqrt{a}; q)_n (-q\sqrt{a}; q)_n (d; q)_n (e; q)_n (-1)^n q^{\frac{1}{2}n(n+1)} a^n}{(q; q)_n (\sqrt{a}; q)_n (-\sqrt{a}; q)_n (aq/d; q)_n (aq/e; q)_n d^n e^n},$$

then $\quad \Phi_\beta = {}_6\Phi_5\left[\begin{matrix} a, q\sqrt{a}, -q\sqrt{a}, & d, & e, & q^{-N}; \\ \sqrt{a}, & -\sqrt{a}, aq/d, aq/e, aq^{N+1}; \end{matrix} q, aq^{n+1}/de \right].$

$$(3.4.1.3)$$

This series can be summed as a well-poised terminating $_6\Phi_5$ series, by (3.3.1.4), to give

$$\Phi_\beta = \frac{(a; q)_n (a/de; q)_n}{(a/d; q)_n (a/e; q)_n}. \qquad (3.4.1.4)$$

This result, when combined with (3.4.1.1), leads to Watson's analogue of Whipple's transform (2.4.1.1),

$$_8\Phi_7\left[\begin{matrix} a, q\sqrt{a}, -q\sqrt{a}, & c, & d, & e, & f, & g; \\ \sqrt{a}, & -\sqrt{a}, aq/c, aq/d, aq/e, aq/f, aq/g; \end{matrix} q, \frac{a^2 q^2}{cdefg} \right]$$

$$= \Pi\left[\begin{matrix} aq, aq/fg, aq/eg, aq/ef; \\ aq/e, aq/f, aq/g, aq/efg; \end{matrix} q \right] {}_4\Phi_3\left[\begin{matrix} aq/cd, & e, & f, & g; \\ & efg/a, aq/c, aq/d; \end{matrix} q, q \right].$$

$$(3.4.1.5)$$

This transforms a well-poised $_8\Phi_7$ series with the special forms of the second and third parameters, into a Saalschutzian $_4\Phi_3$ series. It contains as special cases, very many of the classical identities, (see § 3.5.1).

Next let us suppose that

$$\Phi_\gamma = {}_3\Phi_2\left[\begin{matrix} kq^{2r}, k/a, q^{-N-r}; \\ aq^{2r+1}, k^2/aq^{r-N}; \end{matrix} q, q \right].$$

Then $\Phi_\beta$ can be summed as a well-poised $_8\Phi_7$ series, by Jackson's theorem, and we find finally, after some reduction, that

$$_5\Phi_4\left[\begin{matrix} a, & b, & c, & d, & q^{-N}; \\ aq/b, aq/c, aq/d, a^2q^{-N}/k^2; \end{matrix} q,q\right] = \frac{(kq/a;\,q)_N\,(k^2q/a;\,q)_N}{(kq;\,q)_N\,(k^2q/a^2;\,q)_N}$$

$$\times\,_{12}\Phi_{11}\left[\begin{matrix} a,q\sqrt{k},-q\sqrt{k},kb/a,kc/a,kd/a,\sqrt{a},-\sqrt{a},\sqrt{(aq)}, \\ \sqrt{k},\ -\sqrt{k},aq/b,\,aq/c,\,aq/d,\,k\sqrt{(q/a)},-k\sqrt{(q/a)}, \end{matrix}\right.$$
$$\left.\begin{matrix} -\sqrt{(aq)},k^2q^{N+1}/a,\ q^{-N}; \\ kq/\sqrt{a},-kq/\sqrt{a},aq^{-N}/k,kq^{N+1}; \end{matrix} q,q\right].$$

(3.4.1.6)

This is a transformation between a Saalschutzian nearly-poised $_5\Phi_4(q)$ series and a well-poised $_{12}\Phi_{11}(q)$ series. It is the analogue of (2.4.1.4).

If we take $c = q/a$, and $d = -q/a$, so that $k = -a/bq$ then four of the parameters cancel out and the $_{12}\Phi_{11}(q)$ series reduces to a terminating well-poised $_8\Phi_7(q)$ series which has the special forms of the second and third parameters. This can be summed by Jackson's theorem, and, after some reduction, we find that

$$_5\Phi_4\left[\begin{matrix} a,q\sqrt{a},-q\sqrt{a}, & b, & q^{-N}; \\ \sqrt{a}, & -\sqrt{a},aq/b,\,b^2q^{2-N}; \end{matrix} q,q\right]$$

$$= \frac{(a/b^2;\,q)_{N-1}(1/bq;\,q)_N\,(1-aq^{2N-1}/b^2)}{(aq/b;\,q)_N\,(1/b^2q;\,q)_N}.$$ (3.4.1.7)

This series is nearly-poised and Saalschutzian, and this result corresponds to (2.4.2.6), of which it is a basic analogue.

### 3.4.2 Basic well-poised transforms.
Next, we shall suppose that $\Phi_\gamma$ can be summed by Jackson's theorem (3.3.1.1) as a well-poised $_8\Phi_7(q)$ series. As before, $\Phi_\gamma$ must be well-poised in a $v$ parameter to produce a summable $\Phi_\beta$ series. So let

$$\Phi_\gamma = \,_8\Phi_7\left[\begin{matrix} kq^{2r},q^{r+1}\sqrt{k},-q^{r+1}\sqrt{k},eq^{r-1}, & k/a, & fq^r, \\ q^r\sqrt{k}, & -q^r\sqrt{k},kq^{2+r}/e,aq^{2r+1},kq^{r+1}/f, \end{matrix}\right.$$
$$\left.\begin{matrix} gq^r, & q^{r-N}; \\ kq^{r+1}/g,kq^{r+N+1}; \end{matrix} q,q\right],$$

(3.4.2.1)

where $akq^{N+1} = efg$.

Then we can take $\quad\alpha_r = (a;\,q)_r\,(q/a;\,q)_r,$
so that

$$\Phi_\beta = \,_8\Phi_7\left[\begin{matrix} a,q\sqrt{a},-q\sqrt{a}, & b, & c, & d,kq^n, & q^{-n}; \\ \sqrt{a}, & -\sqrt{a},aq/b,aq/c,aq/d,aq^{1-n}/k,aq^{1+n}; \end{matrix} q,q\right],$$

(3.4.2.2)

where $a^2q = bcdk$. Then summing $\Phi_\beta$ by Jackson's theorem also, we find that

$$\Phi_\beta = \frac{(aq;q)_n\,(aq/cd;q)_n\,(aq/bd;q)_n\,(aq/bc;q)_n}{(aq/b;q)_n\,(aq/c;q)_n\,(aq/d;q)_n\,(aq/bcd;q)_n}. \quad (3.4.2.3)$$

When we substitute these values of $\Phi_\beta$ and $\Phi_\gamma$ in (3.4.7), after some reduction, the theorem leads us to the result

$$_{10}\Phi_9\left[\begin{array}{c} a, q\sqrt{a}, -q\sqrt{a}, \quad c, \quad d, \quad e, \quad f, \quad g, \quad h, \quad j; \\ \sqrt{a}, \ -\sqrt{a}, aq/c, aq/d, aq/e, aq/f, aq/g, aq/h, aq/j; \end{array} q,q\right]$$

$$= \Pi\left[\begin{array}{c} aq, aq/fg, aq/fh, aq/fj, aq/gh, aq/gj, aq/hj, aq/fghj; \\ aq/f, aq/g, aq/h, aq/j, aq/ghj, aq/fhj, aq/fgh, aq/fgj; \end{array} q\right]$$

$$\times\ _{10}\Phi_9\left[\begin{array}{c} k, q\sqrt{k}, -q\sqrt{k}, kc/a, kd/a, ke/a, \quad f, \quad g, \quad h, \quad j; \\ \sqrt{k}, \ -\sqrt{k}, aq/c, aq/d, aq/e, kq/f, \ kq/g, kq/h, kq/j; \end{array} q,q\right],$$

$$(3.4.2.4)$$

where $k = a^2q/cde$, $a^3q^2 = cdefghj$, and $f$, or $g$, or $h$ or $j$ is of the form $q^{-N}$, where $N$ is a positive integer.

This is a relationship between two terminating $_{10}\Phi_9(q)$ series, both well-poised with the special form of the second and third parameters. It is the basic analogue of (2.4.4.1), the relation between two well-poised $_9F_8(1)$ series, and it was first given by Jackson.† It is one of the few results in the special theory which involves as many as six free parameters, and in consequence, it contains many interesting special cases.

Thus, in (3.4.2.4), let us replace $c$ and $k$ by their values in terms of the other parameters, and put $j = q^{-N}$. Then we can let $N$ tend to infinity, and, arguing in the same way as we did for the formula (2.4.4.3), which expresses a well-poised $_7F_6(1)$ series in terms of two Saalschutzian $_4F_3(1)$ series, we can derive the basic analogue of this result, namely,

$$_8\Phi_7\left[\begin{array}{c} a, q\sqrt{a}, -q\sqrt{a}, \quad d, \quad e, \quad f, \quad g, \quad h; \\ \sqrt{a}, \ -\sqrt{a}, aq/d, aq/e, aq/f, aq/g, aq/h; \end{array} q, a^2q^2/defgh\right]$$

$$= \Pi\left[\begin{array}{c} aq, aq/fg, aq/fh, aq/gh; \\ aq/f, aq/g, aq/h, aq/fgh; \end{array} q\right]_4\Phi_3\left[\begin{array}{c} aq/de, f, g, h; \\ aq/d, aq/e, fgh/a; \end{array} q, q\right]$$

$$+ \Pi\left[\begin{array}{c} aq, aq/de, f, g, h, a^2q^2/dfgh, a^2q^2/efgh; \\ aq/d, aq/e, aq/f, aq/g, aq/h, a^2q^2/defgh, fgh/aq; \end{array} q\right]$$

$$\times\ _4\Phi_3\left[\begin{array}{c} aq/gh, aq/fh, aq/fg, a^2q^2/defgh; \\ aq^2/fgh, a^2q^2/dfgh, a^2q^2/efgh; \end{array} q, q\right]. \quad (3.4.2.5)$$

† F. H. Jackson (1921 a).

This is a generalization of (3.4.1.5), and from it, we can see that (3.4.1.5) is true provided only that the series on the right terminates, and the series on the left converges.

We can substitute an $_8\Phi_7(q)$ series for each of the $_4\Phi_3(q)$ series on the right of (3.4.2.5), using (3.4.1.5), and then we obtain an identity between three well-poised $_8\Phi_7(q)$ series.

## 3.5 The Rogers–Ramanujan identities

These two remarkable results are

$$1 + \frac{q}{1-q} + \frac{q^4}{(1-q)(1-q^2)} + \ldots + \frac{q^{n^2}}{(1-q)(1-q^2)(1-q^3)\ldots(1-q^n)} + \ldots$$

$$= 1/\{(1-q)(1-q^4)(1-q^6)(1-q^9)\ldots(1-q^{5n+1})(1-q^{5n+4})\ldots\} \quad (3.5.1)$$

and

$$1 + \frac{q^2}{1-q} + \frac{q^6}{(1-q)(1-q^2)} + \ldots + \frac{q^{n(n+1)}}{(1-q)(1-q^2)\ldots(1-q^n)} + \ldots$$

$$= 1/\{(1-q^2)(1-q^3)(1-q^7)(1-q^8)\ldots(1-q^{5n+2})(1-q^{5n+3})\ldots\}. \quad (3.5.2)$$

When they are restated in the more contracted notation of basic series, these identities become

$$\sum_{n=0}^{\infty} \frac{q^{n^2}}{(q;q)_n} = 1/\Pi(q,q^4;q^5) \quad (3.5.3)$$

and

$$\sum_{n=0}^{\infty} \frac{q^{n(n+1)}}{(q;q)_n} = 1/\Pi(q^2,q^3;q^5). \quad (3.5.4)$$

These identities were first stated by L. J. Rogers[†] in 1893. They were later rediscovered quite independently by S. Ramanujan, and they were stated by him, without proof. Some controversy arose but this was finally settled by G. H. Hardy who arranged for both claimants to publish proofs in 1919[‡]. The simplest of the several proofs given by Rogers and the proof given by Ramanujan both depended on proving the more general formula

$$1 + \sum_{n=1}^{\infty} (-1)^n a^{2n} q^{\frac{1}{2}n(5n-1)} \frac{(1-aq^{2n})(a;q)_n}{(1-a)(q;q)_n} = \Pi(a;q) \sum_{n=0}^{\infty} \frac{a^n q^{n^2}}{(q;q)_n}. \quad (3.5.5)$$

---

† Rogers (1894).          ‡ Ramanujan & Rogers (1919).

The two results (3.5.1) and (3.5.2) can be deduced from this by putting $a = 1$, and $a = q$, respectively as we shall show later.

During the 1920's about six other methods of proving (3.5.1) and (3.5.2) were published. Finally, in 1929, G. N. Watson put forward his elegant proof, using basic series.† This proof depended on Watson's transformation (3.4.1.5) of a well-poised $_8\Phi_7$ series into a Saalschutzian $_4\Phi_3$ series.

In this formula, put $g = q^{-N}$, where $N$ is a positive integer, and then let $c \to \infty$. The part of the $(r+1)$th term of the series involving $c$ is

$$\frac{(c; q)_r c^{-r}}{(aq/c; q)_r} = \frac{(-1)^r c^r q^{\frac{1}{2}r(r-1)}(1 - 1/c)(1 - 1/cq) \dots (1 - 1/cq^{r+1})}{c^r(1 - aq/c)(1 - aq^2/c) \dots (1 - aq^{r-1}/c)}$$

and this expression tends to $(-1)^r q^{\frac{1}{2}r(r-1)}$ as $c \to \infty$.

Similar results follow when we let $d$, $e$, and $f$ tend to infinity in turn, and we find, finally, that (3.4.1.5) has been reduced to

$$1 + \sum_{n=1}^{N} \frac{(aq; q)_{r-1}(1 - aq^{2r}) a^{2r} q^{2r^2} q^{Nr}(q^{-N}; q)_r}{(q; q)_r (aq^{N+1}; q)_r}$$

$$= \prod_{n=1}^{N} (1 - aq^n) \left\{ 1 + \sum_{r=1}^{N} \frac{(-1)^r q^{\frac{1}{2}r(r+1)}(q^{-N}; q)_r a^r q^{Nr}}{(q; q)_r} \right\}. \quad (3.5.6)$$

Now let $N \to \infty$. If $\quad A_r(N) = \dfrac{(q^{-N}; q)_r q^{Nr}}{(aq^{N+1}; q)_r}$,

then $A_r(N) \to (-1)^r q^{\frac{1}{2}r(r-1)}$ as $N \to \infty$, for any fixed value of $r$. For all values of $N$,

$$|A_r(N)| < \left| 1 \Big/ \prod_{n=1}^{\infty} (1 - aq^n) \right| < K, \quad \text{a constant,}$$

since this infinite product is convergent. Hence, in the series on the left of (3.5.6) the modulus of each term is

$$< K \left| \frac{(aq; q)_{r-1}}{(q; q)_r} (1 - aq^{2r}) a^{2r} q^{2r^2} (-1)^r q^{\frac{1}{2}r(r-1)} \right| = K|u_{r+1}|,$$

and $\quad\quad\quad |u_{r+1}| \sim \left| \prod_{n=0}^{\infty} \frac{(1 - aq^n)}{(1 - q^{1+n})} a^{2r} q^{2r^2} q^{\frac{1}{2}r(r-1)} \right|$

$$= C |a^{2r} q^{\frac{1}{2}r(5r-1)}|,$$

which is the term of a convergent series. We can argue in a similar way, about the behaviour of the series on the right-hand side of (3.5.6) as $N \to \infty$, and apply Tannery's theorem, to show that when $N \to \infty$, (3.5.6) becomes (3.5.5).

In order to deduce (3.5.1) from (3.5.5), let us put $a = 1$, in (3.5.5). Then we find that

$$\Pi(q; q) \sum_{r=0}^{\infty} \frac{q^{r^2}}{(q; q)_r} = 1 + \sum_{r=1}^{\infty} (-1)^r \frac{(1 - q^{2r})}{(1 - q^r)} q^{\frac{1}{2}r(5r-1)},$$

$$= 1 + \sum_{r=1}^{\infty} (-1)^r (1 + q^r) q^{\frac{1}{2}r(5r-1)},$$

$$= 1 + \sum_{r=1}^{\infty} (-1)^r \{q^{\frac{1}{2}r(5r-1)} + q^{\frac{1}{2}r(5r+1)}\},$$

$$= \sum_{r=-\infty}^{\infty} (-1)^r q^{\frac{1}{2}r} q^{5r^2/2}. \tag{3.5.7}$$

But Jacobi's formula for the summation of elliptic integrals, (3.1.12), can be re-expressed in the form

$$\sum_{r=-\infty}^{\infty} (-1)^r z^r q^{z^2} = \prod_{n=1}^{\infty} (1 - zq^{2n-1})(1 - q^{2n-1}/z)(1 - q^{2n})$$

$$= \Pi[zq, q/z, q^2; q^2], \tag{3.5.8}$$

if $|q| < 1$.

Hence, if we replace $q$ by $q^{\frac{5}{2}}$ and put $q^{\frac{1}{2}}$ for $z$, we can rewrite (3.5.8) as

$$\sum_{r=-\infty}^{\infty} (-1)^r q^{\frac{1}{2}r} q^{\frac{5}{2}r^2} = \prod_{n=1}^{\infty} (1 - q^{5n-2})(1 - q^{5n-3})(1 - q^{5n}). \tag{3.5.9}$$

So it follows, from (3.5.7), that

$$\sum_{n=0}^{\infty} \frac{q^{n^2}}{(q; q)_n} = \prod_{n=1}^{\infty} \frac{(1 - q^{5n-2})(1 - q^{5n-3})(1 - q^{5n})}{(1 - q^n)}$$

$$= \prod_{n=1}^{\infty} \{1/(1 - q^{5n-1})(1 - q^{5n-4})\}. \tag{3.5.10}$$

The second Rogers–Ramanujan identity follows in a similar way, when we put $a = q$ in (3.5.5). These series, infinite in both directions, which we have used above, are further examples of bilateral basic series, to be discussed more fully in Chapter 7.

### 3.5.1 Some further identities.

If, in (3.4.1.5), we put $cd = aq$, and then let $e, f$ and $g$ tend to infinity, we obtain

$$1 + \sum_{n=1}^{\infty} (-1)^n a^n q^{\frac{1}{2}n(3n-1)} (1 - aq^{2n}) \frac{(aq; q)_n}{(q; q)_n} = \prod_{n=1}^{\infty} (1 - aq^n). \tag{3.5.1.1}$$

When $a = 1$, this gives us the result

$$1 + \sum_{n=1}^{\infty} (-1)^n \{q^{\frac{1}{2}n(3n-1)} + q^{\frac{1}{2}n(3n+1)}\} = \prod_{n=1}^{\infty} (1 - q^n). \qquad (3.5.1.2)$$

This is a classical result, in the theory of partition functions, due originally to Euler.†

Again, if we put $cd = aq, e = \sqrt{(aq)}$ and then let $f$ and $g$ tend to infinity, we obtain, from (3.4.1.5),

$$1 + \sum_{n=1}^{\infty} a^{\frac{1}{2}n} q^{n(n-\frac{1}{2})} (1 - aq^{2n}) \frac{(a;q)_n}{(q;q)_n} = \prod_{n=1}^{\infty} \frac{(1 - aq^n)}{(1 - a^{\frac{1}{2}}q^{n-\frac{1}{2}})}. \tag{3.5.1.3}$$

In particular, if $a = 1$, we have, on writing $q^2$ for $q$,

$$1 + \sum_{n=1}^{\infty} q^{\frac{1}{2}n(n+1)} = \prod_{n=1}^{\infty} \frac{(1 - q^{2n})}{(1 - q^{2n-1})}. \tag{3.5.1.4}$$

This is another classical result, due originally to Gauss.‡

Finally, if we put $c = \sqrt{a}$, $d = -\sqrt{a}$, and then let $e, f$ and $g$ tend to infinity, we obtain the relation

$$1 + \sum_{n=1}^{\infty} \frac{(a;q)_n}{(q;q)_n} a^n q^{\frac{1}{2}n(3n+1)} = \prod_{n=1}^{\infty} (1 - aq^n) \left\{ 1 + \sum_{n=1}^{\infty} \frac{(-q;q)_n a^n q^{n^2}}{(q;q)_n (aq;q)_n} \right\}, \tag{3.5.1.5}$$

from (3.4.1.5). In particular, when $a = 1$,

$$1 + \sum_{n=1}^{\infty} q^{n^2} / \{(q;q)_n\}^2 = \prod_{n=1}^{\infty} \{1/(1 - q^n)\}, \tag{3.5.1.6}$$

and, when $a = q$, we can deduce from (3.5.1.5), by the use of (3.5.1.4) that

$$1 + \sum_{n=1}^{\infty} \frac{(-q;q)_n q^{n(n+1)}}{(q;q)_n (q^3;q^2)_n} = \prod_{n=1}^{\infty} \frac{(1 - q^{6n})}{(1 - q^{6n-3})(1 - q^{n+1})}. \tag{3.5.1.7}$$

A very large number of such results exists. They arise naturally in various branches of number theory, and the theory of elliptic modular functions. A systematic attempt to seek and list them is made in Slater (1951, 1952 a) in which 120 such results are deduced. Some of the most interesting of these will be given in § 7.3.1.

### 3.5.2 Numerical evaluation of infinite products. Since the infinite product

$$\prod_{n=0}^{\infty} (1 - aq^n)$$

† Euler (1748).          ‡ Gauss (1866).

plays a part in the theory of basic functions, similar to that of the Gamma function in the theory of ordinary hypergeometric functions, as we have seen, so it is pertinent to enquire how such products can be evaluated numerically. While the Gamma function has an extensive literature, this type of product has very little.

For real values of $q$ and $a$, the usual method of evaluation is to sum the series in the formula

$$\sum_{n=0}^{\infty} \frac{a^n}{(q;\,q)_n} = 1 \Big/ \Big\{ \prod_{n=0}^{\infty} (1 - aq^n) \Big\}. \qquad (3.5.2.1)$$

Difficulties arise when $0 \cdot 89 < q < 1$, but these can be overcome if logarithms are introduced, thus

$$\log_\epsilon \prod_{n=0}^{\infty} (1 - aq^n) = \sum_{n=0}^{\infty} \log_\epsilon (1 - aq^n). \qquad (3.5.2.2)$$

Two short numerical tables are given in Appendix V and VI. The first of these is of the function

$$1 \Big/ \Big\{ \prod_{n=0}^{\infty} (1 - aq^n) \Big\}$$

over the range

$$a = -0 \cdot 90(0 \cdot 05)\,0 \cdot 95, \quad q = 0 \cdot 0(0 \cdot 05)\,1 \cdot 0,$$

to seven significant figures, and the second table is of the function

$$1 \Big/ \Big\{ \prod_{n=1}^{\infty} (1 - q^n) \Big\}$$

for $\qquad\qquad q = 0 \cdot 0(0 \cdot 005)\,0 \cdot 995,$

also to seven significant figures. Both tables were calculated in the Mathematical Laboratory of Cambridge University, by kind permission of the director of the laboratory, Dr M. V. Wilkes. The original calculations were made by the author, in 1953, who used the electronic computer Edsac I, and later, the calculations were repeated and checked, in 1959, by a research student E. Sugden, who used the electronic computer Edsac II.

# 4

# HYPERGEOMETRIC INTEGRALS

## 4.1 Elementary integral transforms

Many integral representations of the general hypergeometric series $_AF_B(z)$ have been given in the literature. These fall into two main groups, simple integral transforms of the Euler type, first generalized by Pochammer,† and Barnes-type contour integrals. We shall consider the simple Euler-type integrals first. These are very useful in numerical work, but not so fruitful as the Barnes-type integrals as a source of new transformations.

Let

$$I \equiv \int_0^1 t^{c-1}(1-t)^{d-c-1}\,_AF_B[(a);\ (b);\ zt]\,\mathrm{d}t, \qquad (4.1.1)$$

where $|z| < 1$. This integral exists and is convergent if $\mathrm{Rl}\,(\Sigma a - c) > 0$, and $\mathrm{Rl}\,(\Sigma b - \Sigma a + d - c) > 0$. Now, as before, we can exchange the order of summation and integration, and we find that

$$I = \sum_{n=0}^{\infty} \frac{((a))_n z^n}{((b))_n n!} \int_0^1 t^{c+n-1}(1-t)^{d-c-1}\,\mathrm{d}t$$

$$= \Gamma\begin{bmatrix} c, d-c \\ d \end{bmatrix}\,_{A+1}F_{B+1}[c, (a);\ d, (b);\ z]$$

by (1.6.6), so that we have the general Euler transform

$$_{A+1}F_{B+1}[c, (a);\ d, (b);\ z] = \Gamma\begin{bmatrix} d \\ c, d-c \end{bmatrix}$$
$$\times \int_0^1 t^{c-1}(1-t)^{d-c-1}\,_AF_B[(a);\ (b);\ tz]\,\mathrm{d}t. \quad (4.1.2)$$

This integral can itself be expressed as an Euler transform involving an $_{A-1}F_{B-1}(tz)$ series, and so we can go on right back to the simple Euler integral for the Gamma function. In this way, we can produce a multiple Euler integral for the general hypergeometric function

$$_{A+1}F_B[a_0, (a);\ (b);\ z]$$

$$= \Gamma\begin{bmatrix} (b) \\ (a),\ (b-a) \end{bmatrix} \int_0^1 \int_0^1 \cdots \int_0^1 t^{\Sigma a - A}(1-t)^{\Sigma(b-a)-A}\,(1-zt)^{-a_0}\,\mathrm{d}t,$$

$$(4.1.3)$$

where $B = A$.

† Pochammer (1870).

Integrals of this type have been studied by Erdélyi (1939). A closely related integral is

$$_AF_B[(a); d, (b)'; z]$$

$$= \Gamma\begin{bmatrix} d \\ b_\nu, d - b_\nu \end{bmatrix} \int_0^1 t^{b_\nu - 1}(1 - t)^{d - b_\nu - 1} {}_AF_B[(a); (b); tz]\,dt, \quad (4.1.4)$$

where the dash indicates that $b_\nu$ is left out of the $b$ sequence.

## 4.2 Barnes-type integrals

Foremost in the theory of general transforms of the series $_AF_B(z)$ are the contour integrals of the Barnes type. These have a group of Gamma functions as the integrand, and they are evaluated by considering the sum of the residues of the integrand at the sequences of its poles which fall within the contours considered. The integral round a contour is usually split up into integrals round various parts of the contour. Then the proof of the theorem consists of arguments to show that most of these partial integrals become zero, as the contour is expanded towards infinity, leaving, as the result, a single infinite contour integral expressed as some infinite series of residues.

Before we go on to consider the integrals connected with the general transformation theory of the $_AF_B(z)$ series, we shall first study the integral analogues of the special summation theorems connected with the names of Gauss, Dougall, Dixon and Saalschutz. Most of these were proved by Barnes himself in the early years of this century.†

### 4.2.1 Barnes's first lemma.

The first relationship which we shall consider, is usually known as Barnes's first lemma,‡ though it is in fact the integral analogue of Gauss's theorem.

Let

$$I_C = \frac{1}{2\pi i} \int_C \Gamma[a + s, b + s, c - s, d - s]\,ds, \quad (4.2.1.1)$$

where $C$ is the contour of Figure 4.1, which consists of the imaginary axis $Oy$, from $-iR$ to $+iR$, and the semi-circle of radius $R$, centre $O$, which lies to the right of $Oy$. This contour is traversed in a clockwise direction. It is indented along the axis $Oy$, where necessary, to ensure that all the poles of the integrand in the sequences

$$s = c + n, \quad d + n, \quad (n = 0, 1, 2, ..., [R])$$

† Barnes (1907 a, b, c).        ‡ Barnes (1907 b), § 15.

lie within the contour, and that all the poles in the sequences

$$s = -a-n, \quad -b-n, \quad (n = 0, 1, 2, \ldots, [R])$$

lie outside the contour.

The integral $I_C$ can be split up into two parts, the integral from $-iR$ to $+iR$, which tends to

$$I = \frac{1}{2\pi i} \int_{-i\infty}^{i\infty} \Gamma[a+s, b+s, c-s, d-s]\,ds \quad \text{as } R \to \infty, \quad (4.2.1.2)$$

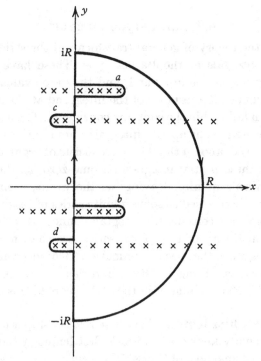

Fig. 4.1

and the integral round the semi-circle,

$$I_S = \frac{-1}{2\pi i} \int_{-\frac{1}{2}\pi}^{\frac{1}{2}\pi} \Gamma[a+R\,e^{i\theta}, b+R\,e^{i\theta}, c-R\,e^{i\theta}, d-R\,e^{i\theta}]\,R\,e^{i\theta}i\,d\theta$$

since $s = R\,e^{i\theta}$ on the boundary of the semi-circle.

Then
$$|I_S| < A\,R\,O(R^{-a-b+1-c+1-d}),$$

where $A$ is a constant, so that, as $R \to \infty$,

$$|I_S| \to 0,$$

provided that $\mathrm{Rl}\,(a+b+c+d) < 1$.

But

$I_C$ = the sum of the residues of the integrand at $s = c + n, d + n$,

$$= \sum_{n=0}^{R} \left\{ \Gamma[a+c+n, b+c+n, d-c-n] \frac{(-1)^{n-1}}{n!} \right.$$
$$\left. + \Gamma[a+d+n, b+d+n, c-d-n] \frac{(-1)^{n-1}}{n!} \right\}.$$

Hence, when $R \to \infty$,

$$I = \Gamma[a+c, b+c, d-c] \, {}_2F_1[a+c, b+c; 1+c-d; 1]$$
$$+ \Gamma[a+d, b+d, c-d] \, {}_2F_1[a+d, b+d; 1+d-c; 1]. \quad (4.2.1.3)$$

We can sum the two series on the right by Gauss's theorem, and we find, after a little reduction, that

$$\frac{1}{2\pi i} \int_{-i\infty}^{i\infty} \Gamma[a+s, b+s, c-s, d-s] \, ds = \Gamma \begin{bmatrix} a+c, a+d, b+c, b+d \\ a+b+c+d \end{bmatrix}$$

$$(4.2.1.4)$$

provided that $\mathrm{Rl}\,(a+b+c+d) < 1$.

This is the integral analogue of Gauss's theorem. It is symmetrical in $a$ and $b$, and in $c$ and $d$. By the theory of analytic continuation, we can remove the restriction $\mathrm{Rl}\,(a+b+c+d) < 1$, and we see that the result is true for all values of $a$, $b$, $c$ and $d$, provided that none of the poles of $\Gamma(a+s)\,\Gamma(b+s)$ coincide with the poles of $\Gamma(c-s)\,\Gamma(d-s)$. Further, if we write $s-k$, $a+k$, $b+k$, $c-k$, $d-k$ for $s$, $a$, $b$, $c$ and $d$, respectively, we see that the result is still true when the limits of integration are $k \pm i\infty$, where $k$ is any real finite constant. This means that we can move our whole contour any distance we wish parallel to the real axis.

**4.2.2 Barnes's second lemma.** This lemma† provides the integral analogue of Saalschutz's theorem (2.3.1.3).

We start with the series

$$\Gamma \begin{bmatrix} a, b, d-a, d-b \\ d \end{bmatrix} \, {}_3F_2[a, b, c; d, e; 1]$$

$$= \Gamma[d-a, d-b] \sum_{n=0}^{\infty} \frac{(c)_n}{(e)_n \, n!} \Gamma \begin{bmatrix} a+n, b+n \\ d+n \end{bmatrix}$$

$$= \sum_{n=0}^{\infty} \frac{(c)_n}{(e)_n \, n!} \frac{1}{2\pi i} \int_{-i\infty}^{i\infty} \Gamma[a+s, b+s, n-s, d-a-b-s] \, ds$$

by Barnes's first lemma,

$$= \frac{1}{2\pi i} \int_{-i\infty}^{i\infty} \Gamma[a+s, b+s, d-a-b-s, -s] \, {}_2F_1[c, -s; e; 1] \, ds$$

† Barnes (1910), § 4(2).

provided that $\mathrm{Rl}\,(e-c+s) > 0$, so that we can justify the interchange in the order of summation and integration. Then we find that

$$\Gamma\!\begin{bmatrix} a,b,d-a,d-b \\ d \end{bmatrix}\, {}_3F_2[a,b,c;\,d,e;\,1]$$

$$= \Gamma\!\begin{bmatrix} e \\ e-c \end{bmatrix} \frac{1}{2\pi i} \int_{-i\infty}^{i\infty} \Gamma\!\begin{bmatrix} a+s,b+s,d-a-b-s,e-c+s,\,-s \\ e+s \end{bmatrix} ds$$

$$(4.2.2.1)$$

when we have summed the inner series by Gauss's theorem.

Now take $d = c$, so that the ${}_3F_2(1)$ series reduces to a ${}_2F_1(1)$ series, which is also summable by Gauss's theorem. Then, if we replace $e-c$ by $c$ and $d-a-b$ by $1-d$, (4.2.2.1) becomes

$$\frac{1}{2\pi i} \int_{-i\infty}^{i\infty} \Gamma\!\begin{bmatrix} a+s,b+s,c+s,1-d-s,\,-s \\ e+s \end{bmatrix} ds$$

$$= \Gamma\!\begin{bmatrix} a,b,c,1-d+a,1-d+b,1-d+c \\ e-a,e-b,e-c \end{bmatrix}, \quad (4.2.2.2)$$

and the condition $c = d$ becomes the Saalschutzian condition that $1+a+b+c = d+e$. If this integral is evaluated by considering the residues at the poles to the right of the imaginary axis, we obtain a relation similar to (2.4.4.4), which reduces to Saalschutz's theorem when one of the parameters $a$, $b$, or $c$ is a negative integer.

### 4.2.3  The integral analogue of Dougall's theorem. Let us consider the integral

$$I_1 = \frac{1}{2\pi i} \int_{-i\infty}^{i\infty} \Gamma\!\begin{bmatrix} a+s,1+\tfrac{1}{2}a+s,b+s,c+s,d+s,b-a-s,\,-s \\ \tfrac{1}{2}a+s,1+a-c+s,1+a-d+s \end{bmatrix} ds.$$

$$(4.2.3.1)$$

This is the integral analogue of the well-poised ${}_5F_4(1)$ series, with the special form of the second parameter. If we consider the residues at the poles to the right of the contour, in the usual way, we find that this integral can be expressed in terms of two well-poised ${}_5F_4(1)$ series which are themselves summable. We can thus evaluate $I_1$ in terms of Gamma functions, and we find that

$$I_1 = \tfrac{1}{2}\Gamma\!\begin{bmatrix} b,c,d,b+c-a,b+d-a \\ 1+a-c-d,b+c+d-a \end{bmatrix}. \quad (4.2.3.2)$$

This is the integral analogue of the ${}_5F_4(1)$ summation theorem.

In a similar way, the more general integral

$$I_2 = \frac{1}{2\pi i}\int_{-i\infty}^{i\infty} \Gamma\left[\begin{array}{l} a+s, 1+\frac{1}{2}a+s, b+s, c+s, d+s, e+s, \\ \qquad\frac{1}{2}a+s, 1+a-c+s, 1+a-d+s, \end{array}\right.$$

$$\left.\begin{array}{l} f+s, b-a-s, -s \\ 1+a-e+s, 1+a-f+s \end{array}\right]ds$$

(4.2.3.3)

can be expressed in terms of two well-poised $_7F_6(1)$ series. But these two series can only be summed by Dougall's theorem when they terminate, and then the contour cannot be drawn to separate the increasing and decreasing sequences of poles. We can, however, get round this difficulty in the following way. From Barnes's second lemma, we have

$$\Gamma\left[\begin{array}{l} d+s, e+s, f+s \\ 1+a-d+s, 1+a-e+s, 1+a-f+s \end{array}\right]$$

$$= \frac{1}{2\pi i}\int_{-i\infty}^{i\infty} \Gamma\left[\begin{array}{l} d+t, e+t, f+t, 1+a-d-e-f-t, s-t \\ 1+a+s+t, 1+a-e-f, 1+a-d-f, 1+a-d-e \end{array}\right]dt.$$

(4.2.3.4)

Thus we can make our single integral into a double integral, and interchange the order of summation and integration, to give us

$$I_2 = \frac{1}{2\pi i}\int_{-i\infty}^{i\infty} \Gamma\left[\begin{array}{l} d+t, e+t, f+t, 1+a-d-e-f-t \\ 1+a-e-f, 1+a-d-f, 1+a-d-e \end{array}\right]$$

$$\times \left\{\frac{1}{2\pi i}\int_{-i\infty}^{i\infty} \Gamma\left[\begin{array}{l} a+s, 1+\frac{1}{2}a+s, b+s, c+s, b-a-s, -s, s-t \\ \frac{1}{2}a+s, 1+a-c+s, 1+a+s+t \end{array}\right]ds\right\}dt.$$

(4.2.3.5)

The lower bound of the distance between the $s$ and $t$ contours is supposed to be positive, the contours being modified if necessary, to ensure that this is so. The integration with respect to $s$ can now be carried out, if we make use of (4.2.3.2) above, to give us

$$I_2 = \frac{1}{2}\Gamma\left[\begin{array}{l} b, c, b+c-a \\ 1+a-e-f, 1+a-d-f, 1+a-d-e \end{array}\right]$$

$$\times \frac{1}{2\pi i}\int_{-i\infty}^{i\infty} \Gamma\left[\begin{array}{l} d+t, e+t, f+t, b-a-t, 1+a-d-e-f-t, -t \\ \qquad 1+a-c+t, b+c-a-t \end{array}\right]dt.$$

(4.2.3.6)

This integral can be evaluated by Barnes's second lemma provided that

$$1+2a = b+c+d+e+f,$$

8

so that

$$I_2 = \tfrac{1}{2}\Gamma\begin{bmatrix} b,c,d,e,f,b+c-a,b+d-a,b+e-a,b+f-a \\ 1+a-d-e,1+a-c-e,1+a-c-d,1+a-c-f, \\ 1+a-d-f,1+a-e-f \end{bmatrix}.$$

(4.2.3.7)

This is the integral analogue of Dougall's theorem.

If further we consider the poles of the integrand of $I_2$ which lie to the right of the contour, we can obtain the formula

$$_7F_6\begin{bmatrix} a,1+\tfrac{1}{2}a, & b, & c, & d, & e, & f; \\ & \tfrac{1}{2}a,1+a-b,1+a-c,1+a-d,1+a-e,1+a-f; \end{bmatrix} 1$$

$$= \Gamma\begin{bmatrix} 1+a-c,1+a-d,1+a-e,1+a-f,b+c-a, \\ 1+a,b-a,1+a-d-e,1+a-c-e,1+a-c-d, \\ \qquad\qquad b+d-a,b+e-a,b+f-a \\ \qquad\qquad 1+a-c-f,1+a-d-f,1+a-e-f \end{bmatrix}$$

$$-\Gamma\begin{bmatrix} 1+2b-a,b+c-a,b+d-a,b+e-a,b+f-a,a-b, \\ \qquad\qquad 1+a-c,1+a-d,1+a-e,1+a-f \\ 1+b-c,1+b-d,1+b-e,1+b-f,b-a,1+a,c,d,e,f \end{bmatrix}$$

$$\times {}_7F_6\begin{bmatrix} 2b-a,1+b-\tfrac{1}{2}a, & b,b+c-a,b+d-a,b+e-a,b+f-a; \\ b-\tfrac{1}{2}a,1+b-a,1+b-c,1+b-d,1+b-e,1+b-f; \end{bmatrix} 1,$$

(4.2.3.8)

where $1+2a = b+c+d+e+f$.

This is the form assumed by Dougall's theorem when we remove the restriction that one of the parameters must be a negative integer.

In this result there is an apparent lack of symmetry in $b, c, d, e$ and $f$, although we know that there must be such symmetry, in fact, on both sides of the equation. This apparent asymmetry arises from the inherent asymmetry in Barnes's lemmas, that is from the terms in $\Gamma(1-b-s)$ and $\Gamma(d-a-b-s)$ respectively.

## 4.3  Relations between $_3F_2(1)$ series

We have already proved, in (2.3.3.7), a fundamental relation between two $_3F_2(1)$ series, which can be written

$$_3F_2[a,b,c;\,d,e;\,1] = \Gamma\begin{bmatrix} d,e,s \\ a,s+b,s+c \end{bmatrix} {}_3F_2\begin{bmatrix} d-a,e-a,s; \\ s+b,s+c; \end{bmatrix} 1 ,$$

(4.3.1)

where $s \equiv d+e-a-b-c$. If one of these two series is well-poised then so is the other one.

A second relation, this time between three $_3F_2(1)$ series, will now be proved by the use of Barnes-type contour integrals. We shall consider the integral

$$I = \frac{1}{2\pi i} \int_C e^{\pm i\pi s} \Gamma \begin{bmatrix} -s, a+s, b+s, c+s \\ d+s, e+s \end{bmatrix} ds, \qquad (4.3.2)$$

taken round the semi-circular contour of Fig. 4.1, which was used in the proof of the Barnes's lemmas. This integral, as we know, is equal to minus the sum of the residues at the poles of the integrand within this semi-circular contour to the right of the imaginary axis $Oy$. But it is also equal to the sum of the residues at the poles of the integrand within a similar semi-circular contour to the left of the imaginary axis $Oy$. Hence we find that

$$I = \Gamma \begin{bmatrix} a, b, c \\ d, e \end{bmatrix} {}_3F_2 \begin{bmatrix} a, b, c; \\ d, e; \end{bmatrix} 1 \end{bmatrix}$$

$$= e^{\pm i\pi a} \Gamma \begin{bmatrix} a, b-a, c-a \\ d-a, e-a \end{bmatrix} {}_3F_2 \begin{bmatrix} a, 1+a-d, 1+a-e; \\ 1+a-b, 1+a-c; \end{bmatrix} 1 \end{bmatrix} + \text{idem} \, (a; b, c).$$

$$(4.3.3)$$

Here the expression 'idem$(a; b, c)$' means the sum of the similar expressions with $b$ and $c$ respectively interchanged with $a$.

If now, we multiply these two relations by $e^{\pm i\pi a}$ and subtract, we find that

$$_3F_2 \begin{bmatrix} a, b, c; \\ d, e; \end{bmatrix} 1 \end{bmatrix} = \Gamma \begin{bmatrix} 1-a, d, e, c-b \\ e-b, d-b, 1+b-a, c \end{bmatrix}$$

$$\times \, _3F_2 \begin{bmatrix} b, 1+b-d, 1+b-e; \\ 1+b-c, 1+b-a; \end{bmatrix} 1 \end{bmatrix} + \Gamma \begin{bmatrix} 1-a, d, e, b-c \\ e-c, d-c, 1+c-a, b \end{bmatrix}$$

$$\times \, _3F_2 \begin{bmatrix} c, 1+c-e, 1+c-d; \\ 1+c-b, 1+c-a; \end{bmatrix} 1 \end{bmatrix}.$$

$$(4.3.4)$$

This is a general relation connecting three general $_3F_2(1)$ series.

The results (4.3.1) and (4.3.4) above are two of many relations due to Thomae,[†] who approached the subject through the calculus of finite differences.

Since there are five unrestricted parameters in the general $_3F_2(1)$ series, we may in general expect to find that there are

$$5! = 120$$

such different series.

In 1923, Whipple investigated the interconnexions between Thomae's many formulae.[‡] He introduced the following notations; let

$$r_0, \quad r_1, \quad r_2, \quad r_3, \quad r_4, \quad r_5$$

---

† Thomae (1879).                    ‡ Whipple (1925).

be six parameters such that

$$r_0 + r_1 + r_2 + r_3 + r_4 + r_5 = 0,$$

and let $\quad \alpha_{lmn} = \tfrac{1}{2} + r_l + r_m + r_n, \quad \beta_{mn} = 1 + r_m - r_n.$

Also, $g$, $h$ and $j$ are used to represent those three numbers out of the six integers 0, 1, 2, 3, 4, 5 not already represented by $l$, $m$ and $n$.

Then we can define the two functions

$$F_p(l; m, n) = \frac{1}{\Gamma[\alpha_{ghj}, \beta_{ml}, \beta_{nl}]}{}_3F_2\left[\begin{matrix} \alpha_{gmn}, \alpha_{hmn}, \alpha_{jmn}; \\ \beta_{ml}, \beta_{nl}; \end{matrix} 1\right], \quad (4.3.5)$$

and $\quad F_n(l; m, n) = \dfrac{1}{\Gamma[\alpha_{lmn}, \beta_{lm}, \beta_{ln}]}{}_3F_2\left[\begin{matrix} \alpha_{lhj}, \alpha_{lgj}, \alpha_{lgh}; \\ \beta_{lm}, \beta_{ln}; \end{matrix} 1\right]. \quad (4.3.6)$

The condition that the series

$$F_p(l; m, n)$$

is convergent is $\quad\quad\quad \mathrm{Rl}(\alpha_{ghj}) > 0,$

and the condition that the series

$$F_n(l; m, n)$$

is convergent is $\quad\quad\quad \mathrm{Rl}(\alpha_{lmn}) > 0.$

Any $F_n$ function can be obtained from any $F_p$ function simply by changing the signs of all the $r$ parameters. By permutation of the suffixes $l$, $m$, $n$ over the six numbers 0, 1, 2, 3, 4, 5, sixty $F_p$ functions and sixty $F_n$ functions can be written down.

Let $\quad \alpha_{145} = a, \quad \alpha_{245} = b, \quad \alpha_{345} = c, \quad \beta_{40} = d, \quad \beta_{50} = e,$

and $\quad\quad\quad \alpha_{123} = s = d + e - a - b - c.$

Then the ${}_3F_2(1)$ series occurring in the definition of $F_p(0; 4, 5)$ is

$${}_3F_2[a, b, c; d, e; 1],$$

and all the $\alpha$'s and $\beta$'s occurring in these one hundred and twenty ${}_3F_2(1)$ functions can be expressed in terms of $a$, $b$, $c$, $d$ and $e$. In table 4.1, these values for $\alpha$ and $\beta$ are given explicitly.

## 4.3.1 Two-term relations.

The fundamental transform (4.3.1) can be rewritten in the above notation as

$$F_p(0; 4, 5) = F_p(0; 2, 3). \quad (4.3.1.1)$$

By the interchange of $r_4$ and $r_1$ in the definition, we find that this implies that $\quad\quad F_p(0; 1, 5) = F_p(0; 2, 3), \quad (4.3.1.2)$

and thus $\quad\quad\quad F_p(0; 4, 5) = F_p(0; 1, 5). \quad (4.3.1.3)$

Accordingly, we find that all the permutations of the indices 1 to 5 are legitimate, and that all the ten expressions $F_p(0; m, n)$ are equal and may be denoted by the symbol $F_p(0)$. Similar results are true for

## Table 4.1

*Expressions for $\alpha$ and $\beta$ in terms of $a$, $b$, $c$, $d$ and $e$*

| | | | | |
|---|---|---|---|---|
| $\alpha_{012} = 1-c$ | $\alpha_{013} = 1-b$ | $\alpha_{014} = 1-e+a$ | $\alpha_{015} = 1-d+a$ | |
| $\alpha_{023} = 1-a$ | $\alpha_{024} = 1-e+b$ | $\alpha_{025} = 1-d+b$ | | |
| $\alpha_{034} = 1-e+c$ | $\alpha_{035} = 1-d+c$ | $\alpha_{045} = 1-s$ | | |
| $\alpha_{123} = s$ | $\alpha_{124} = d-c$ | $\alpha_{125} = e-c$ | | |
| $\alpha_{134} = d-b$ | $\alpha_{135} = e-b$ | $\alpha_{145} = a$ | | |
| $\alpha_{234} = d-a$ | $\alpha_{235} = e-a$ | $\alpha_{245} = b$ | $\alpha_{345} = c$ | |

| | | | | |
|---|---|---|---|---|
| $\beta_{01} = 2-s-a$ | $\beta_{02} = 2-s-b$ | $\beta_{03} = 2-s-c$ | $\beta_{04} = 2-d$ | $\beta_{05} = 2-e$ |
| $\beta_{10} = s+a$ | $\beta_{12} = 1+a-b$ | $\beta_{13} = 1+a-c$ | $\beta_{14} = 1-b-c+e$ | $\beta_{15} = 1-b-c+d$ |
| $\beta_{20} = s+b$ | $\beta_{21} = 1-a+b$ | $\beta_{23} = 1+b-c$ | $\beta_{24} = 1-a-c+e$ | $\beta_{25} = 1-a-c+d$ |
| $\beta_{30} = s+c$ | $\beta_{31} = 1+c-a$ | $\beta_{32} = 1+c-b$ | $\beta_{34} = 1-a-b+e$ | $\beta_{35} = 1-a-b+d$ |
| $\beta_{40} = d$ | $\beta_{41} = 1+b+c-e$ | $\beta_{42} = 1+a+c-e$ | $\beta_{43} = 1+a+b-e$ | $\beta_{45} = 1+d-e$ |
| $\beta_{50} = e$ | $\beta_{51} = 1+b+c-d$ | $\beta_{52} = 1+a+c-d$ | $\beta_{53} = 1+a+b-d$ | $\beta_{54} = 1-d+e$ |

$$s = d+e-a-b-c.$$

all the other $F_p$ and $F_n$ series. Thus the sixty $F_p$ series may be divided into six groups of ten series each; the members of any one group are all equal to one another. A similar remark applies to the sixty $F_n$ series, and the twelve representative series, representing the twelve groups, will be denoted by $F_p(v)$ and $F_n(v)$, $v = 0, 1, 2, 3, 4, 5$.

In tables 4.2 and 4.3, the parameters of the $F_p$ functions and the $F_n$ functions are given in terms of $a$, $b$, $c$, $d$ and $e$. Only the representative forms are given. The permutation of the indices 1, 2 and 3 corresponds to the permutation of the parameters $a$, $b$ and $c$, whilst the permutation of the indices 4 and 5 corresponds to the permutation of the parameters $d$ and $e$. Thus $F_p(2)$ and $F_p(3)$ are of the same general type as $F_p(1)$, in the sense that they can be derived from $F_p(1)$ by the interchange of $b$ or $c$ with $a$ respectively. Similarly, $F_p(5)$ is of the same general type as $F_p(4)$ and can be derived from it by the interchange of $d$ and $e$.

### 4.3.2 Three-term relations.

The fundamental three-term relation (4.3.4) can be written in the present notation as

$$\frac{\sin(\pi\beta_{23})}{\pi\Gamma(\alpha_{023})} F_p(0) = \frac{F_n(2)}{\Gamma[\alpha_{134}, \alpha_{135}, \alpha_{345}]} - \frac{F_n(3)}{\Gamma[\alpha_{124}, \alpha_{125}, \alpha_{245}]}.$$

(4.3.2.1)

By changing the signs of all the $r$'s, we obtain

$$\frac{\sin(\pi\beta_{32})}{\pi\Gamma(\alpha_{145})} F_n(0) = \frac{F_p(2)}{\Gamma[\alpha_{025}, \alpha_{024}, \alpha_{012}]} - \frac{F_p(3)}{\Gamma[\alpha_{035}, \alpha_{034}, \alpha_{013}]}.$$

(4.3.2.2)

Table 4.2

| | $m,n$ | Numerator parameters | | | Denominator parameters | |
|---|---|---|---|---|---|---|
| $F_p(0)$ | 4, 5 | $a,$ | $b,$ | $c,$ | $d,$ | $e,$ |
| | 2, 3 | $s,$ | $d-a,$ | $e-a,$ | $s+b,$ | $s+c,$ |
| | 1, 4 | $a,$ | $d-b,$ | $d-c,$ | $d,$ | $s+a,$ |
| $F_p(1)$ | 0, 2 | $1-d+b,$ | $1-e+b,$ | $1-a,$ | $1+b-a,$ | $2-s-a,$ |
| | 0, 4 | $1-s,$ | $1-e+b,$ | $1-e+c,$ | $1+b+c-e,$ | $2-s-a,$ |
| | 2, 3 | $d-a,$ | $e-a,$ | $1-a,$ | $1-a+b,$ | $1-a+c,$ |
| | 2, 4 | $b,$ | $d-a,$ | $1-e+b,$ | $1+b+c-e,$ | $1+b-a,$ |
| | 4, 5 | $1-s,$ | $b,$ | $c,$ | $1+b+c-e,$ | $1+b+c-d,$ |
| $F_p(2)$ | 0, 1 | $1-d+a,$ | $1-e+a,$ | $1-b,$ | $1+a-b,$ | $2-s-b,$ |
| | 0, 4 | $1-s,$ | $1-e+a,$ | $1-e+c,$ | $1+a+c-e,$ | $2-s-b,$ |
| | 1, 3 | $d-b,$ | $e-b,$ | $1-b,$ | $1-b+a,$ | $1-b+c,$ |
| | 1, 4 | $a,$ | $d-b,$ | $1-e+a,$ | $1+a+c-e,$ | $1+a-b,$ |
| | 4, 5 | $1-s,$ | $a,$ | $c,$ | $1+a+c-e,$ | $1+a+c-d,$ |
| $F_p(3)$ | 0, 2 | $1-d+b,$ | $1-e+b,$ | $1-c,$ | $1+b-c,$ | $2-s-c,$ |
| | 0, 4 | $1-s,$ | $1-e+b,$ | $1-e+a,$ | $1+b+a-e,$ | $2-s-c,$ |
| | 2, 1 | $d-c,$ | $e-c,$ | $1-c,$ | $1-c+b,$ | $1-c+a,$ |
| | 2, 4 | $b,$ | $d-c,$ | $1-e+b,$ | $1+b+a-e,$ | $1+b-c,$ |
| | 4, 5 | $1-s,$ | $b,$ | $a,$ | $1+b+a-e,$ | $1+b+a-d,$ |
| $F_p(4)$ | 0, 1 | $1-d+a,$ | $1-b,$ | $1-c,$ | $2-d,$ | $1+e-b-c,$ |
| | 0, 5 | $1-d+a,$ | $1-d+b,$ | $1-d+c,$ | $2-d,$ | $1-d+e,$ |
| | 1, 2 | $e-c,$ | $1-c,$ | $s,$ | $1+e-a-c,$ | $1+e-b-c,$ |
| | 1, 5 | $1-d+a,$ | $e-c,$ | $e-b,$ | $1+e-d,$ | $1+e-b-c,$ |
| $F_p(5)$ | 0, 1 | $1-e+a,$ | $1-b,$ | $1-c,$ | $2-e,$ | $1+d-b-c,$ |
| | 0, 4 | $1-e+a,$ | $1-e+b,$ | $1-e+c,$ | $2-e,$ | $1-e+d,$ |
| | 1, 2 | $d-c,$ | $1-c,$ | $s,$ | $1+d-a-c,$ | $1+d-b-c,$ |
| | 1, 4 | $1-e+a,$ | $d-c,$ | $d-b,$ | $1+d-e,$ | $1+d-b-c,$ |

$$s = e+f-a-b-c.$$

If we combine three equations of the type (4.3.2.1) we find that

$$\frac{\sin(\pi\beta_{45})\,F_p(0)}{\Gamma[\alpha_{012},\alpha_{013},\alpha_{023}]} + \frac{\sin(\pi\beta_{50})\,F_p(4)}{\Gamma[\alpha_{124},\alpha_{134},\alpha_{234}]} + \frac{\sin(\pi\beta_{04})\,F_p(5)}{\Gamma[\alpha_{125},\alpha_{135},\alpha_{235}]} = 0,$$

(4.3.2.3)

and, again by changing the signs of all the $r$'s, or by combining three equations like (4.3.2.2), we obtain

$$\frac{\sin(\pi\beta_{54})\,F_n(0)}{\Gamma[\alpha_{345},\alpha_{245},\alpha_{145}]} + \frac{\sin(\pi\beta_{05})\,F_n(4)}{\Gamma[\alpha_{035},\alpha_{025},\alpha_{015}]} + \frac{\sin(\pi\beta_{40})\,F_n(5)}{\Gamma[\alpha_{034},\alpha_{024},\alpha_{014}]} = 0.$$

(4.3.2.4)

## Table 4.3

| | $m,n$ | Numerator parameters | | | Denominator parameters | |
|---|---|---|---|---|---|---|
| $F_n(0)$ | 4,5 | $1-a,$ | $1-b,$ | $1-c,$ | $2-d,$ | $2-e,$ |
| | 2,3 | $1-s,$ | $1-d+a,$ | $1-e+a,$ | $2-s-b,$ | $2-s-c,$ |
| | 1,4 | $1-a,$ | $1-d+b,$ | $1-d+c,$ | $2-d,$ | $2-s-a,$ |
| $F_n(1)$ | 0,2 | $d-b,$ | $e-b,$ | $a,$ | $1+a-b,$ | $s+a,$ |
| | 0,4 | $s,$ | $e-b,$ | $e-c,$ | $1-b-c+e,$ | $s+a,$ |
| | 2,3 | $1-d+a,$ | $1-e+a,$ | $a,$ | $1+a-b,$ | $1+a-c,$ |
| | 2,4 | $1-b,$ | $1-d+a,$ | $e-b,$ | $1-b-c+e,$ | $1+a-b,$ |
| | 4,5 | $s,$ | $1-b,$ | $1-c,$ | $1-b-c+e,$ | $1-b-c+d,$ |
| $F_n(2)$ | 0,1 | $d-a,$ | $e-a,$ | $b,$ | $1+b-a,$ | $s+b,$ |
| | 0,4 | $s,$ | $e-a,$ | $e-c,$ | $1-a-c+e,$ | $s+b,$ |
| | 1,3 | $1-d+b,$ | $1-e+b,$ | $b,$ | $1+b-a,$ | $1+b-c,$ |
| | 1,4 | $1-a,$ | $1-d+b,$ | $e-a,$ | $1-a-c+e,$ | $1+b-a,$ |
| | 4,5 | $s,$ | $1-a,$ | $1-c,$ | $1-a-c+e,$ | $1-a-c+d,$ |
| $F_n(3)$ | 0,2 | $d-b,$ | $e-b,$ | $c,$ | $1+c-b,$ | $s+c,$ |
| | 0,4 | $s,$ | $e-b,$ | $e-a,$ | $1-b-a+e,$ | $s+c,$ |
| | 2,1 | $1-d+c,$ | $1-e+c,$ | $c,$ | $1+c-b,$ | $1+c-a,$ |
| | 2,4 | $1-b,$ | $1-d+c,$ | $e-b,$ | $1-b-a+e,$ | $1+c-b,$ |
| | 4,5 | $s,$ | $1-b,$ | $1-a,$ | $1-b-a+e,$ | $1-b-a+d,$ |
| $F_n(4)$ | 0,1 | $d-a,$ | $b,$ | $c,$ | $d,$ | $1-e+b+c,$ |
| | 0,5 | $d-a,$ | $d-b,$ | $d-c,$ | $d,$ | $1+d-e,$ |
| | 1,2 | $1-e+c,$ | $c,$ | $1-s,$ | $1-e+a+c,$ | $1-e+b+c,$ |
| | 1,5 | $d-a,$ | $1-e+c,$ | $1-e+b,$ | $1+d-e,$ | $1-e+b+c,$ |
| $F_n(5)$ | 0,1 | $e-a,$ | $b,$ | $c,$ | $e,$ | $1-d+b+c,$ |
| | 0,4 | $e-a,$ | $e-b,$ | $e-c,$ | $e,$ | $1+e-d,$ |
| | 1,2 | $1-d+c,$ | $c,$ | $1-s,$ | $1-d+a+c,$ | $1-d+b+c,$ |
| | 1,4 | $e-a,$ | $1-d+c,$ | $1-d+b,$ | $1+e-d,$ | $1-d+b+c,$ |

$$s = d+e-a-b-c.$$

Now, if we eliminate $F_n(2)$ from the relation corresponding to (4.3.2.1) which connects $F_p(5)$, $F_n(0)$ and $F_n(2)$, and the relation of type (4.3.2.2) which connects $F_n(2)$, $F_p(0)$ and $F_p(5)$, it follows that

$$\frac{F_p(0)}{\Gamma[\alpha_{120}, \alpha_{130}, \alpha_{230}, \alpha_{240}, \alpha_{140}, \alpha_{340}]} + \frac{\sin(\pi\beta_{05})F_n(0)}{\Gamma[\alpha_{123}, \alpha_{124}, \alpha_{134}, \alpha_{234}]} = K_0 F_p(5),$$

$$(4.3.2.5)$$

where

$$\pi^3 K_0 = \tfrac{1}{4} \sum_{n=1}^{5} \cos\{\pi(r_0 + 2r_n)\} - \tfrac{1}{4}\cos(3\pi r_0),$$

$$= \sin(\pi\alpha_{145})\sin(\pi\alpha_{245})\sin(\pi\alpha_{345}) + \sin(\pi\alpha_{123})\sin(\pi\beta_{40})\sin(\pi\beta_{50}).$$

The corresponding analogue for $F_n(0)$ is

$$\frac{F_n(0)}{\Gamma[\alpha_{345}, \alpha_{245}, \alpha_{145}, \alpha_{135}, \alpha_{235}, \alpha_{125}]} + \frac{\sin(\pi\beta_{50}) F_p(0)}{\Gamma[\alpha_{045}, \alpha_{035}, \alpha_{025}, \alpha_{015}]} = K_0 F_n(5).$$

(4.3.2.6)

All the three-term relations possible between the 120 hypergeometric $_3F_2(1)$ series are summed up in the six relations above. These play a role similar to that of the recurrence relations for the Gauss function (§1.4).

### 4.3.3 Relations between finite series.

When one parameter, say $c$, is a negative integer, $-m$, the series

$$_3F_2[a, b, c; e, f; 1]$$

terminates, and it can be written in the reverse order, as in §2.2.3. From (4.3.2.1), when $\alpha_{345} = c = -m$, we find that

$$(-1)^m \Gamma\begin{bmatrix} \alpha_{124}, \alpha_{125} \\ \alpha_{023}, \alpha_{013} \end{bmatrix} F_p(0) = F_n(3). \qquad (4.3.3.1)$$

There are eighteen such terminating series altogether. Three of these are

$$F_p(0; 4, 5), \quad F_p(0; 3, 5) \quad \text{and} \quad F_p(0; 3, 4).$$

When they are reversed, these three series become

$$F_n(3; 1, 2), \quad F_n(4; 1, 2) \quad \text{and} \quad F_n(5; 1, 2).$$

The six relations between these eighteen series are then

$$\Gamma[\alpha_{123}, \alpha_{124}, \alpha_{125}] F_p(0) = \Gamma[\alpha_{023}, \alpha_{024}, \alpha_{025}] F_p(1), \qquad (4.3.3.2)$$

$$= \Gamma[\alpha_{013}, \alpha_{014}, \alpha_{015}] F_p(2), \qquad (4.3.3.3)$$

$$= (-1)^m \Gamma[\alpha_{123}, \alpha_{023}, \alpha_{013}] F_n(3), \qquad (4.3.3.4)$$

$$= (-1)^m \Gamma[\alpha_{124}, \alpha_{024}, \alpha_{014}] F_n(4), \qquad (4.3.3.5)$$

$$= (-1)^m \Gamma[\alpha_{125}, \alpha_{025}, \alpha_{015}] F_n(5). \qquad (4.3.3.6)$$

The other possible series, such as $F_n(0)$, do not lead to any specially simple results.

### 4.3.4 The non-terminating form of Saalschutz's theorem.

If we interchange the indices 2 and 5 in (4.3.2.1) we obtain

$$F_p(0; 4, 5) = \frac{\pi}{\sin(\pi\beta_{53})} \Gamma\begin{bmatrix} \alpha_{035} \\ \alpha_{234}, \alpha_{134}, \alpha_{123} \end{bmatrix} F_n(5; 0, 3)$$

$$+ \frac{\pi}{\sin(\pi\beta_{35})} \Gamma\begin{bmatrix} \alpha_{035} \\ \alpha_{145}, \alpha_{245}, \alpha_{125} \end{bmatrix} F_n(3; 0, 5). \qquad (4.3.4.1)$$

This can be rewritten in the standard notation as

$$_3F_2\begin{bmatrix} a,b,c; \\ d,e; \end{bmatrix} 1 \end{bmatrix} = \Gamma\begin{bmatrix} d,d-a-b \\ d-a,d-b \end{bmatrix} _3F_2\begin{bmatrix} a,b,e-c; \\ 1+a+b-d,e; \end{bmatrix} 1 \end{bmatrix}$$

$$+\Gamma\begin{bmatrix} d,e,a+b-d,d+e-a-b-c \\ a,b,e-c,d+e-a-b \end{bmatrix}$$

$$\times {}_3F_2\begin{bmatrix} d-a,d-b,d+e-a-b-c; \\ 1+d-a-b,d+e-a-b; \end{bmatrix} 1 \end{bmatrix}. \qquad (4.3.4.2)$$

This formula has an interesting connexion with Saalschutz's theorem. If $d+e = 1+a+b+c$, the first series on the right-hand side reduces to a $_2F_1(1)$ series, which can be summed by Gauss's theorem. We thus obtain

$$_3F_2\begin{bmatrix} a,b,d+e-a-b-1; \\ d,e; \end{bmatrix} 1 \end{bmatrix} = \Gamma\begin{bmatrix} d,e,d-a-b,e-a-b \\ d-a,d-b,e-a,e-b \end{bmatrix}$$

$$+\frac{1}{a+b-d}\Gamma\begin{bmatrix} d,e \\ a,b,d+e-a-b \end{bmatrix} _3F_2\begin{bmatrix} d-a,d-b,1; \\ 1+d-a-b,d+e-a-b; \end{bmatrix} 1 \end{bmatrix}. \qquad (4.3.4.3)$$

If $a$, or $b$ is a negative integer, the second term on the right vanishes and we obtain Saalschutz's theorem. Thus (4.3.4.3) gives us again a form of Saalschutz's theorem, when we have removed the condition that one of the numerator parameters must be a negative integer. This result should be compared with (2.4.4.4).

## 4.3.5 Relations between Saalschutzian $_4F_3(1)$ series.

Just as there were 120 combinations of the five free parameters in the general $_3F_2(1)$ series, so we shall expect to find 120 possible combinations of the five free parameters which occur in the general well-poised $_7F_6(1)$ series, and in the terminating Saalschutzian $_4F_3(1)$ series.

We have already found, in (2.4.1.7), a relation between two $_4F_3(1)$ Saalschutzian series. This result can be rewritten as

$$_4F_3[x,y,z,-n; u,v,w; 1]$$

$$= \frac{(v-z)_n (w-z)_n}{(v)_n (w)_n}$$

$$\times {}_4F_3[u-x,u-y,z,-n; 1-v+z-n, 1-w+z-n, w; 1], \qquad (4.3.5.1)$$

provided that the parameters are subject to the condition

$$u+v+w = 1+x+y+z-n, \qquad (4.3.5.2)$$

which in effect, gives us five free parameters. In this form, we can see that this result can also be proved directly, simply by equating the coefficients of $\zeta^n$ in (1.3.4.1).

Since they are both finite, the series in (4.3.5.1) can be written in the reverse order, and so we can obtain two more Saalschutzian series related to the given $_4F_3(1)$ series. If we interchange the parameters $x, y$ and $z$ or $u, v$ and $w$ in (4.3.5.1) we can obtain nine distinct $_4F_3(1)$ series, each related to the given $_4F_3(1)$ series, as well as the ten equivalent series obtained simply by reversing the order of the terms. Since the number of series involved is fairly large, it is convenient again to use the ideas in notation developed by Whipple.[†]

We shall use $r_1, r_2, \ldots, r_6$ for our six parameters, such that

$$r_1 + r_2 + \ldots + r_6 = 0, \qquad (4.3.5.3)$$

and we shall write
$$\phi \equiv \tfrac{1}{3}(n-1), \qquad (4.3.5.4)$$

$$\epsilon_{ij} = r_i + r_j - \phi, \qquad (4.3.5.5)$$

$$\delta_{ij} = r_i - r_j - n, \qquad (4.3.5.6)$$

so that
$$\epsilon_{12} + \epsilon_{34} + \epsilon_{56} = 1 - n \qquad (4.3.5.7)$$

and
$$\delta_{12} = \epsilon_{13} - \epsilon_{23} - n. \qquad (4.3.5.8)$$

Now let

$$S(k, l, m) = (-1)^n \Gamma \begin{bmatrix} 1 - \epsilon_{LM}, 1 - \epsilon_{KM}, 1 - \epsilon_{KL} \\ 1 - n - \epsilon_{LM}, 1 - n - \epsilon_{KM}, 1 - n - \epsilon_{KL} \end{bmatrix}$$

$$\times \, _4F_3 \begin{bmatrix} \epsilon_{lm}, \epsilon_{km}, \epsilon_{kl}, -n; \\ 1 - n - \epsilon_{LM}, 1 - n - \epsilon_{KM}, 1 - n - \epsilon_{KL}; \end{bmatrix} 1 , \quad (4.3.5.9)$$

where $K, L$ and $M$ are the three suffixes out of $1, 2, \ldots, 6$ which have not already occurred in $k, l, m$.

Then the definition may be rewritten

$$S(k, l, m) = \sum_{p=0}^{n} {}_nC_p (\epsilon_{lm})_p (\epsilon_{km})_p (\epsilon_{kl})_p (\epsilon_{LM})_{n-p} (\epsilon_{KM})_{n-p} (\epsilon_{KL})_{n-p}.$$

$$(4.3.5.10)$$

By reversal of the series, we find immediately that

$$S(k, l, m) = S(K, L, M). \qquad (4.3.5.11)$$

## 4.3.6 Relations between finite $_7F_6(1)$ series.

Whipple's transform (2.4.1.1) connects a well-poised $_7F_6(1)$ series and a finite Saalschutzian $_4F_3(1)$ series. There are two distinct cases, one in which the $_7F_6(1)$ series and the $_4F_3(1)$ series both terminate, and the other case in which the $_7F_6(1)$ series does not terminate although the $_4F_3(1)$ series does. If

$$u + v + w = 1 + x + y + z - n, \qquad (4.3.6.1)$$

† Whipple (1926a), § 3.5.

the formulae in the two cases can be written

$$_4F_3[x,y,z,-n;\ u,v,w;\ 1]$$

$$= \Gamma\begin{bmatrix} v+w-x,\ 1+y-u,\ 1+z-u,\ 1-n-u \\ 1+y-n-u,\ 1+z-n-u,\ 1+y+z-u,\ 1-u \end{bmatrix}$$

$$\times {}_7F_6\begin{bmatrix} a,\ 1+\tfrac{1}{2}a,\ w-x,\ v-x, & y, \\ \tfrac{1}{2}a, & v, & w,\ 1+z-u-n, \end{bmatrix}$$

$$\begin{matrix} z, & -n; \\ 1+y-u-n,\ 1+y+z-u; & 1 \end{matrix}\Bigg],$$

$$(4.3.6.2)$$

in which $\qquad\qquad a = y+z-n-u = w+v-x-1,\qquad$ (4.3.6.3)

and

$$_4F_3[x,y,z,-n;\ u,v,w;\ 1]$$

$$= \Gamma\begin{bmatrix} v+w+n,\ 1+x-u,\ 1+y-u,\ 1+z-u \\ 1+y+z-u,\ 1+z+x-u,\ 1+x+y-u,\ 1-u \end{bmatrix}$$

$$\times {}_7F_6\begin{bmatrix} b,\ 1+\tfrac{1}{2}b,\ w+n,\ v+n, & x, \\ \tfrac{1}{2}b, & v, & w,\ 1+y+z-u, \end{bmatrix}$$

$$\begin{matrix} y, & z; \\ 1+z+x-u,\ 1+x+y-u; & 1 \end{matrix}\Bigg],$$

$$(4.3.6.4)$$

where $\qquad\qquad b = x+y+z-u = v+w+n-1.\qquad$ (4.3.6.5)

Again we shall use Whipple's notation, and write

$$W(k;\ K) = \Gamma\begin{bmatrix} 1+\delta_{kK},\ 1-\epsilon_{mK},\ 1-\epsilon_{lK},\ 1-\epsilon_{KL}, \\ 1+n+\delta_{kK},\ 1-n-\epsilon_{mK},\ 1-n-\epsilon_{lK}, \end{bmatrix}$$

$$\begin{matrix} 1-\epsilon_{KM} \\ 1-n-\epsilon_{KL},\ 1-n-\epsilon_{KM} \end{matrix}\Bigg]$$

$$\times {}_7F_6\begin{bmatrix} \delta_{kK},\ 1+\tfrac{1}{2}\delta_{kK}, & \epsilon_{kM}, & \epsilon_{kL}, \\ \tfrac{1}{2}\delta_{kK},\ 1-n-\epsilon_{KM},\ 1-n-\epsilon_{KL}, \end{bmatrix}$$

$$\begin{matrix} \epsilon_{km}, & \epsilon_{kl}, & -n; \\ 1-n-\epsilon_{mK},\ 1-n-\epsilon_{lK},\ 1+n+\delta_{kK}; & 1 \end{matrix}\Bigg],$$

$$(4.3.6.6)$$

where all the possible permutations of the numbers 1, 2, 3, 4, 5 and 6, are allowed in the suffixes $k, l, m, K, L, M$. This definition can be rewritten

$$W(k;\ K) = (-1)^n \sum_{p=0}^{n} {}_nC_p\ \frac{(r_k-r_K+2p-n)\,(\epsilon_{kl})_p\,(\epsilon_{km})_p\,(\epsilon_{kL})_p}{(r_k-r_K)\qquad (1+r_k-r_K)_p}$$

$$\times \frac{(\epsilon_{kM})_p\,(\epsilon_{Kl})_{n-p}\,(\epsilon_{Km})_{n-p}\,(\epsilon_{KL})_{n-p}\,(\epsilon_{KM})_{n-p}}{(1+r_K-r_k)_{n-p}}.\qquad (4.3.6.7)$$

By the reversal of this series, we have immediately

$$W(k;\,K) = W(K;\,k). \qquad (4.3.6.8)$$

Equation (4.3.6.3) above can now be rewritten

$$S(1,2,3) = W(1;\,4) = S(1,2,5) = W(2;\,6) = S(2,3,5), \qquad (4.3.6.9)$$

and so on.

Formally, we can write down twenty $S$ series and thirty $W$ series, but if we make no distinction between a given series and that obtained from it by a reversal of the terms, each series is counted twice, so that there are in fact, ten distinct $S$ series and fifteen distinct $W$ series, and these twenty-five series are all equal to one another.

### 4.3.7 Relations between non-terminating $_7F_6(1)$ series. In (4.3.6.4), the first parameter on the right is

$$b = v+w+n-1 = -\phi-2r_4-r_5-r_6,$$

so that we shall write $\qquad \lambda_{4;\,5,6} = b,$
and represent the series as

$W(K;\,L,\,M)$

$$= \Gamma\begin{bmatrix}1+\lambda_{K;\,L,\,M}, 1-\epsilon_{kK}, 1-\epsilon_{lK}, 1-\epsilon_{KL}, 1-\epsilon_{KM}, 1-\epsilon_{mK}\\ 1+n+\delta_{kK}, 1+n+\delta_{lK}, 1+n+\delta_{mK}, 1-n-\epsilon_{KL}, 1-n-\epsilon_{KM}, \epsilon_{LM}\end{bmatrix}$$

$$\times {}_7F_6\begin{bmatrix}\lambda_{K;\,L,\,M}, 1+\tfrac{1}{2}\lambda_{K;\,L,\,M}, 1-\epsilon_{KL}, 1-\epsilon_{KM}, \epsilon_{kl}, \epsilon_{km}, \epsilon_{lm};\\ \tfrac{1}{2}\lambda_{K;\,L,\,M}, 1-n-\epsilon_{KM}, 1-n-\epsilon_{KL}, 1+n+\delta_{kK}, \\ 1+n+\delta_{lK}, 1+n+\delta_{mK}; \end{bmatrix}\ 1\ \end{bmatrix}.$$

$$(4.3.7.1)$$

This series $W(K;\,L,M)$ is always convergent if $\mathrm{Rl}(\epsilon_{LM}) > 0$. Then (4.3.6.2) can be rewritten in the form

$$S(1,2,3) = W(4;\,5,6). \qquad (4.3.7.2)$$

Since $S(4,5,6)$ is the same series as $S(1,2,3)$ written in the reverse order, there are six non-terminating well-poised $_7F_6(1)$ series corresponding to each Saalschutzian $_4F_3(1)$ series. It should be noted, however, that not all the six series can be convergent, since the $\epsilon$'s have a negative sum. The total number of non-terminating well-poised $_7F_6(1)$ series derived from the ten equal Saalschutzian $_4F_3(1)$ series is in fact sixty.

By the combination of (4.3.7.2) with the relations between the $S$-series we find the following formulae;

$$
\begin{aligned}
W(4;\,5,6) &= S(1,2,3) = S(4,5,6),\\
&= S(1,2,4) = S(3,5,6),\\
&= S(1,2,5) = S(3,4,6),
\end{aligned}
\qquad (4.3.7.3)
$$

$$
\begin{aligned}
W(4;\,5,6) &= W(1;\,4) = W(4;\,1),\\
&= W(1;\,5) = W(5;\,1),\\
&= W(1;\,2) = W(2;\,1),\\
&= W(4;\,5) = W(5;\,4),\\
&= W(5;\,6) = W(6;\,5),
\end{aligned}
\qquad (4.3.7.4)
$$

$$
\begin{aligned}
W(4;\,5,6) &= W(5;\,4,6),\\
&= W(5;\,1,6),\\
&= W(5;\,1,2),\\
&= W(5;\,1,4),
\end{aligned}
\qquad (4.3.7.5)
$$

$$
\begin{aligned}
W(4;\,5,6) &= W(4;\,1,5),\\
&= W(4;\,1,2),
\end{aligned}
\qquad (4.3.7.6)
$$

$$
\begin{aligned}
W(4;\,5,6) &= W(1;\,2,3),\\
&= W(1;\,2,4),\\
&= W(1;\,2,5),\\
&= W(1;\,5,6),\\
&= W(1;\,4,5).
\end{aligned}
\qquad (4.3.7.7)
$$

Thus, associated with a given non-terminating well-poised series there are three distinct Saalschutzian $_4F_3(1)$ series, five $_7F_6(1)$ terminating well-poised series, and eleven $_7F_6(1)$ non-terminating well-poised series. Also, associated with a given terminating well-poised $_7F_6(1)$ series there are two distinct Saalschutzian $_4F_3(1)$ series, three terminating well-poised $_7F_6(1)$ series, and eight non-terminating well-poised $_7F_6(1)$ series. As before, we tabulate the parameters associated with a given well-poised series, as did Whipple.† Only numerator parameters are given, and the second parameter of the well-poised series is omitted in each case. The parameters tabulated are for

$$
S(1,2,3);\ \epsilon_{23},\epsilon_{13},\epsilon_{12},\ -n;\ 1-n-\epsilon_{56},\, 1-n-\epsilon_{46},\, 1-n-\epsilon_{45}.
$$

† Whipple (1926 b).

Those tabulated for $W(1; 4)$ are

$$\delta_{14}; \ \epsilon_{16}, \epsilon_{15}, \epsilon_{13}, \epsilon_{12}, \ -n,$$

and the parameters shown for $W(4; 5, 6)$ are

$$\lambda_{4; \ 5,6}; \ 1 - \epsilon_{45}, 1 - \epsilon_{46}; \ \epsilon_{23}, \epsilon_{13}, \epsilon_{12}.$$

All these results are due to Whipple.† In a later paper,‡ he goes on to investigate the inter-connexions which hold among groups of three well-poised $_7F_6(1)$ series. In particular, he proves that

$$\frac{\sin\{\pi(d+e+f-a)\}\,\psi[a; \, b,c,d,e,f]}{\Gamma[b+d-a, b+e-a, b+f-a, 1-c]}$$

$$= \frac{\sin\{\pi(b-a)\}\,\psi[e+f-c; \ e,f, 1+a-b-c, 1+a-c-d, e+f-a]}{\Gamma[1+a-e-f, 1+a-d-f, 1+a-d-e, 1-s]}$$

$$+ \frac{\sin\{\pi(c-s)\}\,\psi[2b-a; \ b, b+c-a, b+d-a, b+e-a, b+f-a]}{\Gamma[d, e, f, 1+a-b-c]},$$

$$(4.3.7.8)$$

### Table 4.4

*Parameters of associated series; master series well-poised and terminating*

| | |
|---|---|
| $W(1; 2)$ | $a; c,d,e,f, -n;$ |
| $W(2; 1)$ | $-a-2n; c-a-n, d-a-n, e-a-n, f-a-n, -n;$ |
| | |
| $W(1; 3)$ | $s+a-c; s,d,e,f, -n;$ |
| $W(3; 1)$ | $c-a-s-2n; c-a-n, 1+a-e-f, 1+a-d-f, 1+a-d-e, -n;$ |
| | |
| $W(2; 3)$ | $s-c-n; s, d-a-n, e-a-n, f-a-n, -n;$ |
| $W(3; 2)$ | $c-s-n; c, 1+a-e-f, 1+a-d-f, 1+a-d-e, -n;$ |
| | |
| $W(3; 4)$ | $c-d-n; c, c-a-n, 1+a-d-f, 1+a-d-e, -n;$ |
| | |
| $S(1,2,3)$ | $s, c, c-a-n, -n; \ c+f-a-n, c+e-a-n, c+d-a-n;$ |
| $S(4,5,6)$ | $1+a-c-d, 1+a-c-f, 1+a-c-e, -n; \ 1-n-s, 1-n-c, 1+a-c;$ |
| | |
| $S(1,3,4)$ | $c, d, 1+a-e-f, -n; 1+a-e, 1+a-f, c+d-a-n;$ |
| $S(2,5,6)$ | $e-a-n, f-a-n, 1+a-c-d, -n; \ 1-n-c, 1-n-d, e+f-a-n;$ |
| | |
| $W(1; 2,3)$ | $1-s-n-c; 1-s, 1-c; 1+a-c-d, 1+a-c-e, 1+a-c-f;$ |
| $W(2; 1,3)$ | $1-s+a-c; 1-s, 1+n+a-c; 1+a-c-d, 1+a-c-e, 1+a-c-f;$ |
| $W(3; 1,2)$ | $1+a-2c; 1-c, 1+n+a-c; 1+a-c-d, 1+a-c-e, 1+a-c-f;$ |
| $W(1; 3,4)$ | $1-c-d-n; 1-c, 1-d; e-a-n, f-a-n, 1+a-c-d;$ |
| $W(2; 3,4)$ | $e+f-s; 1-c+a+n, 1-d+a+n; e,f, 1+a-c-d;$ |
| $W(3; 1,4)$ | $e+f-c-a-n; 1-c, e+f-a; e-a-n, f-a-n, 1+a-c-d;$ |
| $W(3; 2,4)$ | $e+f-c; 1-c+a+n, e+f-a; e,f, 1+a-c-d;$ |
| $W(3; 4,5)$ | $s-c; e+f-a, d+f-a; s,f, f-a-n;$ |

$$s = c+d+e+f-2a-n-1.$$

† Whipple (1924, 1926a, b).        ‡ Whipple (1936).

## Table 4.5

*Parameters of associated series; master series well-poised and non-terminating*

$W(4; 5, 6)$    $c+d-n-1; c, d; e, f, g;$
$W(5; 4, 6)$    $c-t-n; c, 1-t; e, f, g;$

$W(5; 1, 6)$    $1+e-t-d; 1-d+e+n, 1-t; e, 1-n-f-t, 1-n-g-t;$
$W(5; 1, 2)$    $1+e+f-2d+n; 1-d+e+n, 1-d+f+n; 1-n-g-t, c-g-n, 1-d;$
$W(5; 1, 4)$    $c-d+e; 1-d+e+n, c; e, c-f-n, c-g-n;$
$W(4; 1, 5)$    $c+e+t-1; e+t+n, c; e, c-f-n, c-g-n;$
$W(4; 1, 2)$    $e+f+2t+n-1; e+t+n, f+t+n; d-g-n, c-g-n, t;$
$W(1; 2, 3)$    $1-f-g-n; 1-f, 1-g; 1-c, 1-d, t;$
$W(1; 2, 4)$    $e-g+t; 1-g, e+t+n; d-g-n, c-g-n, t;$
$W(1; 2, 5)$    $1-d+e-g; 1-g, 1-d+e+n; 1-g-t-n, c-g-n, 1-d;$
$W(1; 5, 6)$    $1-c-d+2e+n; 1-d+e+n, 1-c+e+n; e, 1-n-f-t, 1-n-g-t;$
$W(1; 4, 5)$    $c+e-f-g-n; e+t+n, 1-d+e+n; e, c-f-n, c-g-n;$

$S(1, 2, 3)$    $e, f, g, -n; c-n, d-n, 1-t-n;$
$S(4, 5, 6)$    $1-c, 1-d, t, -n; 1-e-n, 1-f-n, 1-g-n;$
$S(1, 2, 4)$    $1-n-e-t, 1-n-f-t, g, -n; 1-d+g, 1-c+g, 1-t-n;$
$S(3, 5, 6)$    $d-g-n, c-g-n, t, -n; e+t, f+t, 1-g-n;$

$S(1, 2, 5)$    $d-e-n, d-f-n, g, -n; g+t, 1-c+g, d-n;$
$S(3, 4, 6)$    $1-n-g-t, c-g-n, 1-d, -n; 1-d+e, 1-d+f, 1-g-n;$

$W(1; 4)$    $c+d-e-2n-1; f, g, d-e-n, c-e-n, -n;$
$W(4; 1)$    $1-c-d+e; 1-n-f-t, 1-n-g-t, 1-c, 1-d, -n;$

$W(1; 5)$    $f+g-d; f, g, 1-n-e-t, c-e-n, -n;$
$W(5; 1)$    $d-f-g-2n; d-f-n, d-g-n, 1-c, t, -n;$
$W(1; 2)$    $f-e-n; f, 1-n-e-t, d-e-n, c-e-n, -n;$

$W(4; 5)$    $1-d-t-n; 1-n-e-t, 1-n-f-t, 1-n-g-t, 1-d, -n;$
$W(5; 4)$    $d+t-n-1; d-e-n, d-f-n, d-g-n, t, -n;$

$W(5; 6)$    $d-c-n; d-e-n, d-f-n, d-g-n, 1-c, -n;$

$$t = c+d-e-f-g-2n.$$

where

$$\psi[a; b, c, d, e, f]$$

$$\equiv \Gamma\begin{bmatrix} 1+a \\ 1+a-b, 1+a-c, 1+a-d, 1+a-e, 1+a-f, 1-s \end{bmatrix}$$

$$\times {}_7F_6\begin{bmatrix} a, 1+\tfrac{1}{2}a, & b, & c, & d, & e, & f; \\ \tfrac{1}{2}a, 1+a-b, 1+a-c, 1+a-d, 1+a-e, 1+a-f; \end{bmatrix} 1$$

and
$$s \equiv b+c+d+e+f-2a-1.$$

This is a general relation between three well-poised $_7F_6(1)$ series. The proof depends on the result (2.4.4.3) connecting a well-poised $_7F_6(1)$ series and two Saalschutzian $_4F_3(1)$ series. Each of the $_4F_3(1)$ series can be expressed as a well-poised $_7F_6(1)$ series, as in (4.3.7.2) above, and the result follows directly.

If we write

$$
\left.
\begin{aligned}
a &= \tfrac{1}{2} - 3x_0 + x_1 + x_2 + x_3 + x_4 + x_5, \\
b &= \tfrac{1}{2} - x_0 - x_1 + x_2 + x_3 + x_4 + x_5, \\
c &= \tfrac{1}{2} - x_0 + x_1 - x_2 + x_3 + x_4 + x_5, \\
d &= \tfrac{1}{2} - x_0 + x_1 + x_2 - x_3 + x_4 + x_5, \\
e &= \tfrac{1}{2} - x_0 + x_1 + x_2 + x_3 - x_4 + x_5
\end{aligned}
\right\}
\qquad \text{(4.3.7.9)}
$$

and

$$
f = \tfrac{1}{2} - x_0 + x_1 + x_2 + x_3 + x_4 - x_5,
$$

so that

$$
s = \tfrac{1}{2} + x_0 + x_1 + x_2 + x_3 + x_4 + x_5,
$$

we can deduce a large number of equivalent relations between three well-poised $_7F_6(1)$ series simply by permuting the six parameters, $x_0, x_1, x_2, x_3, x_4, x_5$. In this way, Whipple deduced three-term relations of the above type between 196 allied well-poised $_7F_6(1)$ series.

## 4.4 Products of hypergeometric series

We have already seen that, by a simple application of Euler's transform (1.1.3), we can deduce an identity between products of two Gauss functions (§ 2.5), such as

$$
_2F_1[a, b;\ c;\ z]\,_2F_1[1 - a, 1 - b;\ 2 - c; z]
$$

$$
= {}_2F_1[1 + a - c, 1 + b - c;\ 2 - c;\ z]\,_2F_1[c - a, c - b;\ c;\ z]. \quad \text{(4.4.1)}
$$

By the use of a special type of contour integral, we shall now show that relations of this type can be generalized. For series of the type $_3F_2(z)$, the formulae to be proved are

$$
_3F_2[a, b, c;\ d, e;\ z]\,_3F_2[1 - a, 1 - b, 1 - c; 2 - d, 2 - e;\ z]
$$

$$
= \frac{e - 1}{e - d}\,{}_3F_2\!\left[\begin{array}{c} 1 + a - d, 1 + b - d, 1 + c - d;\ \\ 2 - d, 1 + e - d; \end{array} z\right]{}_3F_2\!\left[\begin{array}{c} d - a, d - b, d - c;\ \\ d, 1 + d - e; \end{array} z\right]
$$

$$
+ \frac{d - 1}{d - e}\,{}_3F_2\!\left[\begin{array}{c} 1 + a - e, 1 + b - e, 1 + c - e;\ \\ 2 - e, 1 + d - e; \end{array} z\right]{}_3F_2\!\left[\begin{array}{c} e - a, e - b, e - c;\ \\ e, 1 + e - d; \end{array} z\right]
$$

$$
\text{(4.4.2)}
$$

which reduces to (4.4.1) above when $c = e \to \infty$, and

$$(1-z)^{a+b+c-d-e}{}_3F_2[a,b,c;\,d,e;\,z]$$

$$= \frac{e-1}{e-d}{}_3F_2\left[\begin{matrix}d-a,d-b,d-c;\\d,1+d-e;\end{matrix}\,z\right]{}_3F_2\left[\begin{matrix}e-a,e-b,e-c;\\e-1,1+e-d;\end{matrix}\,z\right]$$

$$+\frac{d-1}{d-e}{}_3F_2\left[\begin{matrix}e-a,e-b,e-c;\\e,1+e-d;\end{matrix}\,z\right]{}_3F_2\left[\begin{matrix}d-a,d-b,d-c;\\d-1,1+d-e;\end{matrix}\,z\right] \quad (4.4.3)$$

which reduces to Euler's transform when $c = e \to \infty$.

First, let us equate to zero the coefficients of $z^n$ on both sides of (4.4.2). Then we shall see that the formula to be proved is

$$\sum_{r=0}^{n}\left\{\frac{(-1)^r(a-r)_n(b-r)_n(c-r)_n}{r!\,(n-r)!\,(d-1-r)_{n+1}(e-1-r)_{n+1}}\right.$$

$$+\frac{(-1)^r(1+a-d-r)_n(1+b-d-r)_n(1+c-d-r)_n}{r!\,(n-r)!\,(1-d-r)_{n+1}(e-d-r)_{n+1}}$$

$$\left.+\frac{(-1)^r(1+a-e-r)_n(1+b-e-r)_n(1+c-e-r)_n}{r!\,(n-r)!\,(1-e-r)_{n+1}(d-e-r)_{n+1}}\right\}=0.$$

$$(4.4.4)$$

The contemplation of this formula suggests to us that it might be proved by a study of the contour integral

$$I_1 = \int_C \frac{(a-s)_n(b-s)_n(c-s)_n}{(-s)_{n+1}(d-1-s)_{n+1}(e-1-s)_{n+1}}\,ds \quad (4.4.5)$$

taken round a large circle $|s| = R$, of centre $O$ and radius $R$. This integral tends to zero as $R \to \infty$. When we equate to zero the sums of the residues of the integrand at the three sequences of poles

$$s = r, \quad s = 1-d-r \quad \text{and} \quad s = 1-e-r, \quad \text{for } r = 0,1,2,...,[R]$$

which lie within this circle, we obtain (4.4.4) immediately.

A proof of (4.4.3) is rather more complicated, and is based on the integral

$$I_2 = \int_C \frac{(a-s)_n(b-s)_n(c-s)_n}{(-s)_{n-m+1}(d-1-s)_{n+1}(e-1-s)_{n+1}}\,ds. \quad (4.4.6)$$

When $m = 0$, this reduces to (4.4.5), and when $m = 1, 2$ or $3$, we can deduce three identities similar to (4.4.4) in form. By the use of some algebraic reduction, finally we can deduce (4.4.3). Outlines of such proofs are given in Bailey (1935) § 10.3, Bailey (1933 a), Burchnall (1932), and Darling (1932).

### 4.5 A general integral

Next we shall discuss a general contour integral of the Barnes type for the series $_AF_B(1)$. Let

$$I_R = \frac{1}{2\pi i} \int_R \Gamma\begin{bmatrix}(a)+s, \ -s \\ (b)+s\end{bmatrix} ds. \qquad (4.5.1)$$

This integral is taken in a clockwise direction round the contour $R$, consisting of a large semi-circle, of centre $O$ and radius $R$, lying to the right of the imaginary axis. This contour is indented in the usual way,

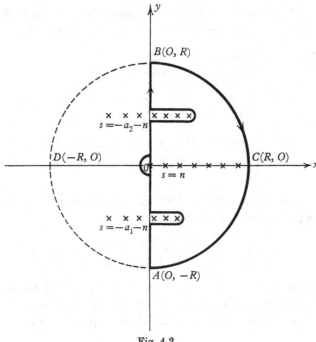

Fig. 4.2

to ensure that all the poles of the integrand in the increasing sequence $s = n$, $n = 0, 1, 2, \ldots, [R]$, lie to the right of the indented imaginary axis, and that all the poles of the integrand in the decreasing sequences, that is $s = -a_1-n, -a_2-n, -a_3-n, \ldots, -a_A-n$, for $n = 0, 1, 2, 3, \ldots$, lie to the left of the indented imaginary axis. This implies that $|R| > \max|a_\nu|$.

Now we can split $I_R$ up into two integrals, $I_{AB}$, the integral taken along the imaginary axis $Oy$, from $A(O, R)$ to $B(O, R)$ and $I_{BCA}$, the integral taken round the semi-circle $BCA$, of radius $R$, that is

$$I_R = I_{AB} + I_{BCA}. \qquad (4.5.2)$$

Also, $I_R$ is equal to the sum of all the residues of the integrand at its poles within the contour, and, since the residue of $\Gamma(-s)$ at $s = n$ is $(-1)^{n-1}/n!$, we find that

$$I_R = \sum_{n=0}^{[R]} \Gamma\begin{bmatrix} (a)+n \\ (b)+n \end{bmatrix} \frac{(-1)^{n-1}}{n!}. \qquad (4.5.3)$$

Now, on this semi-circle, $\qquad s = R\,e^{i\theta}. \qquad (4.5.4)$

Hence $\qquad I_{BCA} = \dfrac{-1}{2\pi i} \displaystyle\int_{-\frac{1}{2}\pi}^{\frac{1}{2}\pi} \Gamma\begin{bmatrix} (a)+R\,e^{i\theta}, \ -R\,e^{i\theta} \\ (b)+R\,e^{i\theta} \end{bmatrix} R\,e^{i\theta}i\,d\theta. \qquad (4.5.5)$

Here the negative sign arises since the integral is taken clockwise, that is from $+\frac{1}{2}\pi$ to $-\frac{1}{2}\pi$. Then

$$|I_{BCA}| = \frac{1}{2\pi} \int_{-\frac{1}{2}\pi}^{\frac{1}{2}\pi} R\,O(R^{-k})\,d\theta, \qquad (4.5.6)$$

where $\qquad k \equiv \mathrm{Rl}\left( \displaystyle\sum_{\nu=1}^{B} b_\nu - \displaystyle\sum_{\nu=1}^{A} a_\nu \right), \qquad (4.5.7)$

so that, as $R \to \infty$, $\qquad I_{BCA} \to 0$,

provided that $k > 0$, and that $A = B+1$.

Also, as $R \to \infty$,

$$I_{AB} \to I = \frac{1}{2\pi i} \int_{-i\infty}^{i\infty} \Gamma\begin{bmatrix} (a)+s, \ -s \\ (b)+s \end{bmatrix} ds. \qquad (4.5.8)$$

When $k > 0$, this integral $I$ exists and is equal to the sum of all the residues of the poles in the increasing sequence, $s = n$, that is

$$I = \lim_{R \to \infty} I_R = \sum_{n=0}^{\infty} \Gamma\begin{bmatrix} (a)+n \\ (b)+n \end{bmatrix} \frac{(-1)^{n-1}}{n!} \qquad (4.5.9)$$

so that we have finally,

$$\frac{1}{2\pi i} \int_{-i\infty}^{i\infty} \Gamma\begin{bmatrix} (a)+s, \ -s \\ (b)+s \end{bmatrix} ds = \Gamma\begin{bmatrix} (a) \\ (b) \end{bmatrix} {}_AF_{A-1}[(a); (b); -1], \qquad (4.5.10)$$

where $\mathrm{Rl}(\Sigma b - \Sigma a) > 0$, and $A = B+1 \geqslant 0$.

We can deduce many well-known results from this integral, by taking special values of $A$ and $B$. Thus, if $A = 1$, and $B = 0$, we have

$$\frac{1}{2\pi i} \int_{-i\infty}^{i\infty} \Gamma(a+s)\,\Gamma(-s)\,ds = \Gamma(a)\,{}_1F_0[a; \ ; -1]$$
$$= \Gamma(a)\,2^{-a}, \qquad (4.5.11)$$

provided that $\mathrm{Rl}(a) < 0$.

If $A = 2$, and $B = 1$,

$$\frac{1}{2\pi i} \int_{-i\infty}^{i\infty} \frac{\Gamma(a+s)\,\Gamma(b+s)\,\Gamma(-s)}{\Gamma(c+s)}\,ds = \frac{\Gamma(a)\,\Gamma(b)}{\Gamma(c)}\,{}_2F_1[a, b; c; -1] \qquad (4.5.12)$$

where $\mathrm{Rl}(c-a-b) > 0$. This is the limiting form of (1.6.1.6) when $|z| \to -1$.

In an exactly similar way, we can consider the integral $I_{R'}$, taken clockwise round an indented contour, consisting of part of the imaginary axis and a semi-circle to the left of $Oy$, that is round the contour $DBA$ of Fig. 4.2. The integral is again split into two parts so that

$$I_{R'} = I_{BA} + I_{ADB}. \qquad (4.5.13)$$

Then $\qquad\qquad I_{ADB} \to 0 \quad \text{as } R \to \infty,$

and $\qquad\qquad I_{BA} \to I \quad \text{as } R \to \infty.$

The sequence of residues in this case, is however, the double sequence

$$s = -a_\nu - n, \quad \text{for } \nu = 1, 2, 3, ..., \nu, ..., A, \quad n = 0, 1, 2, ..., [R],$$

so that we have

$$I = \sum_{\nu=1}^{A} \Gamma\begin{bmatrix} a_\nu, (a)-a_\nu \\ (b)-b_\nu \end{bmatrix} {}_AF_{A-1}\begin{bmatrix} a_\nu, 1+a_\nu-(b); \\ 1+a_\nu-(a)'; \end{bmatrix} -1 \Bigg], \quad (4.5.14)$$

where, for convergence, we must have $A = B+1$, and

$$\mathrm{Rl} \sum_{\nu=1}^{B} (b_\nu - a_\nu) > 0.$$

The dash in the notation $1+a_\nu-(a)'$ indicates that the denominator parameter $1+a_\nu-a_\nu$ has in fact become the factorial element in each term of the series, and so is not written explicitly in the $F$ notation. We can thus state the complete theorem

$$\frac{1}{2\pi i}\int_{-i\infty}^{i\infty} \Gamma\begin{bmatrix} (a)+s, -s \\ (b)+s \end{bmatrix} ds$$

$$= \sum_{\nu=1}^{A} \Gamma\begin{bmatrix} a_\nu, (a)-a_\nu \\ (b)-b_\nu \end{bmatrix} {}_AF_{A-1}\begin{bmatrix} a_\nu, 1+a_\nu-(b); \\ 1+a_\nu-(a)'; \end{bmatrix} -1 \Bigg]$$

$$= \Gamma\begin{bmatrix} (a) \\ (b) \end{bmatrix} {}_AF_{A-1}\begin{bmatrix} (a); \\ (b); \end{bmatrix} -1 \Bigg] \qquad (4.5.15)$$

provided that $\qquad \mathrm{Rl} \sum_{\nu=1}^{B} (b_\nu - a_\nu) > 0,$

and that $A = B+1$.

### 4.5.1 The main theorem for ${}_AF_B(1)$.

Next, we shall consider ways of making this result (4.5.15) more general. If we look at the various results due to Barnes, Whipple and Bailey,† we see that their results involve not only sequences of poles of the type $(a)+s$ in the

† Barnes (1907b); Whipple (1926c); Bailey (1935), § 6.7.

numerator and in the denominator, but also sequences of poles of the type $(a) - s$ in both the numerator and denominator. Hence we are led to study next the general hypergeometric integral

$$I_R = \frac{1}{2\pi i} \int_R \Gamma\left[\begin{matrix}(a)+s, (b)-s\\(c)+s, (d)-s\end{matrix}\right] ds, \qquad (4.5.1.1)$$

where there are $A$ of the $a$ parameters, $B$ of the $b$ parameters, $C$ of the $c$ parameters, and $D$ of the $d$ parameters. Again we consider the integral taken in turn round the two semi-circular contours $ABC$, and $ABD$, where the imaginary axis $Oy$ is now indented so that all the increasing sequences of poles

$$s = b_\nu + n, \quad \text{for } v = 1, 2, 3, \ldots, v, \ldots, B, \quad n = 0, 1, 2, \ldots, [R],$$

lie to the right of $Oy$ and all the decreasing sequences of poles

$$s = -a_\mu - n, \quad \mu = 1, 2, 3, \ldots, \mu, \ldots, A, \quad n = 0, 1, 2, \ldots, [R],$$

lie to its left.

Then we let $R \to \infty$, as before, and find the main theorem

$$\frac{1}{2\pi i} \int_{-i\infty}^{i\infty} \Gamma\left[\begin{matrix}(a)+s, (b)-s\\(c)+s, (d)-s\end{matrix}\right] ds$$

$$= \sum_{\mu=1}^{A} \Gamma\left[\begin{matrix}(a)-a_\mu, (b)+a_\mu\\(c)-a_\mu, (d)+a_\mu\end{matrix}\right]{}_{B+C}F_{A+D-1}\left[\begin{matrix}(b)+a_\mu, 1+a_\mu-(c);\\1+a_\mu-(a)', (d)+a_\mu;\end{matrix}(-1)^{A+C}\right]$$

$$= \sum_{\nu=1}^{B} \Gamma\left[\begin{matrix}(a)+b_\nu, (b)-b_\nu\\(c)+b_\nu, (d)-b_\nu\end{matrix}\right]{}_{A+D}F_{B+C-1}\left[\begin{matrix}(a)+b_\nu, 1+b_\nu-(d);\\(c)+b_\nu, 1+b_\nu-(b)';\end{matrix}(-1)^{B+D}\right],$$

$$(4.5.1.2)$$

where
$$\mathrm{Rl}\,\Sigma(c+d-a-b) > 0,$$

and
$$A - C = B - D \geqslant 0.$$

This is a general theorem of great power, and it contains within itself as special cases, all the Barnes-type integral analogues of §§ 4.2.1–4.2.3.

This theorem was originally due to Whipple. Whipple realized that such general relations must exist when he published the special case for ${}_4F_3(1)$ series.[†] Sears gave the more general form of the result in his thesis in 1949, and later he published four papers containing all his results.[‡] In these papers Sears also restated Whipple's corresponding results for well-poised series which we shall consider in § 4.5.2.

[†] Whipple (1936).                    [‡] Sears (1951 a, b, c, d).

### 4.5.2 General theorems for well-poised series.

We shall now state the four general theorems for well-poised series given by Sears. First let us consider the integral

$$I_1 = \int_{-i\infty}^{i\infty} \Gamma\begin{bmatrix} (a)+s, (a)-a_0-s \\ 1+a_0-(b)+s, 1-(b)-s \end{bmatrix} e^{2\pi is}\, ds, \qquad (4.5.2.1)$$

where there are $A+1$ of the $a$ parameters, including $a_0$, and $A-1$ of the $b$ parameters, such that

$$\mathrm{Rl}\,(Aa_0 + A - 1 - \Sigma a - \Sigma b) > 0.$$

The usual contour integration round the two semicircular contours of Fig. 4.2 leads us to the following result,†

$$\Gamma\begin{bmatrix} (a), (a)-a_0 \\ 1+a_0-(b), 1-(b), \tfrac{1}{2}a_0, -\tfrac{1}{2}a_0, 1-\tfrac{1}{2}a_0, 1+\tfrac{1}{2}a_0 \end{bmatrix}$$

$$\times {}_{2A}F_{2A-1}\begin{bmatrix} a_0, (a), (b); \\ 1+a_0-(a), 1+a_0-(b); \end{bmatrix} 1 \end{bmatrix}$$

$$= \sum_{\nu=1}^{A} a_\nu \Gamma\begin{bmatrix} a_0-a_\nu, (a)-a_\nu, a_\nu, a_\nu-a_0+(a) \\ 1+a_\nu-(b), 1+a_0-a_\nu-(b), \tfrac{1}{2}a_0-a_\nu, \\ \tfrac{1}{2}a_0+a_\nu, 1+a_\nu-\tfrac{1}{2}a_0, 1+a_\nu+\tfrac{1}{2}a_0 \end{bmatrix}$$

$$\times {}_{2A}F_{2A-1}\begin{bmatrix} a_\nu-a_0+(a), a_\nu-a_0+(b), a_\nu; \\ 1+a_\nu-(a)', 1+a_\nu-(b), 1+a_\nu-a_0; \end{bmatrix} 1 \end{bmatrix}. \qquad (4.5.2.2)$$

This theorem expresses a ${}_{2A}F_{2A-1}(1)$ series, well-poised in $a_0$, in terms of $A$ other ${}_{2A}F_{2A-1}(1)$ series, each well-poised in $2a_\nu - a_0$.

In a similar way, the integral

$$I_2 = \int_{-i\infty}^{i\infty} \Gamma\begin{bmatrix} (a)+s, (a)-a_0-s \\ 1+a_0-(b)+s, 1-(b)-s \end{bmatrix} e^{\pi is}\, ds \qquad (4.5.2.3)$$

leads us to the result‡

$$\Gamma\begin{bmatrix} (a), (a)-a_0 \\ 1+a_0-(b), 1-(b), \tfrac{1}{2}a_0, 1-\tfrac{1}{2}a_0 \end{bmatrix} {}_{2A}F_{2A-1}\begin{bmatrix} a_0, (a), (b); \\ 1+a_0-(a), 1+a_0-(b); \end{bmatrix} -1 \end{bmatrix}$$

$$= \sum_{\nu=1}^{A} a_\nu \Gamma\begin{bmatrix} a_0-a_\nu, a_\nu, (a)-a_\nu, a_\nu-a_0+(a) \\ 1+a_\nu-(b), \tfrac{1}{2}+\tfrac{1}{2}a_\nu-\tfrac{1}{2}a_0, 1+a_0-a_\nu-(b), 1+\tfrac{1}{2}a_\nu-\tfrac{1}{2}a_0 \end{bmatrix}$$

$$\times {}_{2A}F_{2A-1}\begin{bmatrix} a_\nu-a_0+(a), a_\nu-a_0+(b), a_\nu; \\ 1+a_\nu-(a)', 1+a_\nu-(b), a_\nu-a_0+1; \end{bmatrix} -1 \end{bmatrix} \qquad (4.5.2.4)$$

provided that        $\mathrm{Rl}\,(Aa_0 + A - 1 - \Sigma a - \Sigma b) > 0.$

† Sears (1951 a), 11.11.                    ‡ Sears (1951 a), 11.12.

This theorem expresses a $_{2A}F_{2A-1}(-1)$ series, well-poised in $a_0$, in terms of $A$ other $_{2A}F_{2A-1}(-1)$ series, well-poised in $2a_\nu - a_0$.

If, next, we assume that there are $A+1$ of the $a$ parameters, including $a_0$, and $A$ of the $b$ parameters, we get the integral

$$I_3 = \int_{-i\infty}^{i\infty} \Gamma\begin{bmatrix} (a)+s, (a)-a_0-s \\ 1+a_0-(b)+s, 1-(b)-s \end{bmatrix} e^{\pi i s}\, ds. \qquad (4.5.2.5)$$

This differs only by a factor $e^{\pi i s}$ from the integral $I_1$, and it leads us to the result†

$$\Gamma\begin{bmatrix} a_0, (a), (a)-a_0 \\ 1+a_0-(b), 1-(b), \tfrac{1}{2}a_0, \tfrac{1}{2}-\tfrac{1}{2}a_0, 1-\tfrac{1}{2}a_0 \end{bmatrix}$$

$$\times\,_{2A+1}F_{2A}\begin{bmatrix} a_0, (a), (b); \\ 1+a_0-(a), 1+a_0-(b); \end{bmatrix} 1 \end{bmatrix}$$

$$= \sum_{\nu=1}^{A} a_\nu \Gamma\begin{bmatrix} a_0-a_\nu, (a)-a_\nu, a_\nu-a_0-(a) \\ 1+a_\nu-(b), \tfrac{1}{2}+\tfrac{1}{2}a_\nu, 1+a_0-a_\nu-(b), \tfrac{1}{2}+\tfrac{1}{2}a_\nu-a_0 \end{bmatrix}$$

$$\times\,_{2A+1}F_{2A}\begin{bmatrix} a_\nu, a_\nu-a_0+(a), a_\nu-a_0+(b); \\ 1+a_\nu-a_0, 1+a_\nu-(a)', 1+a_\nu-(b); \end{bmatrix} 1 \end{bmatrix} \qquad (4.5.2.6)$$

provided that $\quad \mathrm{Rl}\,(Aa_0 + \tfrac{1}{2}a_0 + A - \Sigma a - \Sigma b) > 0.$

This theorem states a relation between a $_{2A+1}F_{2A}(1)$ series, well-poised in $a_0$ and $A$ other $_{2A+1}F_{2A}(1)$ series, each well-poised in $2a_\nu - a_0$.

Finally, if we assume that there are $A+2$ of the $a$ parameters, including $a_0$, and $A-1$ of the $b$ parameters, the integral

$$I_4 = \int_{-i\infty}^{i\infty} \Gamma\begin{bmatrix} (a)+s, (a)-a_0-s \\ 1+a_0-(b)+s, 1-(b)-s \end{bmatrix} e^{2\pi i s}\, ds \qquad (4.5.2.7)$$

which differs from (4.5.2.3) above by a factor of $e^{\pi i s}$ also, leads us to this relation‡ between a $_{2A+1}F_{2A}(-1)$ series, well-poised in $a_0$, and $A+1$ other $_{2A+1}F_{2A}(-1)$ series, each well-poised in $2a_\nu - a_0$;

$$\Gamma\begin{bmatrix} a_0, (a), (a)-a_0 \\ 1+a_0-(b), 1-(b), \tfrac{1}{2}a_0, 1-\tfrac{1}{2}a_0 \end{bmatrix}$$

$$\times\,_{2A+1}F_{2A}\begin{bmatrix} a_0, (a), & (b); \\ 1+a_0-(a), 1+a_0-(b); \end{bmatrix} -1 \end{bmatrix}$$

$$= \sum_{\nu=1}^{A} a_\nu \Gamma\begin{bmatrix} (a)-a_\nu, a_\nu, a_\nu-a_0+(a) \\ 1+a_\nu-(b), 1+a_0-a_\nu-(b), \tfrac{1}{2}a_\nu-a_0, 1+a_\nu-a_0 \end{bmatrix}$$

$$\times\,_{2A+1}F_{2A}\begin{bmatrix} a_\nu, a_\nu-a_0+(a), a_\nu-a_0+(b); \\ 1+a_\nu-a_0, 1+a_\nu-(a)', 1+a_\nu-(b); \end{bmatrix} -1 \end{bmatrix} \qquad (4.5.2.8)$$

provided that $\quad \mathrm{Rl}\,(Aa_0 + \tfrac{1}{2}a_0 + A - \tfrac{1}{2} - \Sigma a - \Sigma b) > 0.$

† Sears (1951a), 11.13.      ‡ Sears (1951a), 11.14.

## 4.6 The general integral for $_AF_B(z)$

The result (4.5.1.2), although very general, still does not contain, as special cases, all the Barnes-type integrals that we have used in the preceding chapters. In particular, it makes no provision for the inclusion of the variable $z$ in our general series. But the Barnes-type integral for $_2F_1(a, b; c; z)$ of § 1.6.1 gives us the clue as to how this may be done. This time we consider

$$I_C = \frac{1}{2\pi i} \int_R \Gamma\begin{bmatrix}(a)+s, (b)-s\\(c)+s, (d)-s\end{bmatrix} z^s \, ds, \qquad (4.6.1)$$

taken round the rectangular contour of Fig. 1.2, in a clockwise direction. It is now necessary that $N$ be an integer such that

$$N > \max\{|\mathrm{Im}\, a_\mu|,\ |\mathrm{Im}\, b_\nu|\},$$

in order that all the sequences of poles may fall between the lines $BC$ and $DA$.

Again, we indent the imaginary axis between $A$ and $B$ so that all the poles in the ascending sequences

$$s = b_\nu + n, \quad (\nu = 1, 2, 3, \ldots, B;\ n = 0, 1, 2, \ldots, N),$$

fall within the contour, and all the poles in the descending sequences

$$s = -a_\mu - n, \quad (\mu = 1, 2, \ldots, A;\ n = 0, 1, 2, \ldots, N),$$

fall outside the contour.

Again $I_C$ splits up into four integrals,

$$I_N, J_1, J_2 \text{ and } J_3,$$

and, by arguments exactly parallel to those of § 1.6.1, we can show that as $N \to \infty$,

$$J_1 \to 0, \quad J_2 \to 0, \quad J_3 \to 0,$$

and

$$I_C \to I(z) = \frac{1}{2\pi i} \int_{-i\infty}^{i\infty} \Gamma\begin{bmatrix}(a)+s, (b)-s\\(c)+s, (d)-s\end{bmatrix} z^s \, ds \qquad (4.6.2)$$

provided that $\frac{1}{2}\pi |A + B - C - D| > |\arg z|$, and that $B + C > A + D$.

But the sum of the residues of $I_C$ at the poles of the integrand within the contour $C$, becomes, as $N \to \infty$,

$$\sum_B(z) \equiv \sum_{\nu=1}^{B} z^{b_\nu} \Gamma\begin{bmatrix}(a)+b_\nu, (b)-b_\nu\\(c)+b_\nu, (d)-b_\nu\end{bmatrix}$$

$$\times {}_{A+D}F_{B+C-1}\begin{bmatrix}(a)+b_\nu, 1+b_\nu-(d);\\1+b_\nu-(b)', (c)+b_\nu;\end{bmatrix} (-1)^{B+D} z\end{bmatrix}. \qquad (4.6.3)$$

Hence we have now shown that

$$I(z) = \underset{B}{\Sigma}(z), \qquad (4.6.4)$$

when $B + C > A + D$.

If we consider the similar contour $FEBA$, symmetrical about $Oy$, to $ABCD$, we can show that, when $A + D > B + C$,

$$I(z) = \underset{A}{\Sigma}(z), \qquad (4.6.5)$$

provided that $\tfrac{1}{2}\pi\,|A + B - C - D| > |\arg z|$, where

$$\underset{A}{\Sigma}(z) \equiv \sum_{\mu=1}^{A} z^{-a_\mu}\Gamma\begin{bmatrix}(a) - a_\mu, (b) + a_\mu \\ (c) - a_\mu, (d) + a_\mu\end{bmatrix}$$
$$\times {}_{B+C}F_{A+D-1}\begin{bmatrix}(b) + a_\mu, 1 + a_\mu - (c); & (-1)^{A+C} \\ 1 + a_\mu - (a)', (d) + a_\mu; & z\end{bmatrix} \quad (4.6.6)$$

since the '$a$' sequences of poles are now included within the contour instead of the '$b$' sequences of poles, as before.

**4.6.1 Asymptotic integrals.** As yet we have not deduced any asymptotic transforms from our integrals. But the series of (4.6.6) above which involves powers of $1/z$, coupled with the arguments of §1.8.1, for the analytic continuation of the Gauss function, lead us to the required theorem.

Exactly similar arguments to those of §1.8.1, give us the results

$$I(z) \sim \underset{A}{\Sigma}(z) \quad \text{when} \quad A + D < B + C, \qquad (4.6.1.1)$$

and

$$I(z) \sim \underset{B}{\Sigma}(z) \quad \text{when} \quad B + C < A + D. \qquad (4.6.1.2)$$

Thus we can combine all our results into one theorem, which will include all the known transformations of the Gauss-type series within itself.

**4.6.2 The main theorem for $_AF_B(z)$.** We shall now state the general theorem in full. *If*

$$I(z) = \frac{1}{2\pi i}\int_{-i\infty}^{i\infty}\Gamma\begin{bmatrix}(a) + s, (b) - s \\ (c) + s, (d) - s\end{bmatrix} z^s\,ds, \qquad (4.6.2.1)$$

$$\underset{A}{\Sigma}(z) = \sum_{\mu=1}^{A} z^{-a_\mu}\Gamma\begin{bmatrix}(a)' - a_\mu, (b) + a_\mu \\ (c) - a_\mu, (d) + a_\mu\end{bmatrix}$$
$$\times {}_{B+C}F_{A+D-1}\begin{bmatrix}(b) + a_\mu, & 1 + a_\mu - (c); \\ 1 + a_\mu - (a)', (d) + a_\mu; \end{bmatrix}(-1)^{A+C}z^{-1}\end{bmatrix}$$

$$(4.6.2.2)$$

*and*
$$\sum_B(z) = \sum_{\nu=1}^{B} z^{b_\nu} \Gamma\begin{bmatrix}(a)+b_\nu, (b)'-b_\nu\\(c)+b_\nu, (d)-b_\nu\end{bmatrix}$$

$$\times {}_{A+D}F_{B+C-1}\begin{bmatrix}(a)+b_\nu, & 1+b_\nu-(d);\\1+b_\nu-(b)', (c)+b_\nu;\end{bmatrix}(-1)^{B+D}z\Bigg], \quad (4.6.2.3)$$

*then, provided that* $\frac{1}{2}\pi|A+B-C-D| > |\arg z|$,

(i) $I(z) = \sum_A(z) \sim \sum_B(z)$   *when*   $B+C < A+D$,     $(4.6.2.4)$

(ii) $I(z) = \sum_B(z) \sim \sum_A(z)$   *when*   $B+C > A+D$,     $(4.6.2.5)$

*and* (iii), *if we put* $z = 1$,

$$I(1) = \sum_A(1) = \sum_B(1) \quad \text{when} \quad A-C = B-D \geqslant 0, \quad (4.6.2.6)$$

*provided that*          $\mathrm{Rl}\,\Sigma(c+d-a-b) > 0$.

The proof of this theorem follows on exactly the same lines as the proofs of the previous theorems, $(4.5.15)$ and $(4.5.1.2)$. We consider the integration round semi-circular contours to the left and to the right of the imaginary axis $Oy$; then we evaluate the various sequences of residues. Finally we let the radius of the contour become infinite, and establish the necessary conditions for convergence, or for the existence of an asymptotic expansion. This theorem was implicit in some of Whipple's later work, but it was developed fully and stated by Meijer (1941 $b$–$f$).

## 4.7 Contour integrals of hypergeometric functions

The theorem stated in §4.6.2, general as it is, still does not represent the fullest possible extent of the generalizing process yet attained. The next step is to introduce the concept of summation into the integrand of the contour integral. Thus we consider the effect of replacing $I(z)$, $\Sigma_A(z)$, and $\Sigma_B(z)$ in the theorem of §4.6.2 by

$$I_M(z) = \frac{1}{2\pi i}\int_{-i\infty}^{i\infty}\sum_{m=0}^{M}\frac{((e))_m x^m}{((f))_m m!}\Gamma\begin{bmatrix}(a)+m+s, (b)+m-s\\(c)+m+s, (d)+m-s\end{bmatrix}z^s\,\mathrm{d}s,$$
$$(4.7.1)$$

$$\sum_{AM}(z) = \sum_{\mu=1}^{A}\sum_{m=0}^{M}\sum_{n=0}^{\infty}\Gamma\begin{bmatrix}(a)'-a_\mu-n, (b)+a_\mu+2m+n\\(c)-a_\mu-n, (d)+a_\mu+2m+n\end{bmatrix}$$

$$\times\frac{((e))_m x^m(-1)^n}{((f))_m m!\,n!\,z^{a+m+n}} \quad (4.7.2)$$

and
$$\sum_{BM} (z) = \sum_{\nu=1}^{B} \sum_{m=0}^{M} \sum_{n=0}^{\infty} \Gamma\begin{bmatrix} (a)+b_\nu+2m+n, (b)'-b_\nu-n \\ (c)+b_\nu+2m+n, (d)-b_\nu-n \end{bmatrix}$$

$$\times \frac{((e))_m x^m z^{b+m+n}(-1)^n}{((f))_m\, m!\, n!}. \quad (4.7.3)$$

Under conditions which make the convergence of the various series, absolute and uniform, we can let $M \to \infty$, and prove the following theorem:

*if*
$$\tfrac{1}{2}\pi\, |A+B-C-D| > |\arg z|,$$

*and the series*
$$_{A+B+E}F_{C+D+F}\begin{bmatrix} (a), (b), (e); \\ (c), (d), (f); \end{bmatrix} x$$

*is absolutely and uniformly convergent in x, then*

$$\text{(i)} \quad I(z) = \sum_{A,\infty} (z) \sim \sum_{B,\infty} (z), \quad (4.7.4)$$

*when $B+C < A+D$, or when $B+C = A+D$ and $|z| > 1$; or*

$$\text{(ii)} \quad I(z) = \sum_{B,\infty} (z) \sim \sum_{A,\infty} (z) \quad (4.7.5)$$

*when $B+C > A+D$, or when $B+C = A+D$ and $|z| < 1$;*

*where*
$$I_\infty(z) = \frac{1}{2\pi i} \int_{-i\infty}^{i\infty} \Gamma\begin{bmatrix} (a)+s, (b)-s \\ (c)+s, (d)-s \end{bmatrix}$$

$$\times {}_{A+B+E}F_{C+D+F}\begin{bmatrix} (a)+s, (b)-s, (e); \\ (c)+s, (d)-s, (f); \end{bmatrix} x\, z^s\, ds. \quad (4.7.6)$$

$$\sum_{A,\infty} (z) = \sum_{\mu=1}^{A} z^{-a_\mu}\Gamma\begin{bmatrix} (a)'-a_\mu, (b)+a_\mu \\ (c)-a_\mu, (d)+a_\mu \end{bmatrix}$$

$$\times \sum_{m=0}^{\infty}\sum_{n=0}^{\infty} \frac{((b)+a_\mu)_{2m+n}(1+a_\mu-(c))_n ((e))_m x^m(-1)^{(A+C)n}}{((d)+a_\mu)_{2m+n}(1+a_\mu-(a)')_n ((f))_m\, m!\, n!\, z^{m+n}},$$

$$(4.7.7)$$

*and*
$$\sum_{B,\infty} (z) = \sum_{\nu=1}^{B} \Gamma\begin{bmatrix} (a)+b_\nu, (b)'-b_\nu \\ (c)+b_\nu, (d)-b_\nu \end{bmatrix} z^{b_\nu}$$

$$\times \sum_{m=0}^{\infty}\sum_{n=0}^{\infty} \frac{((a)+b_\nu)_{2m+n}(1+b_\nu-(d))_n ((e))_m x^m(-1)^{(B+D)n} z^{m+n}}{((c)+b_\nu)_{2m+n}(1+b_\nu-(b)')_n ((f))_m\, m!\, n!}.$$

$$(4.7.8)$$

*In particular,*

    (iii) *when $A-C = B-D \geqslant 0$, $z = 1$, and $\mathrm{Rl}\,\Sigma(c+d-a-b) > 0$, we have*

$$I_\infty(1) = \sum_{A,\infty} (1) = \sum_{B,\infty} (1). \quad (4.7.9)$$

The conditions for the series $_{A+B+E}F_{C+D+F}(x)$ to be absolutely and uniformly convergent in $x$, are either

   (i)   $A+B+E < C+D+F+1$,

or    (ii)   $A+B+E = C+D+F+1$, and either

          (a)   $|x| < 1$,

      or    (b)   $x = 1$, and $\mathrm{Rl}\,\Sigma(c+d+f-a-b-e) > 0$.

The proof of this theorem follows on exactly the same lines as the proofs of the previous theorems (4.5.1.2) and (4.6.2.4, 5, 6) of which this is a generalization. We simply carry out the contour integration round the two finite semi-circular contours of Fig. 4.2 and evaluate the sequences of residues. Then we investigate the conditions under which the radius of the contours and the number of terms in the series under the integrand can both become infinite at the same time.

As an example on this theorem, let $A = B = C = D = 1$, $z = 1$, and $E = F = 0$. Then

$$I(1) = \frac{1}{2\pi i} \int_{-i\infty}^{i\infty} \Gamma\begin{bmatrix} a+s, & -s \\ c+s, d-s \end{bmatrix} {}_2F_2\begin{bmatrix} a+s, & -s; \\ c+s, d-s; \end{bmatrix} x \Bigg]\, ds \quad (4.7.10)$$

and we have   $$I(1) = \Gamma\begin{bmatrix} a \\ c, d \end{bmatrix} \sum_{m=0}^{\infty} \sum_{n=0}^{\infty} \frac{(a)_{2m+n}(1-d)_n x^m}{(c)_{2m+n}\, m!\, n!} \quad (4.7.11)$$

provided that          $\mathrm{Rl}\,(c+d-a) > 0$.

The series in $n$ is summable by Gauss's theorem, (1.1.5), and so

$$I(1) = \Gamma\begin{bmatrix} a, c+d-a-1 \\ c-a, c+d-1, d \end{bmatrix} {}_2F_2\begin{bmatrix} \frac{1}{2}a, \frac{1}{2}a+\frac{1}{2}; \\ \frac{1}{2}c+\frac{1}{2}d-\frac{1}{2}, \frac{1}{2}c+\frac{1}{2}d; \end{bmatrix} x \Bigg].$$
$$(4.7.12)$$

If now we introduce one $e$ parameter, we find that

$$\frac{1}{2\pi i} \int_{-i\infty}^{i\infty} \Gamma\begin{bmatrix} a+s, & -s \\ c+s, d-s \end{bmatrix} {}_3F_2\begin{bmatrix} a+s, e, & -s; \\ c+s, & d-s; \end{bmatrix} x \Bigg]\, ds$$

$$= \Gamma\begin{bmatrix} a, c+d-a-1 \\ c-a, c+d-1, d \end{bmatrix} {}_3F_2\begin{bmatrix} \frac{1}{2}a, \frac{1}{2}a+\frac{1}{2}, e; \\ \frac{1}{2}c+\frac{1}{2}d, \frac{1}{2}c+\frac{1}{2}d+\frac{1}{2}; \end{bmatrix} x \Bigg] \quad (4.7.13)$$

and, in general, when there are $E$ of the $e$ parameters, and $F$ of the $f$ parameters, we find that

$$\frac{1}{2\pi i} \int_{-i\infty}^{i\infty} \Gamma\begin{bmatrix} a+s, & -s \\ c+s, d-s \end{bmatrix} {}_{E+2}F_{F+2}\begin{bmatrix} a+s, (e), -s; \\ c+s, d-s, (f); \end{bmatrix} x \Bigg]\, ds$$

$$= \Gamma\begin{bmatrix} a, c+d-a-1 \\ c-a, c+d-1, d \end{bmatrix} {}_{E+2}F_{F+2}\begin{bmatrix} \frac{1}{2}a, \frac{1}{2}a+\frac{1}{2}, (e); \\ \frac{1}{2}c+\frac{1}{2}d-\frac{1}{2}, \frac{1}{2}c+\frac{1}{2}d, (f); \end{bmatrix} x \Bigg],$$
$$(4.7.14)$$

provided that either

    (i) $E < F+1$,

or   (ii) $E = F+1$, and either

      $(a)$ $|x| < 1$,

      or  $(b)$ $x = 1$, and $\mathrm{Rl}\,(c+d-a+\Sigma f-\Sigma e) > 0$.

In particular,

$$\frac{1}{2\pi i}\int_{-i\infty}^{i\infty}\Gamma[a+s,\,-s]\,_2F_1[a+s,\,-s;\,b;\,1]\,ds$$

$$= 2^{-a}\Gamma(a)\,_2F_1[\tfrac{1}{2}a,\,\tfrac{1}{2}+\tfrac{1}{2}a;\,b;\,1]$$

$$= 2^{-a}\Gamma\begin{bmatrix}a,b,b-a-\tfrac{1}{2}\\ b-\tfrac{1}{2}a,b-\tfrac{1}{2}-\tfrac{1}{2}a\end{bmatrix},\qquad (4.7.15)$$

if we sum by the binomial theorem, and then by Gauss's theorem.

### 4.7.1 Mixed integrals of Gamma functions and hypergeometric functions.

We have assumed, so far, that all the parameters, which occur in the Gamma functions must also occur in the $F$-functions, but this is an unnecessary restriction, and in Bailey (1929$b$), §6.3(1), we find the result

$$_{R+4}F_{S+3}\begin{bmatrix}a,a_1,\,a_2,a_3,(\rho);\\ k-a_1,k-a_2,k-a_3,(\sigma);\end{bmatrix}1$$

$$= \Gamma\begin{bmatrix}k-a_1,k-a_2,k-a_3\\ a_1,a_2,a_3,k-a_2-a_3,k-a_3-a_1,k-a_1-a_2\end{bmatrix}$$

$$\times \frac{1}{2\pi i}\int_{-i\infty}^{i\infty}\Gamma\begin{bmatrix}a_1+s,a_2+s,a_3+s,k-a_1-a_2-a_3-s,\,-s\\ k+s\end{bmatrix}$$

$$\times\,_{R+2}F_{S+1}\begin{bmatrix}a,(\rho),\,-s;\\ (\sigma),k+s;\end{bmatrix}1\,ds.\qquad (4.7.1.1)$$

The proof follows, by expansion and the interchange of the order of summation and integration, from Barnes's second lemma, (4.2.2.2) written in the form,

$$\Gamma\begin{bmatrix}a_1+n,a_2+n,a_3+n,k-a_2-a_3,k-a_3-a_1,k-a_1-a_2\\ k-a_1+n,k-a_2+n,k-a_3+n\end{bmatrix}$$

$$= \frac{1}{2\pi i}\int_{-i\infty}^{i\infty}\Gamma\begin{bmatrix}a_1+s,a_2+s,a_3+s,k-a_1-a_2-a_3-s,n-s\\ k+n+s\end{bmatrix}ds.$$

$$(4.7.1.2)$$

Bailey goes on to seek those special cases of the $_{R+2}F_{S+1}(1)$ series under the integrand on the right of (4.7.1.1) above, which can be summed, in order to find a pure Barnes-type integral to represent special cases of the $_{R+4}F_{S+3}(1)$ series on the left of (4.7.1.1). Thus, if the series on the right can be summed by Dixon's theorem, we can obtain an integral representing the series

$$_5F_4[a, b, c, d, e; 1+a-b, 1+a-c, 1+a-d, 1+a-e; 1].$$

When, further, $b = 1 + \tfrac{1}{2}a$, the integral can be evaluated by Barnes's second lemma (4.2.2.2), to give the formula (4.2.3.2) which is the integral analogue of the $_5F_4(1)$ summation theorem.

Again, we can adjust the parameters in (4.7.1.1) above, so that the series on the right can be summed by Dougall's theorem, (2.3.4.1), and we obtain

$$_7F_6\left[\begin{matrix} a, 1+\tfrac{1}{2}a, & b, & c, & d, & e, & f; \\ \tfrac{1}{2}a, 1+a-b, 1+a-c, 1+a-d, 1+a-e, 1+a-f; \end{matrix}\, 1\right]$$

$$= \Gamma\left[\begin{matrix} 1+a-b, 1+a-c, 1+a-d, 1+a-e, 1+a-f \\ 1+a, b, c, d, 1+a-c-d, 1+a-b-d, 1+a-e-f, 1+a-b-c \end{matrix}\right]$$

$$\times \frac{1}{2\pi i}\int_{-i\infty}^{i\infty} \Gamma\left[\begin{matrix} b+s, c+s, d+s, 1+a-e-f+s, \\ \phantom{xxxxxxxxxxxxxxxx} 1+a-b-c-d-s, -s \\ 1+a-e+s, 1+a-f+s \end{matrix}\right] ds.$$

$$(4.7.1.3)$$

In particular, when $f = -n$, and

$$1 + 2a = b + c + d + e - n,$$

the integral on the right can be evaluated by Barnes's second lemma, (4.2.2.2), and we obtain Dougall's theorem again (2.3.4.1). We cannot use this result on the right of (4.7.1.1) to get further transforms, since we would now have to have the condition

$$1 + 2a = r_1 + r_2 + r_3 + r_4 - s,$$

where $r_1, r_2, r_3, r_4$ and $a$ are all supposed to be independent of $s$, and so the process comes to an end, for well-poised series. It can be noted that, if we evaluate the integral on the right of (4.7.1.1) by considering the residues at poles on the right on the contour of Fig. 4.2, we can obtain the transformation (2.4.4.3) of a well-poised $_7F_6(1)$ series, in terms of two Saalschutzian $_4F_3(1)$ series.

This paper,† contains the earliest examples, which I have found in

† Bailey (1929 b).

the literature, of integrals of this type (4.7.1.1), and this integral suggests immediately that the theorem of §4·7 can be generalized, as in the following paragraph, by the introduction of sequences of Gamma functions in the integrand which do not occur also in the $F$-functions.

## 4.7.2  The mixed integral theorem. If

$$
I(z) = \frac{1}{2\pi i} \int_{-i\infty}^{i\infty} \Gamma\left[\begin{matrix} (a)+s, (b)-s, (g)+s, (h)-s \\ (c)+s, (d)-s, (j)+s, (k)-s \end{matrix}\right]
$$

$$
\times {}_{A+B+E}F_{C+D+F}\left[\begin{matrix} (a)+s, (b)-s, (e); \\ (c)+s, (d)-s, (f); \end{matrix} x \right] z^s \, ds, \qquad (4.7.2.1)
$$

$$
\sum_{A,\,\infty} (z) = \sum_{\mu=1}^{A} \Gamma\left[\begin{matrix} (a)'-a_\mu, (b)+a_\mu, (g)-a_\mu, (h)+a_\mu \\ (c)-a_\mu, (d)+a_\mu, (j)-a_\mu, (k)+a_\mu \end{matrix}\right]
$$

$$
\times \sum_{m=0}^{\infty}\sum_{n=0}^{\infty} \frac{((b)+a_\mu)_{2m+n}\,((h)+a_\mu)_{m+n}\,((e))_m\,(1+a_\mu-(c))_n}{(1+a_\mu-(j))_{m+n}\,x^m z^{-a_\mu-m-n}(-1)^{m(A+G-C-J)}}{(1+a_\mu-(a)')_n\,(1+a_\mu-(g))_{m+n}\,((f))_m} 
$$
$$
{((d)+a_\mu)_{2m+n}\,((k)+a_\mu)_{m+n}\,m!\,n!}
$$

$$
+ \sum_{\mu=1}^{G} \Gamma\left[\begin{matrix} (a)-g_\mu, (b)+g_\mu, (g)'-g_\mu, (h)+g_\mu \\ (c)-g_\mu, (d)+g_\mu, (j)-g_\mu, (k)+g_\mu \end{matrix}\right]
$$

$$
\times \sum_{m=0}^{\infty}\sum_{n=0}^{\infty} \frac{((a)-g_\mu)_{m-n}((b)+g_\mu)_{m+n}\,(1+g_\mu-(j))_n}{((h)+g_\mu)_n\,((e))_m\,x^m z^{-g_\mu-n}(-1)^{n(G-J)}}{((c)-g_\mu)_{m-n}\,((d)+g_\mu)_{m+n}\,(1+g_\mu-(g)')_n},
$$
$$
{((k)+g_\mu)_n\,((f))_m\,m!\,n!}
$$

$$
(4.7.2.2)
$$

$$
\sum_{B,\,\infty} (z) = \sum_{\nu=1}^{B} \Gamma\left[\begin{matrix} (a)+b_\nu, (b)'-b_\nu, (g)+b_\nu, (h)-b_\nu \\ (c)+b_\nu, (d)-b_\nu, (j)+b_\nu, (k)-b_\nu \end{matrix}\right]
$$

$$
\times \sum_{m=0}^{\infty}\sum_{n=0}^{\infty} \frac{((a)+b_\nu)_{2m+n}\,((g)+b_\nu)_{m+n}\,(1+b_\nu-(d))_n}{(1+b_\nu-(k))_{m+n}((e))_m\,x^m z^{b_\nu+m+n}(-1)^{n(B+H-D-K)}}{(1+b_\nu-(b)')_n\,(1+b_\nu-(k))_{m+n}\,((c)+b_\nu)_{2m+n}}
$$
$$
{((j)+b_\nu)_{m+n}\,((f))_m\,m!\,n!}
$$

$$
+ \sum_{\nu=1}^{H} \Gamma\left[\begin{matrix} (a)+h_\nu, (b)-h_\nu, (g)+h_\nu, (h)'-h_\nu \\ (c)+h_\nu, (d)-h_\nu, (j)+h_\nu, (k)-h_\nu \end{matrix}\right]
$$

$$
\times \sum_{m=0}^{\infty}\sum_{n=0}^{\infty} \frac{((a)+h_\nu)_{m+n}\,((b)-h_\nu)_{m-n}\,((g)+h_\nu)_n}{((e))_m\,(1+h_\nu-(k))_n\,x^m z^{h_\nu+n}(-1)^{n(H-K)}}{(1+h_\nu-(h)')_n\,((c)+h_\nu)_{m+n}\,((d)-h_\nu)_{m-n}},
$$
$$
{((f))_m\,((j)+h_\nu)_n\,m!\,n!}
$$

$$
(4.7.2.3)
$$

where the series $_{A+B+E}F_{C+D+F}(x)$ is absolutely and uniformly convergent in $x$, then, provided that

$$\tfrac{1}{2}\pi(A+G+B+H-C-D-J-K) > |\arg z|,$$

we have

(i) $$I(z) = \sum_{A,\infty} (z) \sim \sum_{B,\infty} (z), \qquad (4.7.2.4)$$

either $(a)$ when $\quad A+G+D+K > B+H+C+J,$
or $(b)$ when

$$A+G+D+K = B+H+C+J \quad \text{and} \quad |z| > 1,$$

and (ii) $$I(z) = \sum_{B,\infty} (z) \sim \sum_{A,\infty} (z), \qquad (4.7.2.5)$$

either $(a)$ when $\quad A+G+D+K < B+H+C+J,$
or $(b)$ when

$$A+G+D+K = B+H+C+J \quad \text{and} \quad |z| < 1.$$

Also, provided that $z = 1$, and

$$\text{Rl } \Sigma(c+d+j+k-a-b-g-h) > 0,$$

(iii) $$I(1) = \sum_{A,\infty} (1) = \sum_{B,\infty} (1), \qquad (4.7.2.6)$$

when $\qquad A+G-C-J = B+H-D-K \geqslant 0.$

Since this theorem is the most general of the series of theorems we have been considering, we shall give more details of the proof. So, let us write the integral (4.7.2.1) above as

$$I(z) = \int_{-i\infty}^{i\infty} f(s)\,ds, \qquad (4.7.2.7)$$

and consider the two integrals

$$I_{ABCD} = \int_{ABCD} f(s)\,ds, \qquad (4.7.2.8)$$

and $$I_{ADEF} = \int_{ADEF} f(s)\,ds, \qquad (4.7.2.9)$$

taken round the two rectangular contours of Fig. 4.3,

$$A(-iN), \quad B(L-iN), \quad C(L+iM), \quad D(iM),$$

and $\quad A(-iN), \quad D(iM), \quad E(-K+iM), \quad F(-K-iN),$

indented so that the first $L$ poles of each of the ascending sequences of poles
$\qquad (b)+m+n \quad \text{and} \quad (h)+n \quad \text{for } n = 0,1,2,\dots,[L],$

fall within the right-hand contour $ABCD$, and the first $K$ poles of each of the decreasing sequences of poles,

$$-(a)-m-n \quad \text{and} \quad -(g)-n \quad \text{for } n = 0,1,2,...,[K],$$

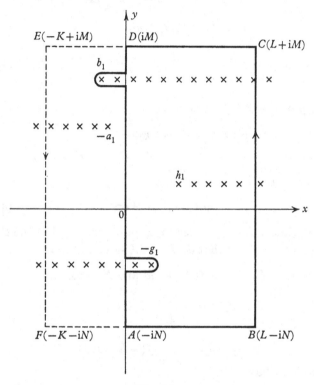

Fig. 4.3

fall within the left-hand contour $ADEF$. Then, by considering the residues at these poles, we find that

$$\sum_{B,L}(z) = I_{ABCD}$$

$$= -\int_{-iN}^{iM} f(s)\,ds + \int_0^L f(x-iN)\,dx$$

$$+ \int_{-iN}^{iM} f(L+s)\,ds - \int_0^L f(x+iM)\,dx,$$

$$= -I_{M,N}(z) + J_1 + J_2 - J_3 \text{ say,} \qquad (4.7.2.10)$$

and $\quad \underset{A,\,K}{\Sigma}(z) = I_{ADEF}$,

$$= \int_{-iN}^{iM} f(s)\,ds - \int_{-K}^{0} f(x+iM)\,dx$$

$$- \int_{-iN}^{iM} f(-K+s)\,ds + \int_{-K}^{0} f(x-iN)\,dx,$$

$$= I_{M,\,N}(z) - J_4 - J_5 + J_6 \quad \text{say.} \tag{4.7.2.11}$$

Now $\quad |J_1| \sim e^{N\alpha} N^\beta \sum_{m=0}^{L} A_m N^{(A+B-C-D)m} \int_0^L N^{l\gamma} |z|^t\,dt, \tag{4.7.2.12}$

where

$$\alpha = |\arg z| - \tfrac{1}{2}\pi(A+B+G+H-C-D-J-K),$$

$$\beta = \mathrm{Rl}\{\Sigma(a+b+g+h-c-d-j-k)\}$$
$$- \tfrac{1}{2}(A+B+G+H-C-D-J-K),$$

$$\gamma = A-B+G-H-C+D-J+K.$$

Hence $\qquad\qquad |J_1| \sim |z|^L N^{-L} e^{-N}, \tag{4.7.2.13}$

as $N \to \infty$, that is $J_1 \to 0$, under the conditions stated in the theorem. Similarly, we can show that $J_3, J_4$ and $J_6 \to 0$, and that

$$I_{M,\,N}(z) \to I(z) \quad \text{as } M \text{ and } N \to \infty.$$

Further $\qquad\qquad J_2 \to J_L = \int_{-i\infty}^{i\infty} f(L+s)\,ds \tag{4.7.2.14}$

and $\qquad\qquad J_5 \to J_K = \int_{-i\infty}^{i\infty} f(-K+s)\,ds, \tag{4.7.2.15}$

so that we now have,

$$I(z) = \underset{A,\,K}{\Sigma}(z) + J_K = \underset{B,\,L}{\Sigma}(z) + J_L. \tag{4.7.2.16}$$

Now, as $K \to \infty$, $J_K \to 0$, and $J_L$ is bounded for finite values of $L$, provided that $\alpha < 0$, and either (a) $\gamma < 0$, or (b) $\gamma = 0$, and $|z| > 1$.

Hence
$$I(z) = \underset{A,\,\infty}{\Sigma}(z) \sim \underset{B,\,\infty}{\Sigma}(z). \tag{4.7.2.17}$$

Similarly, as $L \to \infty$, $J_L \to 0$, and $J_K$ is bounded for finite values of $K$, provided that $\alpha < 0$, and either (a) $\gamma < 0$, or (b) $\gamma = 0$, and $|z| < 1$.

Hence
$$I(z) = \underset{B,\,\infty}{\Sigma}(z) \sim \underset{A,\,\infty}{\Sigma}(z). \tag{4.7.2.18}$$

Finally, if $z = 1$,

$$\mathrm{Rl}\{\Sigma(c+d+j+k-a-b-g-h)\} > 0,$$

and $\qquad A + G - C - J = B + H - D - K \geqslant 0,$

then $\qquad J_K \to 0$ and $J_L \to 0$ as $K \to \infty$ and $L \to \infty,$

and we have $\qquad I(1) = \underset{A,\,\infty}{\sum}(1) = \underset{B,\,\infty}{\sum}(1).$ $\qquad$ (4.7.2.19)

## 4.7.3 Some examples on the mixed theorem.

The theorem, which we have just proved is one of the central theorems of the general theory, so we shall now give some examples of its application. Let us consider first, the integral

$$I = \frac{1}{2\pi i}\int_{-i\infty}^{i\infty}\Gamma\left[\begin{matrix}a_1+s,\,a_2+s,\,a_3+s,\,b-s,\,-s\\c+s\end{matrix}\right]$$

$$\times\,_{E+2}F_{F+1}\left[\begin{matrix}d,\,(e),\,-s;\\(f),\,c+s;\end{matrix}1\right]ds. \qquad (4.7.3.1)$$

Here

$$\underset{A,\,\infty}{\sum} = \Gamma\left[\begin{matrix}a_1,\,a_2,\,a_3,\,b\\c\end{matrix}\right]$$

$$\times\sum_{m=0}^{\infty}\frac{(a_1)_m\,(a_2)_m\,(a_3)_m\,((e))_m\,(d)_m\,(-1)^m}{(c)_{2m}\,((f))_m\,(1-b)_m\,m!}\,_3F_2\left[\begin{matrix}a_1+m,\,a_2+m,\,a_3+m;\\c+2m,\,1-b+m;\end{matrix}1\right]$$

$$+\Gamma\left[\begin{matrix}a_1+b,\,a_2+b,\,a_3+b,\,-b\\c+b\end{matrix}\right]$$

$$\times\sum_{m=0}^{\infty}\frac{(d)_m\,((e))_m\,(-b)_m}{(c+b)_m\,((f))_m\,m!}\,_3F_2\left[\begin{matrix}a_1+b,\,a_2+b,\,a_3+b;\\c+b+m,\,1+b-m;\end{matrix}1\right] \qquad (4.7.3.2)$$

and

$$\underset{B,\,\infty}{\sum} = \sum_{\mu=1}^{3}\Gamma\left[\begin{matrix}b+a_\mu,\,(a)'-a_\mu,\,a_\mu\\c-a_\mu\end{matrix}\right]$$

$$\times\sum_{m=0}^{\infty}\frac{(d)_m\,((e))_m\,(a_\mu)_m}{((f))_m\,(c-a_\mu)_m\,m!}\,_3F_2\left[\begin{matrix}b+a_\mu,\,a_\mu+m,\,1+a_\mu-c-m;\\1+a_\mu-(a)';\end{matrix}1\right].$$

$$(4.7.3.3)$$

The conditions of the third case of the theorem are satisfied, so that

$$I = \underset{A,\,\infty}{\sum} = \underset{B,\,\infty}{\sum}. \qquad (4.7.3.4)$$

When we impose the condition that

$$c - a_1 - a_2 - a_3 = b, \qquad (4.7.3.5)$$

we find that in (4.7.3.2) and (4.7.3.3), the five series in $m$ are all $_3F_2(1)$ series of the Saalschutzian type. If now we make use of the non-terminating form of Saalschutz's theorem (2.4.4.4), we find that the first two series in $m$ in (4.7.3.2), combine into a single product of Gamma functions, and we have

$$I = \Gamma\begin{bmatrix} a_1, a_2, a_3, b+a_1, b+a_2, b+a_3 \\ c-a_1, c-a_2, c-a_3 \end{bmatrix}$$

$$\times {}_{E+4}F_{F+3}\begin{bmatrix} d, a_1, a_2, a_3, (e); \\ c-a_1, c-a_2, c-a_3, (f); \end{bmatrix} 1 \end{bmatrix}. \quad (4.7.3.6)$$

When, further, $E = F$, this is Bailey's original result, (4.7.1.1).

## 4.8 Mellin transforms

The next possibility which we shall consider is that of deducing still more general relations between hypergeometric series, by the introduction of the concept of Mellin integral transforms. We state the Mellin transform theorem thus:

if

$$f(x) = \frac{1}{2\pi i} \int_{c-i\infty}^{c+i\infty} x^{-s} g(s) \, ds, \quad (4.8.1)$$

then

$$g(s) = \int_0^\infty x^{s-1} f(x) \, dx, \quad (4.8.2)$$

provided that $g(s)$ exists in the Lebesgue sense, over the range 0 to infinity. The Mellin transform can be translated into a Laplace transform by putting $e^{1/x}$ for $x$ in both integrals, and into a Fourier transform by putting $e^{1/ix}$ for $x$ in both integrals. See Erdélyi (1954), §7, or Titchmarsh (1948a), for detailed proofs of this theorem, and its various transformations.

If we now apply the Mellin transform to the four theorems of §4.5.1, §4.6, §4.7 and §4.7.2 on the evaluation of contour integrals with hypergeometric or Gamma functions in the integrand, we can deduce immediately four further general theorems on hypergeometric transforms.

**Theorem I.** *If* $\quad g(s) = \Gamma\begin{bmatrix} (a)+s, (b)-s \\ (c)+s, (d)-s \end{bmatrix},$ $\quad\quad (4.8.3)$

$$\sum_A (x) = \sum_{\mu=1}^A x^{a_\mu} \Gamma\begin{bmatrix} (a)'-a_\mu, (b)+a_\mu \\ (c)-a_\mu, (d)+a_\mu \end{bmatrix}$$

$$\times {}_{B+C}F_{A+D-1}\begin{bmatrix} (b)+a_\mu, & 1+a_\mu-(c); \\ 1+a_\mu-(a)', (d)+a_\mu; \end{bmatrix} (-1)^{A+C} x \end{bmatrix} \quad (4.8.4)$$

and
$$\sum_B (1/x) = \sum_{\nu=1}^B x^{-b_\nu} \Gamma \begin{bmatrix} (a)+b_\nu, (b)'-b_\nu \\ (c)+b_\nu, (d)-b_\nu \end{bmatrix}$$

$$\times {}_{A+D}F_{B+C-1} \begin{bmatrix} (a)+b_\nu, 1+b_\nu-(d); \\ (c)+b_\nu, 1+b_\nu-(b)'; \end{bmatrix} (-1)^{B+D}/x \end{bmatrix}, \quad (4.8.5)$$

*then*      (i)    $g(s) = \displaystyle\int_0^\infty x^{s-1} \sum_A (x)\,dx$            (4.8.6)

*if* $A+D > B+C$,

     (ii)    $g(s) = \displaystyle\int_0^1 x^{s-1} \sum_A (x)\,dx + \int_1^\infty x^{s-1} \sum_B (1/x)\,dx$     (4.8.7)

*if* $A+D = B+C$,

*and*

     (iii)    $g(s) = \displaystyle\int_0^\infty x^{s-1} \sum_B (1/x)\,dx$          (4.8.8)

*if* $A+D < B+C$, *provided that, in all three cases,*

$$A+B \geqslant C+D,$$

$$-\mathrm{Rl}\,(a_\mu) < \mathrm{Rl}\,(s) < \mathrm{Rl}\,(b_\nu)$$

*and* $\mathrm{Rl}\,(c_\nu)$, $\mathrm{Rl}\,(d_\nu) \neq -N$, *for N a positive integer or zero, and for all integer values of* $\nu$.

The proof follows from the application of the Mellin transform to the theorem of §4.6. We have, from (4.6.4) and (4.6.5),

$$f(x) = \sum_A (x), \quad (4.8.9)$$

if $A+D > B+C$, or if $A+D = B+C$ and $0 < x < 1$. Also

$$f(x) = \sum_B (1/x) \quad (4.8.10)$$

if $A+D < B+C$, or if $A+D = B+C$, and $1 < x < \infty$.

We must always have $A+B \geqslant C+D$ for $f(x)$ to exist at all, under the condition
$$\tfrac{1}{2}\pi(A+B-C-D) > |\arg x|,$$

and the theorem follows at once, under the stated conditions.

It should be noted that

$$g(-s) \to g(s) \quad \text{when} \quad f(1/x) \to f(x),$$

and so, given any $g(s)$ which falls under the case (iii) above, it is possible to regard it as equivalent to a function $g(-s)$ which can be treated under case (i) simply by the substitution of $b$'s for $a$'s, $d$'s for $c$'s, and *vice versa*. Also, when $A+D = B+C$, $\sum_B (1/x)$ is the analytic continuation of $\sum_A (x)$ for $1 < x < \infty$.

If $B = 0$ and $A = C$, we must also have $D = 0$, since always $A + B \geqslant C + D$. In this particular case, the integral over the range one to infinity is zero, and

$$g(s) = \int_0^1 x^{s-1} \sum_A (x) \, dx. \qquad (4.8.11)$$

If we have $f(x)$ expressed as the contour integral of (4.8.1) above, without giving its evaluation in terms of hypergeometric series, we obtain the Mellin transform of Meijer's $G$-function.

### 4.8.1 The most elementary cases.
If $g(s)$ contains only one Gamma function in $s$, there are two possibilities arising out of Theorem I,    $g(s) = \Gamma(a+s)$ and $g(s) = \Gamma(b-s)$.

These two cases transform into one another under the rule for $g(s)$ into $g(-s)$, and they both lead to the Euler integral of the second kind

$$\Gamma(a+s) = \int_0^\infty x^{a+s-1} e^{-x} \, dx. \qquad (4.8.1.1)$$

The possibilities which arise when $g(s)$ contains two Gamma functions only in $s$, reduce in the same way, to four distinct cases†; two Euler integrals of the first kind,

$$\Gamma \begin{bmatrix} a+s, b-s \\ a+b \end{bmatrix} = \int_0^1 x^{s+a-1} {}_1F_0[a+b; \; ; -x] \, dx$$

$$+ \int_1^\infty x^{s-b-1} {}_1F_0[a+b; \; ; -1/x] \, dx, \qquad (4.8.1.2)$$

and    $$\Gamma \begin{bmatrix} a+s, c-a \\ c+s \end{bmatrix} = \int_0^1 x^{s+a-1} {}_1F_0[1+a-c; \; ; x] \, dx, \qquad (4.8.1.3)$$

and two integrals of Bessel functions,

$$\Gamma \begin{bmatrix} a+s, a+d \\ d-s \end{bmatrix} = \int_0^\infty x^{s+a-1} {}_0F_1[\; ; a+d; -x] \, dx, \qquad (4.8.1.4)$$

and    $$\Gamma[a+s, b+s] = \int_0^\infty x^{s+a-1} \Gamma(b-a) {}_0F_1[\; ; 1+a-b; x] \, dx$$

$$+ \int_0^\infty x^{s+b-1} \Gamma(a-b) {}_0F_1[\; ; 1+b-a; x] \, dx. \qquad (4.8.1.5)$$

† See Erdélyi (1954), § 7.3, (15), (17), (20), (22) and (23).

When $g(s)$ contains three Gamma functions in $s$ we find that there are five distinct possibilities. The first of these is

$$\Gamma\begin{bmatrix} a+s, b-s \\ d-s \end{bmatrix} = \int_0^\infty x^{s+a-1}\Gamma\begin{bmatrix} a+b \\ a+d \end{bmatrix} {}_1F_1[a+b; a+d; -x]\,dx.$$

(4.8.1.6)

This is an integral of a confluent hypergeometric function, which in its turn can be expressed as an integral of a Whittaker function.

The special case $b = \frac{1}{2}$, $d = 1+a$, leads us to

$$\pi^{\frac{1}{2}}\Gamma\begin{bmatrix} s+\nu, s-\nu \\ s+\frac{1}{2} \end{bmatrix} = \int_0^\infty x^{s-1}e^{-\frac{1}{2}x}K_\nu(\tfrac{1}{2}x)\,dx,$$

(4.8.1.7)

where $|\mathrm{Rl}(\nu)| < \mathrm{Rl}(s)$, and the corresponding case† for $g(-s)$ leads us to

$$\pi^{-\frac{1}{2}}\Gamma\begin{bmatrix} \frac{1}{2}-s, s+\nu \\ 1+\nu-s \end{bmatrix} = \int_0^\infty x^{s-1}e^{-\frac{1}{2}x}I_\nu(\tfrac{1}{2}x)\,dx,$$

(4.8.1.8)

where $-\mathrm{Rl}(\nu) < \mathrm{Rl}(s) < \frac{1}{2}$.

The second case is

$$\Gamma[a_1+s, a_2+s, b-s]$$

$$= \int_0^\infty x^{s+a_1-1}\Gamma[a_2-a_1, a_1+b]\,{}_1F_1\begin{bmatrix} a_1+b; \\ 1+a_1-a_2; \end{bmatrix} x \end{bmatrix}\,dx$$

$$+ \int_0^\infty x^{s+a_2-1}\,\Gamma[a_1-a_2, a_2+b]\,{}_1F_1\begin{bmatrix} a_2+b; \\ 1+a_2-a_1; \end{bmatrix} x \end{bmatrix}\,dx.$$

(4.8.1.9)

Here, if we write $\mu-\frac{1}{2} = a_1$, $\frac{1}{2}-\mu = a_2$, $-\kappa = b$, (4.8.1.9) leads us to‡

$$\Gamma\begin{bmatrix} s+\mu-\frac{1}{2}, s-\mu+\frac{1}{2}, -\kappa-s \\ \frac{1}{2}+\mu-\kappa, \frac{1}{2}-\mu-\kappa \end{bmatrix} = \int_0^\infty x^{s-1}e^{\frac{1}{2}x}W_{\kappa,\mu}(x)\,dx,$$

(4.8.1.10)

where $|\mathrm{Rl}\,\mu-\frac{1}{2}| < \mathrm{Rl}(s) < -\mathrm{Rl}(\kappa)$.

The remaining three results§ in this group are

$$\Gamma\begin{bmatrix} a+s, b+s \\ c+s \end{bmatrix} = \int_0^\infty x^{s+a-1}\Gamma\begin{bmatrix} b-a \\ c-a \end{bmatrix}{}_1F_1\begin{bmatrix} 1+a-c; \\ 1+a-d; \end{bmatrix} -x \end{bmatrix}\,dx$$

$$+ \int_0^\infty x^{s+b-1}\Gamma\begin{bmatrix} a-b \\ c-b \end{bmatrix}{}_1F_1\begin{bmatrix} 1+b-c; \\ 1+b-a; \end{bmatrix} -x \end{bmatrix}\,dx, \quad (4.8.1.11)$$

$$\Gamma\begin{bmatrix} a+s, b+s \\ d-s \end{bmatrix} = \int_0^\infty x^{s+a-1}\Gamma\begin{bmatrix} b-a \\ d+a \end{bmatrix}{}_0F_2[\,; 1+a-b, d+a; x]\,dx$$

$$+ \int_0^\infty x^{s+b-1}\Gamma\begin{bmatrix} a-b \\ d+b \end{bmatrix}{}_0F_2[\,; 1+b-a, d+b; x]\,dx, \quad (4.8.1.12)$$

† See Erdélyi (1954), § 7.3 (25) and (26).
‡ See Erdélyi (1954), § 6.8 (7).
§ See Slater (1955e), eqs. (3.11), (3.12) and (3.13).

and     $\Gamma[a_1+s, a_2+s, a_3+s]$

$$= \sum_{\nu=0}^{3} \int_0^\infty x^{s+a_\nu-1} \Gamma[(a)'-a_\nu] {}_0F_2[\ ; 1+a_\nu-(a)';\ -x]\,dx.$$

(4.8.1.13)

Finally, when $g(s)$ contains four Gamma functions in $s$ we can find ten distinct cases, each of which can be extended to the general hypergeometric series. Thus the extension of (4.8.1.13) above is

$$\Gamma[(a)+s] = \sum_{\mu=1}^{A} \Gamma[(a)'-a_\mu]$$

$$\times \int_0^\infty x^{s+a_\mu-1} {}_0F_{A-1}[\ ; 1+a_\mu-(a)';\ (-1)^A x]\,dx.$$

(4.8.1.14)

When sequences of $a$ and $c$ parameters only occur, we have

$$\Gamma\begin{bmatrix}(a)+s \\ (c)+s\end{bmatrix} = \sum_{\mu=1}^{A} \Gamma\begin{bmatrix}(a)'-a_\mu \\ (c)-a_\mu\end{bmatrix}$$

$$\times \int_0^\infty x^{s+a_\mu-1} {}_cF_{A-1}\begin{bmatrix}1+a_\mu-(c); \\ 1+a_\mu-(a)';\end{bmatrix} (-1)^{A+C} x\begin{bmatrix}\\\end{bmatrix}dx,$$

(4.8.1.15)

when $A > C$, and

$$\Gamma\begin{bmatrix}(a)+s \\ (c)+s\end{bmatrix} = \sum_{\mu=1}^{A} \Gamma\begin{bmatrix}(a)'-a_\mu \\ (c)-a_\mu\end{bmatrix} \int_0^1 x^{s+a_\mu-1} {}_AF_{A-1}\begin{bmatrix}1+a_\mu-(c); \\ 1+a_\mu-(a)';\end{bmatrix} x\begin{bmatrix}\\\end{bmatrix}dx,$$

(4.8.1.16)

when $A = C$.

When only sequences of $a$ and $d$ parameters occur, we find that

$$\Gamma\begin{bmatrix}(a)+s \\ (d)-s\end{bmatrix} = \sum_{\mu=1}^{A} \Gamma\begin{bmatrix}(a)'-a_\mu \\ (d)+a_\mu\end{bmatrix} \int_0^\infty x^{s+a_\mu-1}$$

$$\times {}_0F_{A+D-1}[\ ; 1+a_\mu-(a)', (d)+a_\mu;\ (-1)^A x]\,dx,$$     (4.8.1.17)

for $A \geqslant D$, and

$$\Gamma\begin{bmatrix}(a)+s \\ (c)+s, (d)-s\end{bmatrix} = \sum_{\mu=1}^{A} \Gamma\begin{bmatrix}(a)'-a_\mu \\ (c)+s, (d)-s\end{bmatrix} \int_0^\infty x^{s+a_\mu-1}$$

$$\times {}_cF_{A+D-1}\begin{bmatrix}1+a_\mu-(c); \\ 1+a_\mu-(a)', (d)+a_\mu;\end{bmatrix} (-1)^{A+C} x\begin{bmatrix}\\\end{bmatrix}dx,$$     (4.8.1.18)

for $A \geqslant C+D$.

When only sequences of $a$ and $b$ parameters occur, we have

$$\Gamma[(a)+s,(b)-s] = \sum_{\mu=1}^{A} \Gamma[(a)'-a_\mu,(b)+a_\mu]$$

$$\times \int_0^\infty x^{s+a_\mu-1} {}_BF_{A-1}\begin{bmatrix}(b)+a_\mu; \\ 1+a_\mu-(a)';\end{bmatrix} (-1)^A x \Big] dx, \quad (4.8.1.19)$$

for $A > B$, and

$$\Gamma[(a)+s,(b)-s] = \sum_{\mu=1}^{A} \{\Gamma[(a)'-a_\mu,(b)+a_\mu]$$

$$\times \int_0^1 x^{s+a_\mu-1} {}_AF_{A-1}\begin{bmatrix}(b)+a_\mu; \\ 1+a_\mu-(a)';\end{bmatrix} (-1)^A x \Big] dx$$

$$+ \Gamma[(a)+b_\mu,(b)'-b_\mu]\int_1^\infty x^{s-b_\mu-1}$$

$$\times {}_AF_{A-1}[(a)+b_\mu; 1+b_\mu-(b)'; (-1)^A x] dx\}, \quad (4.8.1.20)$$

for $A = B$.

Further,

$$\Gamma\begin{bmatrix}(a)+s,(b)-s \\ (c)+s\end{bmatrix} = \sum_{\mu=1}^{A}\Gamma\begin{bmatrix}(a)'-a_\mu,(b)+a_\mu \\ (c)-a_\mu\end{bmatrix}\int_0^\infty x^{s+a_\mu-1}$$

$$\times {}_{B+C}F_{A-1}\begin{bmatrix}(b)+a_\mu,1+a_\mu-(c); \\ 1+a_\mu-(a)';\end{bmatrix} (-1)^{A+C} x \Big] dx, \quad (4.8.1.21)$$

for $A > B+C$, and

$$\Gamma\begin{bmatrix}(a)+s,(b)-s \\ (c)+s\end{bmatrix} = \sum_{\mu=1}^{A}\Gamma\begin{bmatrix}(a)'-a_\mu,(b)+a_\mu \\ (c)-a_\mu\end{bmatrix}\int_0^1 x^{s+a_\mu-1}$$

$$\times {}_AF_{A-1}\begin{bmatrix}(b)+a_\mu,1+a_\mu-(c); \\ 1+a_\mu-(a)';\end{bmatrix} (-1)^{A+C} x \Big] dx$$

$$+ \sum_{\nu=1}^{B}\Gamma\begin{bmatrix}(a)+b_\nu,(b)'-b_\nu \\ (c)+b_\nu\end{bmatrix}\int_1^\infty x^{s-b_\nu-1}$$

$$\times {}_AF_{A-1}\begin{bmatrix}(a)+b_\nu; \\ (c)+b_\nu,1+b_\nu-(b)';\end{bmatrix} (-1)^B/x \Big] dx, \quad (4.8.1.22)$$

for $A = B+C$, and finally,

$$\Gamma\begin{bmatrix}(a)+s,(b)-s \\ (d)-s\end{bmatrix} = \sum_{\mu=1}^{A}\Gamma\begin{bmatrix}(a)'-a_\mu,(b)+a_\mu \\ (d)+a_\mu\end{bmatrix}\int_0^\infty x^{s+a_\mu-1}$$

$$\times {}_BF_{A+D-1}\begin{bmatrix}(b)+a_\mu; \\ 1+a_\mu-(a)',(d)+a_\mu;\end{bmatrix} (-1)^A x \Big] dx, \quad (4.8.1.23)$$

for $B = D$.

Many special cases of these results are well-known. In particular when $A = 2$, we have, from (4.8.1.16),

$$\Gamma\begin{bmatrix} a+s, b+s \\ c+s, d+s \end{bmatrix} = \Gamma\begin{bmatrix} b-a \\ c-a, d-a \end{bmatrix}$$

$$\times \int_0^1 x^{s+a-1} \, {}_2F_1\begin{bmatrix} 1+a-c, 1+a-d; \\ 1+a-b; \end{bmatrix} x \end{bmatrix} dx$$

$$+ \Gamma\begin{bmatrix} a-b \\ c-b, d-b \end{bmatrix} \int_0^1 x^{s+b-1} \, {}_2F_1\begin{bmatrix} 1+b-c, 1+b-d; \\ 1+b-a; \end{bmatrix} x \end{bmatrix} dx.$$

$$(4.8.1.24)$$

From (4.8.1.20), we have

$$\Gamma[a+s, b+s, c-s, d-s]$$

$$= \Gamma[b-a, c+a, d+a] \int_0^1 x^{s+a-1} \, {}_2F_1\begin{bmatrix} c+a, d+a; \\ 1+a-b; \end{bmatrix} x \end{bmatrix} dx$$

$$+ \Gamma[a-b, c+b, d+b] \int_0^1 x^{s+b-1} \, {}_2F_1\begin{bmatrix} c+b, d+b; \\ 1+b-a; \end{bmatrix} x \end{bmatrix} dx$$

$$+ \Gamma[a+c, b+c, d-c] \int_1^\infty x^{s-c-1} \, {}_2F_1\begin{bmatrix} a+c, b+c; \\ 1+c-d; \end{bmatrix} 1/x \end{bmatrix} dx$$

$$+ \Gamma[a+d, b+d, c-d] \int_1^\infty x^{s-d-1} \, {}_2F_1\begin{bmatrix} a+d, b+d; \\ 1+d-c; \end{bmatrix} 1/x \end{bmatrix} dx.$$

$$(4.8.1.25)$$

Similar results involving the Gauss ${}_2F_1(x)$ functions follow from (4.8.1.22) and (4.8.1.23). Results of this kind are well known, but frequently they are quoted incompletely in the literature.

**4.8.2 The second theorem.** We consider next the Mellin transform of a general hypergeometric function. Suppose that

$$g(s) = \Gamma\begin{bmatrix} (a)+s, (b)-s \\ (c)+s, (d)-s \end{bmatrix}_{A+B+E}F_{C+D+F}\begin{bmatrix} (a)+s, (b)-s, (e); \\ (c)+s, (d)-s, (f); \end{bmatrix} y \end{bmatrix},$$

$$(4.8.2.1)$$

$$\sum_A (x) = \sum_{\mu=1}^A x^{a_\mu} \Gamma\begin{bmatrix} (a)'-a_\mu, (b)+a_\mu \\ (c)-a_\mu, (d)+a_\mu \end{bmatrix} \sum_{m=0}^\infty \frac{((b)+a_\mu)_{2m}}{((d)+a_\mu)_{2m}}$$

$$\times \frac{((e))_m x^m y^m}{((f))_m m!} {}_{B+C}F_{A+D-1}\begin{bmatrix} (b)+a_\mu+2m, 1+a_\mu-(c); \\ (d)+a_\mu+2m, 1+a_\mu-(a)'; \end{bmatrix} (-1)^{A+C} x \end{bmatrix}$$

$$(4.8.2.2)$$

and

$$\sum_{B}(1/x) = \sum_{\nu=1}^{B} x^{-b_\nu}\Gamma\begin{bmatrix}(a)+b_\nu, (b)'-b_\nu\\(c)+b_\nu, (d)-b_\nu\end{bmatrix}\sum_{m=0}^{\infty}\frac{((a)+b_\nu)_{2m}}{((c)+b_\nu)_{2m}}$$

$$\times\frac{((e))_m\,x^{-m}y^m}{((f))_m\,m!}\,_{A+D}F_{B+C-1}\begin{bmatrix}(a)+b_\nu+2m, 1+b_\nu-(d);\\(c)+b_\nu+2m, 1+b_\nu-(b)';\end{bmatrix}(-1)^{B+D}/x\end{bmatrix}.$$

$$(4.8.2.3)$$

Now, let us apply the Mellin transform of §4.8, to the results of the theorem of §4.7. We find **Theorem II.**

*If $A+B \geqslant C+D$, $A+B+E \leqslant C+D+F+1$,*

$$-\mathrm{Rl}\,(a_\mu) < \mathrm{Rl}\,(s) < \mathrm{Rl}\,(b_\nu),$$

*and $\mathrm{Rl}\,(c_\nu)$, $\mathrm{Rl}\,(d_\nu)$, $\mathrm{Rl}\,(f_\nu) \neq -N$, for all values of $\nu$, then*

$$\text{(i)}\quad g(s) = \int_0^\infty x^{s-1}\sum_A(x)\,dx \qquad (4.8.2.4)$$

*for $A+D > B+C$,*

$$\text{(ii)}\quad g(s) = \int_0^1 x^{s-1}\sum_A(x)\,dx + \int_1^\infty x^{s-1}\sum_B(1/x)\,dx \qquad (4.8.2.5)$$

*for $A+D = B+C$, and*

$$\text{(iii)}\quad g(s) = \int_0^\infty x^{s-1}\sum_B(1/x)\,dx \qquad (4.8.2.6)$$

*for $A+D < B+C$.*

Again, we can see that (i) and (iii) are equivalent under the transform

$$g(s) \to g(-s),$$

and that in (ii) $\sum_B(1/x)$ is the analytic continuation of $\sum_A(x)$ for $1 < x < \infty$.

In particular, if $B = D = 0$, then we must have $A = C$. The second integral in (ii) is then zero, and

$$g(s) = \int_0^1 x^{s-1}\sum_A(x)\,dx. \qquad (4.8.2.7)$$

In order to justify the application of the Mellin transform here, we need only notice that, under the conditions of this theorem, $g(s)$ is a function of bounded variation, on this part of the real axis.[†]

† See Titchmarsh (1948a), Theorem 29.

### 4.8.3 Simple cases of the second theorem. The simplest possibility is to consider a function containing one $a$ parameter only. This gives us

$$\Gamma(a+s)\,{}_1F_0[a+s;\ ;y] = \int_0^\infty x^{s+a-1}\,\mathrm{e}^{xy-x}\,\mathrm{d}x, \qquad (4.8.3.1)$$

for $|y| < 1$. This is another variation of Euler's integral of the second kind.

When an $f$ parameter is added we find that

$$\Gamma(a+s)\,{}_1F_1[a+s;f;y] = \int_0^\infty x^{s+a-1}\,\mathrm{e}^{-x}\,{}_0F_1[\ ;f;xy]\,\mathrm{d}x, \qquad (4.8.3.2)$$

and when sequences of $e$ and $f$ parameters are added also, we have†

$$\Gamma(a+s)\,{}_{E+1}F_F[a+s,(e);(f);y] = \int_0^\infty x^{s+a-1}\,\mathrm{e}^{-x}\,{}_EF_F[(e);(f);xy]\,\mathrm{d}x, \qquad (4.8.3.3)$$

where $E < F$ or $E = F$ and $|y| < 1$.

When $g(s)$ has two gamma functions involving $s$ in it, there are four possibilities, which arise. The first of these is

$$\Gamma[a+s,b+s]\,{}_2F_1[a+s,b+s;f;y]$$

$$= \Gamma(b-a)\int_0^\infty x^{s+a-1}\,{}_0F_1[\ ;1+a-b;x]\,{}_0F_1[\ ;f;xy]\,\mathrm{d}x$$

$$+ \Gamma(a-b)\int_0^\infty x^{s+b-1}\,{}_0F_1[\ ;1+b-a;x]\,{}_0F_1[\ ;f;xy]\,\mathrm{d}x. \qquad (4.8.3.4)$$

From this result, we can deduce various integrals of products of Bessel functions. When sequences of $e$ and $f$ parameters are added to $g(s)$, we have

$$\Gamma[a+s,b+s]\,{}_{E+2}F_F[a+s,b+s,(e);(f);y]$$

$$= \Gamma(b-a)\int_0^\infty x^{s+a-1}\,{}_0F_1[\ ;1+a-b;x]\,{}_EF_F[(e);(f);xy]\,\mathrm{d}x$$

$$+ \Gamma(a-b)\int_0^\infty x^{s+b-1}\,{}_0F_1[\ ;1+b-a;z]\,{}_EF_F[(e);(f);xy]\,\mathrm{d}x, \qquad (4.8.3.5)$$

where $E \leqslant F$.

---

† Erdélyi (1954), §§ 6.3 (1), 7.5 (24) and 7.6 (27).

This is a general type of Hankel transform. The extension to a general number of $a$ parameters is

$$\Gamma[(a)+s]_{A+E}F_F[(a)+s, (e); (f); y]$$

$$= \sum_{\mu=1}^{A} \Gamma[(a)' - a_\mu] \int_0^\infty x^{s+a_\mu-1} {}_0F_A[\ ; 1 + a_\mu - (a)'; x] {}_EF_F[(e); (f); xy]\,dx,$$

$$(4.8.3.6)$$

for $E \leqslant F$.

The second possibility is to introduce an $a + s$ parameter and a $b - s$ parameter into $g(s)$. This gives us

$$\Gamma[a+s, b-s] {}_2F_1[a+s, b-s; f; y] = \Gamma(a+b)\int_0^\infty x^{s+a-1}(1+x)^{-a-b}$$

$$\times {}_2F_1[\tfrac{1}{2}a + \tfrac{1}{2}b, \tfrac{1}{2} + \tfrac{1}{2}a + \tfrac{1}{2}b; f; 4xy/(1+x)^2]\,dx,$$

$$(4.8.3.7)$$

and when sequences of $e$ and $f$ parameters are added to $g(s)$, we find that

$$\Gamma[a+s, b-s] {}_{E+2}F_F[a+s, b-s, (e); (f); y]$$

$$= \Gamma(a+b)\int_0^\infty x^{s+a-1}(1+x)^{-a-b}$$

$$\times {}_{E+2}F_F\left[\begin{array}{c} \tfrac{1}{2}a + \tfrac{1}{2}b, \tfrac{1}{2} + \tfrac{1}{2}a + \tfrac{1}{2}b, (e); \\ (f); \end{array} \frac{4xy}{(1+x)^2}\right]dx, \quad (4.8.3.8)$$

for $E < F$. When $A > 1$, or $B > 1$, the double series under the integrand is neither separable nor summable, and so in general it cannot be reduced to a single summation.

In the third case, when an $a + s$ parameter, and a $c + s$ parameter are both present in $g(s)$, we have

$$\Gamma[a+s, c-a; c+s] {}_1F_1[a+s; c+s; y] = \int_0^1 x^{s+a-1} e^{xy}(1-x)^{c-a-1}\,dx.$$

$$(4.8.3.9)$$

When we add an $e$ parameter, we find that

$$\Gamma[a+s, c-a; c+s] {}_2F_1[a+s, e; c+s; y]$$

$$= \int_0^1 x^{s+a-1}(1-x)^{c-a-1}(1-xy)^{-e}\,dx, \quad (4.8.3.10)$$

for $|y| < 1$, and when we add an $f$ parameter also, we have

$$\Gamma[a+s, c-a; c+s] {}_2F_2[a+s, e; c+s, f; y]$$

$$= \int_0^1 x^{s+a-1}(1-x)^{c-a-1} {}_1F_1[e; f; xy]\,dx. \quad (4.8.3.11)$$

For general values of $E$ and $F$, when we have added sequences of both $e$ and $f$ parameters to $g(s)$, we find that[†]

$$\Gamma[a+s, c-a; c+s]\,_{E+1}F_{F+1}[a+s, (e); c+s, (f); y]$$

$$= \int_0^1 x^{s+a-1}(1-x)^{c-a-1}\,_{E}F_{F}[(e); (f); xy]\,dx, \quad (4.8.3.12)$$

for $E < F+1$, or for $E = F+1$ and $|y| < 1$.

For $A = C > 1$, the double series under the integrand is separable, but not summable, in general, and the theorem leads to the result that

$$\Gamma[(a)+s; (c)+s]\,_{A+E}F_{A+F}[(a)+s, (e); (c)+s, (f); y]$$

$$= \sum_{\mu=1}^{A} \Gamma[(a)'-a_\mu; (c)-a_\mu]\int_0^1 x^{s+a_\mu-1}$$

$$\times\,_{A}F_{A-1}[1+a_\mu-(c); 1+a_\mu-(a)'; x]\,_{E}F_{F}[(e); (f); xy]\,dx,$$

$$(4.8.3.13)$$

for $E < F$ or for $E = F$ and $|y| < 1$.

Here the most interesting particular case is probably $A = 2$, which leads us to a $_2F_1(x)$ Gauss series under the integral. In particular, when $E = F = 0$, we have

$$\Gamma[a+s, b+s; c+s, d+s]\,_2F_2[a+s, b+s; c+s, d+s; y]$$

$$= \Gamma[b-a; c-a, d-a]\int_0^1 x^{s+a-1}e^{-xy}$$

$$\times\,_2F_1[1+a-c, 1+a-d; 1+a-b; x]\,dx$$

$$+\Gamma[a-b; c-b, d-b]\int_0^1 x^{s+b-1}e^{-xy}$$

$$\times\,_2F_1[1+b-c, 1+b-d; 1+b-a; x]\,dx. \quad (4.8.3.14)$$

For $A > C \geqslant 1$, the corresponding general result is

$$\Gamma[(a)+s; (c)+s]\,_{A+E}F_{C+F}[(a)+s, (e); (c)+s, (f); y]$$

$$= \sum_{\mu=1}^{A} \Gamma[(a)'-a_\mu; (c)-a_\mu]\int_0^\infty x^{s+a_\mu-1}$$

$$\times\,_{C}F_{A-1}[1+a_\mu-(c); 1+a_\mu-(a)'; (-1)^{A+C}x]$$

$$\times\,_{E}F_{F}[(e); (f); xy]\,dx, \quad (4.8.3.15)$$

† Erdélyi (1954), §§ 7.5 (25), 7.5 (16) and 6.9 (10).

and, in particular, for $A = 2$, $C = 1$, $E = F = 0$, we find that

$$\Gamma[a+s, b+s; c+s]\,_2F_1[a+s, b+s; c+s; y]$$

$$= \Gamma[b-a; c-a]\int_0^\infty x^{s+a-1}e^{-xy}\,_1F_1[1+a-c; 1+a-b; -x]\,dx$$

$$+ \Gamma[a-b; c-b]\int_0^\infty x^{s+b-1}e^{-xy}\,_1F_1[1+b-c; 1+b-a; -x]\,dx,$$

(4.8.3.16)

for $|y| < 1$.

The fourth case is

$$\Gamma[a+s, a+d; d-s]\,_1F_1[a+s; d-s; x]$$

$$= \int_0^\infty x^{s+a-1} \sum_{m=0}^\infty \frac{x^m y^m}{(d+a)_{2m}\, m!}\,_0F_1[\,; a+d+2m; -x]\,dx. \quad (4.8.3.17)$$

Here, the double series under the integral sign is neither separable, nor summable, and the introduction of further $a$, $d$, $e$ or $f$ parameters does not alter this state of affairs.

When $g(s)$ contains three different parameters involving $s$, three further groups of cases arise. First, we have

$$g(s) = \Gamma[a+s, b-s; d-s]\,_2F_1[a+s, b-s; d-s; y], \quad (4.8.3.18)$$

where $|y| < 1$, and its extensions

$$g(s) = \Gamma[a+s, b-s; d-s]\,_{E+2}F_{F+1}[a+s, b-s, (e); d-s, (f); y],$$

(4.8.3.19)

where $E \leqslant F$, and

$$g(s) = \Gamma[(a)+s, (b)-s; (d)-s]\,_{A+B+E}F_{D+F}\begin{bmatrix}(a)+s, (b)-s, (e); \\ (d)-s, (f); \end{bmatrix} y\end{bmatrix},$$

(4.8.3.20)

where $A+B+E \leqslant D+F+1$, and $A+B \geqslant D$.

Secondly, we have

$$g(s) = \Gamma[a+s, b-s; c+s]\,_2F_1[a+s, b+s; c+s; y], \quad (4.8.3.21)$$

where $|y| < 1$, and its extensions,

$$g(s) = \Gamma[a+s, b-s; c+s]\,_{E+2}F_{F+1}[a+s, b-s, (e); c+s, (f); y],$$

(4.8.3.22)

where $E \leqslant F$, and

$$g(s) = \Gamma[(a)+s, (b)-s; (c)+s]\,_{A+B+E}F_{C+F}\begin{bmatrix}(a)+s, (b)-s, (e); \\ (c)+s, (f); \end{bmatrix} y\end{bmatrix},$$

(4.8.3.23)

where $A+B+E \leqslant C+F+1$, and $A+B \geqslant C$.

Thirdly, we have

$$g(s) = \Gamma[a_1+s, a_2+s, a_3+s; c+s, d-s]$$

$$\times {}_3F_2[a_1+s, a_2+s, a_3+s; c+s, d-s; y], \quad (4.8.3.24)$$

and its extensions,

$$g(s) = \Gamma[a_1+s, a_2+s, a_3+s; c+s, d-s]$$

$$\times {}_{E+3}F_{F+2}\begin{bmatrix} a_1+s, a_2+s, a_3+s, (e); \\ c+s, d-s, (f); \end{bmatrix} y \Big], \quad (4.8.3.25)$$

where $E \leqslant F+1$, and

$$g(s) = \Gamma[(a)+s; (c)+s, (d)-s] {}_{A+E}F_{C+D+F}\begin{bmatrix} (a)+s, (e); \\ (c)+s, (d)-s, (f); \end{bmatrix} y \Big],$$

$$(4.8.3.26)$$

where $A \geqslant C+D$, and $A+E \leqslant C+D+F+1$.

In these three groups, however, the inner double sum is neither separable nor summable, owing to the presence of an $a$ and a $d$ parameter, or a $b$ and a $c$ parameter at one and the same time.

# 5

# BASIC HYPERGEOMETRIC INTEGRALS

## 5.1 Basic contour integrals

All the results of the previous chapter can be generalized even further in terms of basic hypergeometric functions. But, before we can do this successfully, it is necessary to develop the concept of a basic contour integral. The earliest attempt to produce a basic analogue of the Barnes-type integral seems to have been due to G. N. Watson (1910a).

This attempt was not very successful, since the poles were taken in the q-plane, so that no direct analogues could be found. The subject was neglected for nearly forty years. In the meantime, the general transformation theory of basic series was being developed by other methods, but these methods of proofs were often long and unsatisfactory. The desire to produce simple, unified and elegant proofs on the lines of the Barnes-type contour integrals led to a renewed interest in the possibility of defining a basic contour integral.

In the first place, we shall consider in detail the proof, by the use of such a contour integral, of a particular relation between four $_3\Phi_2$ basic series, to illustrate the general method. Suppose that $a$, $b$, $c$, $d$, $e$ and $z$ are complex numbers such that none of the members of the sequences

$$-\log a - n, \quad -\log b - n, \quad 1 - \log e + n \quad \text{and} \quad n,$$

for $n = 0, 1, 2, \ldots$, coincide. Let

$$P_N(s) \equiv \frac{(q^s zc/e; q)_N (q^{1-s}/c; q)_N (dq^s; q)_N (q^{1+s}e/zc; q)_N}{(aq^s; q)_N (q^{1-s}/e; q)_N (bq^s; q)_N (q^{-s}; q)_N}$$

where $0 < q < 1$, that is, $q = e^{-t}$, and $t > 0$.

This restriction, that $q$ is real, is not such a drastic one as it appears, for, when the result has been proved for $q$ real, it can easily be extended to complex values of $q$. The restriction is only introduced to simplify and shorten the proof, by making sure that the strip contours to be considered will all lie parallel to the real axis.

Thus,

$$aq^s = e^{-(\log a + st)},$$

and has the period $2\pi i/t$. Hence $1/(aq^s; q)_N$ has poles in the $s$-plane, at all the points $s = -\log a + n + 2k\pi i/t$, one set of poles in every strip of width $2\pi/t$.

Thus, we are led to consider the integral

$$I_N = \int P_N(s)\,ds, \tag{5.1.1}$$

taken in a clockwise direction round the rectangular contour of Fig. 5.1,

$$A(-i\pi/t), \quad B(i\pi/t), \quad C(2N+i\pi/t), \quad D(2N-i\pi/t),$$

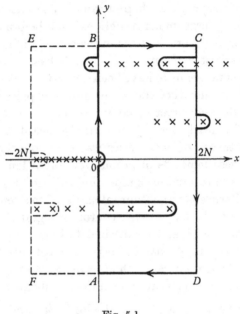

Fig. 5.1

for large enough values of $N$. This contour is indented, if necessary, to ensure that the first $N$ of the points in the increasing sequences

$$1-\log e+n \quad \text{and} \quad n, \quad \text{for } n = 0, 1, 2, ..., N-1,$$

lie within $ABCD$, and that all the points in the decreasing sequences

$$-\log a-n \quad \text{and} \quad -\log b-n, \quad \text{for } n = 0, 1, 2, ..., N-1,$$

lie to the left of $AB$.

Now, by the periodicity of $P_N(s)$, in $2\pi i/t$,

$$\int_{BC} P_N(s)\,ds + \int_{DA} P_N(s)\,ds = 0. \tag{5.1.2}$$

On $CD$, $s = 2N+ir$, where $-\pi/t \leqslant r \leqslant \pi/t$. Hence,

$$I_N - \int_{-i\pi/t}^{i\pi/t} P_N(s)\,ds = i\int_0^{\pi/t} [P_N(2N-ir) - P_N(2N+ir)]\,dr. \tag{5.1.3}$$

Now $\qquad 1-e^{\mathrm{Rl}\,(\log a)}q^N \leqslant |1-aq^{2N+ir}| \leqslant 1+e^{\mathrm{Rl}\,(\log a)}q^{2N},$

and

$$\mathrm{Rl}\,(a)\,q^{-2N}\{1-q^{2N}/\mathrm{Rl}\,(a)\} \leqslant |1-aq^{-2N+ir}| \leqslant \mathrm{Rl}\,(a)\,q^{-2N}\{1+\mathrm{Rl}\,(a)\,q^{2N}\}.$$

Hence

$$\left|\frac{(aq^{2N+ir};\,q)_N\,(bq^{-2N-ir};\,q)_N}{(cq^{2N+ir};\,q)_N\,(dq^{-2N-ir};\,q)_N}\right|$$

$$\leqslant \frac{(\mathrm{Rl}\,(a)\,q^{2N+\pi i/t};\,q)_N\,(-\mathrm{Rl}\,(b)\,q^{2N+\pi i/t};\,q)_N}{(\mathrm{Rl}\,(c)\,q^{2N};\,q)_N\,(-\mathrm{Rl}\,(d)\,q^{2N};\,q)_N}\,q^{\mathrm{Rl}\,(\log b-\log d)},$$

$$\leqslant O(q^{N\,\mathrm{Rl}\,(\log b-\log d)}).$$

Accordingly,

$$\left|I_N-\int_{-i\pi/t}^{i\pi/t}P_N(s)\,ds\right| < O(q^{N\,\mathrm{Rl}\,(\log z)})\int_0^{\pi/t}dr,$$

$$< O(q^{N\,\mathrm{Rl}\,(\log z)}),$$

and this expression $\to 0$, as $N \to \infty$, provided that $\mathrm{Rl}\,z > 0$. Hence

$$\frac{t}{2\pi i}I_N \to I \equiv \frac{t}{2\pi i}\int_{-i\pi/t}^{i\pi/t}\Pi\left[\begin{matrix}q^{1-s}/c,\,dq^s,\,q^{1+s}e/cz,\,q^{-s}zc/e;\\aq^s,\,q^{1-s}/e,\,bq^s,\,q^{-s};\end{matrix}\;q\right]ds, \quad (5.1.4)$$

as $N \to \infty$, for $q$ real and $0 < q < 1$.

But $P_N(s)$ has poles at all the points which fall within the strip $ABCD$ in the $s$-plane, that is at

$$1-\log e+n \quad \text{and} \quad n \quad \text{for } n = 0, 1, 2, \ldots,$$

or at $1-\log e+n+2\pi iK/t$ and $n+2\pi iK/t$, for some integer $K$, if the original points do not fall within this strip. Of each set of poles, only one can occur in any one strip. From this result (5.1.4) and a consideration of the residues of $I_N$, we have

$$\frac{t}{2\pi i}I_N = \frac{(q/c;\,q)_N\,(eq/cz;\,q)_N\,(zc/e;\,q)_N}{(a;\,q)_N\,(q/e;\,q)_N\,(b;\,q)_N\,(q;\,q)_N}$$

$$\times \sum_{n=0}^N \frac{(a;\,q)_n\,(b;\,q)_n\,(c;\,q)_n}{(q;\,q)_n\,(d;\,q)_n\,(e;\,q)_n}z^n[1+O(nq^N)]$$

$$+\frac{(e/c;\,q)_N\,(qd/e;\,q)_N\,(q^2/cz;\,q)_N\,(zc/q;\,q)_N}{(qa/e;\,q)_N\,(qb/e;\,q)_N\,(e/q;\,q)_N\,(q;\,q)_N}$$

$$\times \sum_{n=0}^N \frac{(aq/e;\,q)_n\,(qb/e;\,q)_n\,(qc/e;\,q)_n}{(q;\,q)_n\,(qd/e;\,q)_n\,(q^2/e;\,q)_n}z^n[1+O(nq^N)]. \quad (5.1.5)$$

When $N \to \infty$, using Tannery's theorem, we find that

$$\frac{t}{2\pi i} I = \Pi\left[\begin{matrix} q/c, d, qe/cz, cz/e; \\ a, q/e, b, q; \end{matrix} q\right] {}_3\Phi_2\left[\begin{matrix} a, b, c; \\ d, e; \end{matrix} q, z\right]$$

$$+ \Pi\left[\begin{matrix} e/c, qd/e, q^2/cz, cz/q; \\ qa/e, qb/e, e/q, q; \end{matrix} q\right] {}_3\Phi_2\left[\begin{matrix} qa/e, qb/e, qc/e; \\ qd/e, q^2/e; \end{matrix} q, z\right]. \quad (5.1.6)$$

In the same way, from a consideration of the poles of $P_N(s)$ in a similar strip to the left of $AB$, we find that

$$\frac{t}{2\pi i} I = \Pi\left[\begin{matrix} qa/c, d/a, qe/caz, caz/e; \\ b/a, qa/e, a, q; \end{matrix} q\right] {}_3\Phi_2\left[\begin{matrix} qa/d, qa/e, a; \\ qa/b, qa/c; \end{matrix} q, y/z\right]$$

$$+ \Pi\left[\begin{matrix} qb/c, d/b, qe/cbz, cbz/e; \\ a/b, qb/e, b, q; \end{matrix} q\right] {}_3\Phi_2\left[\begin{matrix} qb/d, qb/e, b; \\ qb/a, qb/c; \end{matrix} q, y/z\right], \quad (5.1.7)$$

where $y \equiv qde/abc$, and $\mathrm{Rl}\,(y) > 0$.

By analytic continuation, these results (5.1.6) and (5.1.7) will hold also for $|q| < 1$, and when we equate (5.1.6) and (5.1.7) our final result connecting four ${}_3\Phi_2$ series follows under the conditions

$$|de/abc| < |z| < 1, \quad |q| < 1.$$

When $q \to 1$, that is when $t \to 0$, (5.1.6) and (5.1.7) reduce formally to the special cases of (4.6.3) and (4.6.6) when $A = B = 2$, and $C = D = 1$.

## 5.2  General basic integral theorems

For an immediate extension of the result (5.1.6), let us consider the integral

$$I_N = \int P_N(s)\,ds, \quad (5.2.1)$$

where

$$P_N(s) = \frac{(q^{1-s}/(c); q)_N\,(q^s(b); q)_N\,(q^{1+s}/\alpha x; q)_N\,(\alpha x q^{-s}; q)_N}{((a)\,q^s; q)_N\,(q^{1-s}/(d); q)_N\,(q^{-s}; q)_N}, \quad (5.2.2)$$

$\alpha = \dfrac{c_1 c_2 \dots c_C}{d_1 d_2 \dots d_C}$, and there are $A+1$ of the $a$ parameters, $A$ of the $b$ parameters, and $C$ of the $c$ and $d$ parameters. If this integral is taken round the same two strip contours of § 5.1, we have

$$\left| I_N - \int_{-i\pi/t}^{i\pi/t} P_N(s)\,ds \right| = O(q^{N\,\mathrm{Rl}\,(\log x)})$$

to the right of $AB$, and

$$\left| I_N - \int_{-i\pi/t}^{i\pi/t} P_N(s)\, ds \right| = O(q^{N\,\mathrm{Rl}\,(\log\,(\beta - x))})$$

to the left of $AB$, where

$$\beta = \frac{b_1 b_2 \ldots b_A d_1 d_2 \ldots d_C}{a_1 a_2 \ldots a_{A+1} c_1 c_2 \ldots c_C}.$$

Thus, if $\mathrm{Rl}\,(\log \beta) > \mathrm{Rl}\,(\log x) > 0$, $I_N$ exists and tends to $I$ as $N \to \infty$, where

$$I = \int_{-i\pi/t}^{i\pi/t} \Pi \left[ \begin{matrix} q^{1-s}/(c),\, (b)\, q^s,\, q^{1+s}/x\alpha,\, x\alpha q^{-s}; \\ (a)\, q^s,\, q^{1-s}/(d),\, q^{-s}; \end{matrix}\; q \right] ds. \qquad (5.2.3)$$

From a consideration of the residues of the integrand within the same strips

$$(\pm 2N - i\pi/t,\, i\pi/t,\, -i\pi/t,\, \pm 2N + i\pi/t)$$

we deduce that

$$\frac{t}{2\pi i} I = \Pi \left[ \begin{matrix} q/(c),\, (b),\, q/x\alpha,\, x\alpha; \\ (a),\, q/(d),\, q; \end{matrix}\; q \right]{}_{A+C+1}\Phi_{A+C} \left[ \begin{matrix} (a),\, (c); \\ (b),\, (d); \end{matrix}\; q, x \right]$$

$$+ \sum_{\nu=1}^{C} \Pi \left[ \begin{matrix} d_\nu/(c),\, q(b)/d_\nu,\, q^2/\alpha x d_\nu,\, x\alpha q^{-1} d_\nu; \\ (a)q/d_\nu,\, d_\nu/(d)',\, d_\nu^2/q,\, q; \end{matrix}\; q \right]$$

$$\times\,{}_{A+C+1}\Phi_{A+C} \left[ \begin{matrix} q(a)/d_\nu,\, q(c)/d_\nu; \\ q(b)/d_\nu,\, q(d)'/d_\nu,\, q^2/d_\nu; \end{matrix}\; q, x \right]$$

$$= \sum_{\mu=1}^{A+1} \Pi \left[ \begin{matrix} qa_\mu/(c),\, (b)/a_\mu,\, q/(x\alpha a_\mu),\, x\alpha a_\mu; \\ (a)'/a_\mu,\, qa_\mu/(d),\, a_\mu,\, q; \end{matrix}\; q \right]$$

$$\times\,{}_{A+C+1}\Phi_{A+C} \left[ \begin{matrix} qa_\mu/(b),\, qa_\mu/(d),\, a_\mu; \\ qa_\mu/(a),\, qa_\mu/(c); \end{matrix}\; q, \beta/x \right], \qquad (5.2.4)$$

where $\quad \alpha \equiv \dfrac{c_1 c_2 \ldots c_C}{d_1 d_2 \ldots d_D}\quad$ and $\quad \beta \equiv \dfrac{qb_1 b_2 \ldots b_A d_1 d_2 \ldots d_C}{a_1 a_2 \ldots a_{A+1} c_1 c_2 \ldots c_C},$

under the convergence conditions

$$|q| < 1,\, |x| < 1,\, |qb_1 b_2 \ldots b_A/(a_1 a_2 \ldots a_{A+1})| < 1.$$

This is Sears's general basic theorem.† As we shall see now, it can itself be generalized still further.

The first step is to try to improve the symmetry in the integrand, in the hope that we might be able to remove the restrictions on the number of parameters, that is that the number of $a$ parameters should have to be equal to the number of $b$ parameters plus one, and similarly

† Sears (1951$d$), §4.3.

that the number of $c$ parameters should have to be equal to the number of $d$ parameters. With this end in view, we consider the integral

$$I = \int_{-i\pi/t}^{i\pi/t} \prod \left[ \begin{matrix} (a)\, q^s, (b)\, q^{-s}; \\ (c)\, q^s, (d)\, q^{-s}; \end{matrix} q \right] ds \qquad (5.2.5)$$

in which there are $A$ of the $a$ parameters, $B$ of the $b$ parameters, $C$ of the $c$ parameters and $D$ of the $d$ parameters. Again we assume, for the moment, that $q$ is real, only to ensure that our strip contours will all lie parallel to the real axis. All the other parameters in (5.2.5) are assumed to be complex, and they can take any values for which the resulting series do not become infinite or undefined numerically. Thus

$$c_1, c_2, ..., c_C \quad \text{and} \quad d_1, d_2, ..., d_D$$

are complex numbers such that none of the members of the sequences

$$-\log c_\nu - n \quad \text{and} \quad \log d_\mu + n,$$

coincide for $n = 0, 1, 2, ..., \nu = 1, 2, ..., C, \mu = 1, 2, ..., D.$

Let
$$\prod^N(s) = \frac{((a)\, q^s; q)_N \, ((b)\, q^{-s}; q)_N}{((c)\, q^s; q)_N \, ((d)\, q^{-s}; q)_N}. \qquad (5.2.6)$$

Then, by considering the integral

$$I_{N,R} = \int \prod^N(s)\, ds \qquad (5.2.7)$$

taken round the contour $ABCD$ of Fig. 5.1, we find, as before, that

$$2\pi i I_{N,R} = \int_{AB} \prod^N(s)\, ds + \int_{CD} \prod^N(s)\, ds,$$

that is

$$I_{N,R} - \frac{t}{2\pi i} \int_{-i\pi/t}^{i\pi/t} \prod^N(s)\, ds = \frac{1}{2\pi} \int_0^{\pi/t} \{\prod^N(2R+ir) - \prod^N(2R-ir)\}\, dr. \qquad (5.2.8)$$

Let
$$\prod^N(R)' = \frac{(-\mathrm{Rl}\,(a)\, q^{2R}; q)_N \, (-q^{2R}/\mathrm{Rl}\,(b); q)_N}{(\mathrm{Rl}\,(c)\, q^{2N}; q)_N \, (-q^{2N}/\mathrm{Rl}\,(d); q)_N}. \qquad (5.2.9)$$

Then, as before,

$$\prod^N(R)' \to 1 \quad \text{as} \quad R \to \infty,$$

and
$$\prod^N(2R \pm ir) \leqslant \prod^N(R)'\, q^k,$$

where
$$k = N(\Sigma\, \mathrm{Rl}\, b - \Sigma\, \mathrm{Rl}\, d) + 2(D - B)\, NR.$$

Hence
$$\left| I_{N,R} - \frac{t}{2\pi i} \int_{-i\pi/t}^{i\pi/t} \prod^N(s)\, ds \right|$$
$$< \frac{1}{\pi} \int_0^{\pi/t} dr\, O(q^k) \prod^N(R)',$$
$$< O(q^k) \prod^N(R)'/t. \qquad (5.2.10)$$

Let
$$I_N = \lim_{N \to \infty} I_{N,R}.$$

Then, provided that $D > B$,

$$I_N = \frac{1}{2\pi i} \int_{-i\pi/t}^{i\pi/t} \prod^N (s)\, ds, \tag{5.2.11}$$

and, if $D = B$,

$$\left| I_N - \frac{t}{2\pi i} \int_{-i\pi/t}^{i\pi/t} \prod^N (s)\, ds \right| \leqslant O\{(\Sigma \operatorname{Rl} b - \Sigma \operatorname{Rl} d)\, q^N\}. \tag{5.2.12}$$

But $I_N$ has residues within the strip $ABCD$ at the poles of $\prod^N (s)$. Hence

$$I_N = \frac{1}{t} \sum_{\nu=1}^{D} \prod^N \left[ \begin{matrix} (a)\, d_\nu,\ (b)/d_\nu; \\ (c)\, d_\nu,\ (d)'/d_\nu; \end{matrix}\ q \right]$$
$$\times \sum_{n=0}^{N} \frac{((c)\, d_\nu;\ q)_n\, (qd_\nu/(b);\ q)_n}{((a)\, d_\nu;\ q)_n\, (qd_\nu/(d);\ q)_n}\, Q_\nu^n, \tag{5.2.13}$$

where
$$Q_\nu = (-q^{\frac{1}{2}n+\frac{1}{2}}d_\nu)^{D-B}\frac{b_1 b_2 \ldots b_B}{d_1 d_2 \ldots d_D}.$$

Now let $N \to \infty$, and we find, if $D > B$, or if $D = B$ and
$$\operatorname{Rl}(b_1 b_2 \ldots b_B) < \operatorname{Rl}(d_1 d_2 \ldots d_D),$$
that

$$I = \frac{1}{t} \sum_{\nu=1}^{D} \prod \left[ \begin{matrix} (a)\, d_\nu,\ (b)/d_\nu; \\ (c)\, d_\nu,\ (d)'/d_\nu; \end{matrix}\ q \right]_{B+C} \Phi_{A+D-1} \left[ \begin{matrix} (c)\, d_\nu,\ qd_\nu/(b); \\ (a)\, d_\nu,\ qd_\nu/(d)'; \end{matrix}\ q, Q_\nu \right]. \tag{5.2.14}$$

Similarly, by considering the residues of $\prod^N (s)$ in a similar strip to the left of $AB$, we have, if $C > A$ or if $C = A$ and
$$\operatorname{Rl}(a_1 a_2 \ldots a_A) < \operatorname{Rl}(c_1 c_2 \ldots c_C),$$
then

$$I = \frac{1}{t} \sum_{\mu=1}^{C} \prod \left[ \begin{matrix} (b)\, c_\mu,\ (a)/c_\mu; \\ (d)\, c_\mu,\ (c)'/c_\mu; \end{matrix}\ q \right]_{A+D} \Phi_{B+C-1} \left[ \begin{matrix} (d)\, c_\mu,\ qc_\mu/(a); \\ (b)\, c_\mu,\ qc_\mu/(c)'; \end{matrix}\ q, Q_\mu' \right], \tag{5.2.15}$$

where
$$Q_\mu' = (-q^{\frac{1}{2}n-\frac{1}{2}}c_\mu)^{C-A}\frac{a_1 a_2 \ldots a_A}{c_1 c_2 \ldots c_C}.$$

Hence we have the main theorem,

$$\sum_{\nu=1}^{D} \prod \left[ \begin{matrix} (a)\, d_\nu,\ (b)/d_\nu; \\ (c)\, d_\nu,\ (d)'/d_\nu; \end{matrix}\ q \right]_{B+C} \Phi_{A+D-1} \left[ \begin{matrix} (c)\, d_\nu,\ qd_\nu/(b); \\ (a)\, d_\nu,\ qd_\nu/(d)'; \end{matrix}\ q, Q_\nu \right]$$
$$= \sum_{\mu=1}^{C} \prod \left[ \begin{matrix} (b)\, c_\mu,\ (a)/c_\mu; \\ (d)\, c_\mu,\ (c)'/c_\mu; \end{matrix}\ q \right]_{A+D} \Phi_{B+C-1} \left[ \begin{matrix} (d)\, c_\mu,\ qc_\mu/(a); \\ (b)\, c_\mu,\ qc_\mu/(c)'; \end{matrix}\ q, Q_\mu' \right], \tag{5.2.16}$$

provided that

(i) $D > B$  or  $D = B$  and  $|b_1 b_2 \ldots b_B| < |d_1 d_2 \ldots d_D|$,

and (ii) $C > A$  or  $C = A$  and  $|a_1 a_2 \ldots a_A| < |c_1 c_2 \ldots c_C|$.

It should be noted that the results corresponding to (5.2.14) and (5.2.15) for ordinary hypergeometric series, (4.6.2.4) and (4.6.2.5) cannot be deduced simply by letting $q \to 1$, in (5.2.14) and (5.2.15), since this would require that $t \to 0$, in the integrals of (5.2.7). However, these results can still be proved directly by the use of analogous Barnes-type contour integrals, as we saw in § 4.6.2.

We give here a few examples of the use of this theorem (5.2.16). First, if $A = C$ and $B = D$, it reduces to the previous theorem, (5.2.4). If $A = B = C = D$, we have

$$\sum_{\nu=1}^{A} \Pi \begin{bmatrix} (a)\,d_\nu,\ (b)/d_\nu; \\ (c)\,d_\nu,\ (d)'/d_\nu; \end{bmatrix} {}_{2A}\Phi_{2A-1} \begin{bmatrix} (c)\,d_\nu,\ qd_\nu/(b); \\ (a)\,d_\nu,\ qd_\nu/(d)'; \end{bmatrix} q, \frac{b_1 b_2 \ldots b_A}{d_1 d_2 \ldots d_A} \end{bmatrix}$$

$$= \sum_{\nu=1}^{A} \Pi \begin{bmatrix} (a)\,c_\nu,\ (b)/c_\nu; \\ (d)\,c_\nu,\ (c)'/c_\nu; \end{bmatrix} {}_{2A}\Phi_{2A-1} \begin{bmatrix} (d)\,c_\nu,\ qc_\nu/(a); \\ (b)\,c_\nu,\ qc_\nu/(c)'; \end{bmatrix} q, \frac{a_1 a_2 \ldots a_A}{c_1 c_2 \ldots c_A} \end{bmatrix},$$

$$(5.2.17)$$

where

$$|b_1 b_2 \ldots b_A| < |d_1 d_2 \ldots d_A|, \quad |a_1 a_2 \ldots a_A| < |c_1 c_2 \ldots c_A| \text{ and } |q| < 1.$$

In particular, if $A = 1$, then

$$\Pi \begin{bmatrix} ad, b/d; \\ bc,\ a/c; \end{bmatrix} {}_2\Phi_1 \begin{bmatrix} cd,\ qd/b; \\ ad; \end{bmatrix} q, b/d \end{bmatrix} = {}_2\Phi_1 \begin{bmatrix} cd,\ qc/a; \\ bc; \end{bmatrix} q, a/c \end{bmatrix} \quad (5.2.18)$$

provided that $|b| < |d|$, $|a| < |c|$ and that $|q| < 1$.

Similarly, if $C = 2$, $D = 1$, $A = B = 0$, then

$${}_2\Phi_0[cd, de;\ ;q, -q^{\frac{1}{2}n+\frac{1}{2}}]$$
$$= \Pi[de;\ d/c;\ q]_1\Phi_1[ce;\ qc/d;\ q, q^{n+1}c/d]$$
$$\quad + \Pi[ce;\ c/d;\ q]_1\Phi_1[de;\ qd/c;\ q, q^{n+1}d/c], \quad (5.2.19)$$

for all values of $c, d$ and $e$, and $|q| < 1$. This is a basic analogue of a well-known result in the theory of Whittaker functions.†

The theorem (5.2.16) can be generalized further by the introduction of a free parameter $z$. We can achieve this by writing $A + 1$ for $A$, $B + 1$ for $B$, $z$ for $b_{B+1}$, and $q/z$ for $a_{A+1}$. Then the theorem can be rewritten

$$\sum_{\mu=1}^{D} \Pi \begin{bmatrix} (a)\,d_\mu,\ (b)/d_\mu,\ qd_\mu/z,\ z/d_\mu; \\ (c)\,d_\mu,\ (d)'/d_\mu; \end{bmatrix} {}_{B+C}\Phi_{A+D-1} \begin{bmatrix} (c)\,d_\mu,\ qd_\mu/(b); \\ (a)\,d_\mu,\ qd_\mu/(d)'; \end{bmatrix} q, Q_\mu \end{bmatrix}$$

$$= \sum_{\nu=1}^{C} \Pi \begin{bmatrix} (b)\,c_\nu,\ (a)/c_\nu,\ zc_\nu,\ q/zc_\nu; \\ (d)\,c_\nu,\ (c)'/c_\nu; \end{bmatrix} {}_{A+D}\Phi_{B+C-1} \begin{bmatrix} (d)\,c_\nu,\ qc_\nu/(a); \\ (b)\,c_\nu,\ qc_\nu/(c)'; \end{bmatrix} q, Q'_\nu \end{bmatrix},$$

$$(5.2.20)$$

where $\quad |q| < 1, \quad Q_\mu \equiv (-q^{\frac{1}{2}n+\frac{1}{2}}d_\mu)^{D-B-1} \dfrac{b_1 b_2 \ldots b_B z}{d_1 d_2 \ldots d_D},$

† Slater (1960), (1.9.4).

and
$$Q'_\nu \equiv (-q^{\frac{1}{2}n+\frac{1}{2}}c_\nu)^{C-A-1} \frac{a_1 a_2 \dots a_A q}{c_1 c_2 \dots c_C z}.$$

The conditions for convergence now become

(i) $D > B+1$, or $D = B+1$

and $|b_1 b_2 \dots b_B z| < |d_1 d_2 \dots d_{B+1}|$,

and (ii) $C > A+1$, or $C = A+1$

and $|a_1 a_2 \dots a_A q| < |z c_1 c_2 \dots c_{A+1}|$.

In particular, if $A = B = 0$, $C = D = 1$, we have

$$\Pi \begin{bmatrix} z/d, qd/z; \\ zc, q/zc; \end{bmatrix} q \Big]_1 \Phi_0[cd; ; q, z/d] = {}_1\Phi_0[cd; ; q, q/zc], \quad (5.2.21)$$

provided that $|z/d| < 1$ and that $|q/zc| < 1$.

If $A = B = 0$, $C = 2$, and $D = 1$, we find that

$$\Pi \begin{bmatrix} qd/z, z/d; \\ c_1 d, c_2 d; \end{bmatrix} q \Big]_2 \Phi_0[c_1 d, c_2 d; ; q, z/d]$$

$$= \Pi \begin{bmatrix} zc_1, q/zc_1; \\ c_1 d, c_2/c_1; \end{bmatrix} q \Big]_1 \Phi_1 \begin{bmatrix} c_1 d; \\ qc_1/d; \end{bmatrix} q, -q^{\frac{1}{2}n+\frac{3}{2}}/zc_1 \Big]$$

$$+ \Pi \begin{bmatrix} zc_2, q/zc_2; \\ dc_2, c_1/c_2; \end{bmatrix} q \Big]_1 \Phi_1 \begin{bmatrix} c_2 d; \\ qc_2/d; \end{bmatrix} q, -q^{\frac{1}{2}n+\frac{3}{2}}/zc_2 \Big], \quad (5.2.22)$$

provided that $|z/d| < 1$, and, finally, if $A = B = 1$, $C = 3$, and $D = 2$, we have

$$\sum_{\nu=1}^{2} \Pi \begin{bmatrix} ad_\nu, b/d_\nu, z/d_\nu, qd_\nu/z; \\ c_1 d_\nu, c_2 d_\nu, c_3 d_\nu, (d)'/d_\nu; \end{bmatrix} q \Big]_4 \Phi_3 \begin{bmatrix} c_1 d_\nu, c_2 d_\nu, c_3 d_\nu, qd_\nu/b; \\ ad_\nu, qd_\nu/(d)'; \end{bmatrix} q, \frac{bz}{d_1 d_2} \Big]$$

$$= \sum_{\mu=1}^{3} \Pi \begin{bmatrix} bc_\mu, a/c_\mu, zc_\mu, q/zc_\mu; \\ d_1 c_\mu, d_2 c_\mu, (c)'/c_\mu; \end{bmatrix} q \Big]_2 \Phi_3 \begin{bmatrix} d_1 c_\mu, d_2 c_\mu; \\ bc_\mu, qc_\mu/(c)'; \end{bmatrix} q, \frac{-q^{\frac{1}{2}n+\frac{3}{2}}ac_\mu}{zc_1 c_2 c_3} \Big],$$

$$(5.2.23)$$

provided that $|bz| < |d_1 d_2|$.

## 5.3 Well-poised basic integrals

There are four integrals which provide the basic analogues of § 4.5.2. These lead to the four general transformations of basic well-poised series first given by Sears.† The first of these integrals is

$$I_1 = \int_{-i\pi/t}^{i\pi/t} \Pi \begin{bmatrix} q^{1-s}/(b), q^{1+s}a_0^{\frac{1}{2}}, q^{1-s}a_0^{-\frac{1}{2}}, q^{1+s+\pi i/t}a_0^{\frac{1}{2}}, q^{1-s+\pi i/t}a_0^{-\frac{1}{2}}; \\ (a) q^s, \qquad\qquad\qquad (a) q^{-s}/a_0; \end{bmatrix} q \Big] q^s \, ds,$$

$$(5.3.1)$$

where there are $M$ of the $a$ parameters, and $M-1$ of the $b$ parameters.

† Sears (1951d), (7.2), (7.3), (7.4) and (7.5).

This integral leads us in the usual way to the result

$$\Pi\left[\begin{matrix} qa_0/(b), q/(b), a_0^{\frac{1}{2}}, -a_0^{\frac{1}{2}}, qa_0^{-\frac{1}{2}}, -qa_0^{-\frac{1}{2}}; \\ a_0, (a), (a)/a_0; \end{matrix} q\right]$$

$$\times {}_{2M}\Phi_{2M-1}\left[\begin{matrix} a_0, (a), (b); \\ qa_0/(a), qa_0/(b); \end{matrix} q, y\right]$$

$$= \sum_{\nu=1}^{M} a_\nu \Pi\left[\begin{matrix} qa_\nu/(b), qa_0/a_\nu(b), a_0^{\frac{1}{2}}/a_\nu, -a_0^{\frac{1}{2}}/a_\nu, qa_\nu a_0^{-\frac{1}{2}}, -qa_\nu a_0^{-\frac{1}{2}}; \\ (a)/a_\nu, a_\nu, a_\nu(a)/a_0; \end{matrix} q\right]$$

$$\times {}_{2M}\Phi_{2M-1}\left[\begin{matrix} a_\nu, a_\nu(a)/a_0, a_\nu(b)/a_0; \\ qa_\nu/a_0, qa_\nu/(a)', qa_\nu/(b); \end{matrix} q, y\right], \quad (5.3.2)$$

where $\quad y \equiv -q^M a_0^{M-1}/(a_1 a_2 \ldots a_M b_1 b_2 \ldots b_{M-1})$.

This theorem expresses a well-poised ${}_{2M}\Phi_{2M-1}$ series in terms of $M$ other well-poised ${}_{2M}\Phi_{2M-1}$ series. It can be thought of† as the process of 'pivoting on $a_0$'.

If, in the integral (5.3.1), we add an extra term $q^{\frac{1}{2}}a_0^{\frac{1}{2}}$, we obtain the integral

$$I_2 = \int_{-i\pi/t}^{i\pi/t} \Pi\left[\begin{matrix} q^{1-s}/(b), q^{1+s}a_0/(b), q^{1+s}a_0^{\frac{1}{2}}, q^{1-s}a_0^{-\frac{1}{2}}, q^{1+s+\pi i/t}a_0^{\frac{1}{2}}, \\ a_0 q^s, (a) q^s, q^{-s}, q^{-s}(a)/a_0; \\ q^{1-s+\pi i/t}a_0^{-\frac{1}{2}}, q^{\frac{1}{2}-s}a_0^{-\frac{1}{2}}, q^{\frac{1}{2}+s}a_0^{\frac{1}{2}}; \\ \phantom{x} \end{matrix} q\right] q^s \, ds. \quad (5.3.3)$$

From this integral we can deduce the second well-poised theorem,‡

$$\Pi\left[\begin{matrix} qa_0/(b), q/(b), a_0^{\frac{1}{2}}, -a_0^{\frac{1}{2}}, qa_0^{-\frac{1}{2}}, -qa_0^{-\frac{1}{2}}, q^{\frac{1}{2}}a_0^{-\frac{1}{2}}, q^{\frac{1}{2}}a_0^{\frac{1}{2}}; \\ (a), (a)/a_0; \end{matrix} q\right]$$

$$\times {}_{2M}\Phi_{2M-1}\left[\begin{matrix} a_0, (a), (b); \\ qa_0/(a), qa_0/(b); \end{matrix} q, y'\right]$$

$$= \sum_{\nu=1}^{M} a_\nu \Pi\left[\begin{matrix} qa_\nu/(b), qa_0/a_\nu(b), q^{\frac{1}{2}}a_\nu^{\frac{1}{2}}a_0^{-\frac{1}{2}}, q^{\frac{1}{2}}a_0^{\frac{1}{2}}a_\nu^{-\frac{1}{2}}, a_0^{\frac{1}{2}}/a_\nu, \\ a_0/a_\nu, (a)'/a_\nu, a_\nu, a_\nu(a)/a_0; \end{matrix}\right.$$

$$\left.\begin{matrix} -a_0^{\frac{1}{2}}/a_\nu, qa_\nu a_0^{-\frac{1}{2}}, -qa_\nu a_0^{-\frac{1}{2}}; \\ \phantom{x} \end{matrix} q\right]$$

$$\times {}_{2M}\Phi_{2M-1}\left[\begin{matrix} a_\nu, a_\nu(a)/a_0, a_\nu(b)/a_0; \\ qa_\nu/a_0, qa_\nu/(a)', qa_\nu/(b); \end{matrix} q, y'\right], \quad (5.3.4)$$

where $\quad y' \equiv -q^{M-\frac{1}{2}}a_0^{M-\frac{1}{2}}/(a_0 a_1 a_2 \ldots a_M b_1 b_2 \ldots b_M)$.

Similarly,§ by putting $M+1$ for $M$ and $b_{2M+2} = q^{\frac{1}{2}}a_0^{\frac{1}{2}}$, we obtain

$$I_3 = \int_{-i\pi/t}^{i\pi/t} \Pi\left[\begin{matrix} q^{1-s}/(b), q^{\frac{1}{2}-s}a_0^{-\frac{1}{2}}, q^{1+s}a_0^{\frac{1}{2}}, q^{1-s}a_0^{-\frac{1}{2}}, q^{1+s+\pi i/t}a_0^{\frac{1}{2}}, \\ (a) q^s, (a) q^{-s}/a_0; \\ q^{1-s+\pi i/t}a_0^{-\frac{1}{2}}; \\ \phantom{x} \end{matrix} q\right] q^s \, ds, \quad (5.3.5)$$

---

† Sears (1951 d), (7.2).       ‡ Sears (1951 d), (7.3).
§ Sears (1951 d), (7.4).

from which we can deduce that

$$\Pi\begin{bmatrix} qa_0/(b), q/(b), q^{\frac{1}{2}}a_0^{-\frac{1}{2}}, a_0^{\frac{1}{2}}, -a_0^{\frac{1}{2}}, qa_0^{-\frac{1}{2}}, -qa_0^{-\frac{1}{2}}; \\ a_0, (a), (a)/a_0; \end{bmatrix} q \end{bmatrix}$$

$$\times {}_{2M+1}\Phi_{2M}\begin{bmatrix} a_0, (a), (b); \\ qa_0/(a), qa_0/(b); \end{bmatrix} q, y'\end{bmatrix}$$

$$= \sum_{\nu=1}^{M+1} a_\nu \Pi\begin{bmatrix} qa_\nu/(b), q^{\frac{1}{2}}a_\nu, qa_0/a_\nu(b), q^{\frac{1}{2}}a_0^{\frac{1}{2}}/a_\nu, a_0^{\frac{1}{2}}/a_\nu, \\ a_0/a_\nu, (a)'/a_\nu, a_\nu, (a)\,a_\nu/a_0; \end{bmatrix}$$

$$-a_0^{\frac{1}{2}}/a_\nu, qa_\nu a_0^{-\frac{1}{2}}, -qa_\nu a_0^{-\frac{1}{2}}; \\ q \end{bmatrix}$$

$$\times {}_{2M+1}\Phi_{2M}\begin{bmatrix} a_\nu, a_\nu(a)/a_0, a_\nu(b)/a_0; \\ qa_\nu/a_0, qa_\nu/(a)', qa_\nu/(b); \end{bmatrix} q, y'\end{bmatrix}, \quad (5.3.6)$$

where $\quad y' \equiv -q^{M+\frac{1}{2}}a_0^{M+\frac{1}{2}}/(a_0 a_1 \ldots a_M b_1 \ldots b_M).$

Finally,† if we have $M$ of the $b$ parameters, we find the integral

$$I_4 = \int_{-i\pi/t}^{i\pi/t} \Pi\begin{bmatrix} q^{1-s}/(b), q^{1+s}a_0/(b); \\ a_0 q^s, (a)\,q^s, q^{-s}, (a)\,q^{-s}/a_0; \end{bmatrix} q \end{bmatrix} q^s \, ds, \quad (5.3.7)$$

from which it follows that

$$\Pi\begin{bmatrix} qa_0/(b), q/(b), a_0^{\frac{1}{2}}, -a_0^{\frac{1}{2}}, qa_0^{-\frac{1}{2}}, -qa_0^{-\frac{1}{2}}; \\ a_0, (a), (a)/a_0; \end{bmatrix} q \end{bmatrix}$$

$$\times {}_{2M+1}\Phi_{2M}\begin{bmatrix} a_0, (a), (b); \\ qa_0/(a), qa_0/(b); \end{bmatrix} q, y''\end{bmatrix}$$

$$= \sum_{\nu=1}^{M} a_\nu \Pi\begin{bmatrix} qa_\nu/(b), qa_0/a_\nu(b), a_0^{\frac{1}{2}}/a_\nu, -a_0^{\frac{1}{2}}/a_\nu, qa_\nu a_0^{-\frac{1}{2}}, -qa_\nu a_0^{-\frac{1}{2}}; \\ a_0/a_\nu, (a)'/a_\nu, a_\nu, a_\nu(a)/a_0; \end{bmatrix} q \end{bmatrix}$$

$$\times {}_{2M+1}\Phi_{2M}\begin{bmatrix} a_\nu, a_\nu(a)/a_0, a_\nu(b)/a_0; \\ qa_\nu/a_0, qa_\nu/(a)', qa_\nu/(b); \end{bmatrix} q, y''\end{bmatrix}, \quad (5.3.8)$$

where $\quad y'' = -q^M a_0^{M-1}/(a_1 \ldots a_M b_1 \ldots b_M).$

## 5.4 Asymptotic forms for basic integrals

We can extend the main result of §5.2, (5.2.16), still further, if we introduce the concepts of asymptotic expansions, and analytic continuation into our basic integrals. In order to do this, we are going to

† Sears (1951 d), (7.5).

investigate next the form that (5.2.16) would assume if the restrictions $D \geqslant B+1$ and $C \geqslant A+1$ were removed. We shall restate this previous theorem shortly thus;

$$\int \Pi(s)\,ds = \Sigma\Pi\Phi(d) \quad if \quad D = B \ and \ \mathrm{Rl}\,\Sigma(b-d) > 0, \ or \ if \ D > B,$$

and

$$\int \Pi(s)\,ds = \Sigma\Pi\Phi(c) \quad if \ C = A \ and \ \mathrm{Rl}\,\Sigma(a-c) > 0, \ or \ if \ C > A.$$

Now, in all cases, even when $C < A$ or when $D < B$, we have, for $R$ fixed,

$$\left| \int_{DC} \overset{N}{\Pi}(s)\,ds \right| \leqslant \frac{1}{t} \frac{|1+(a)\,q^{R+n}|}{|1-(c)\,q^{R+n}|} \frac{|1+(b)\,q^{n-R}|}{|1-(d)\,q^{n-R}|} \; |1+q^{1+n+R}/z| \; |1+zq^{n-R}|.$$

The next term of the series $\overset{D\;N}{\Sigma\Pi}\Phi_R(d)$ is of the same order in $R$ as $\int_{DC} \overset{N}{\Pi}(s)\,ds$; similarly, for $R'$ fixed, $\int_{FE} \overset{N}{\Pi}(s)\,ds$ is also bounded above as $N \to \infty$, and this integral is of the same order in $R'$ as the $(R'+1)$th term in the series

$$\overset{C\;N}{\Sigma\Pi}\Phi_{R'}(c).$$

Hence, we can say that when $D < B$,

$$\frac{1}{2\pi i} \int_{-i\pi/t}^{i\pi/t} \Pi \begin{bmatrix} (a)\,q^s, q^{1+s}/z, (b)\,q^{-s}, zq^{-s}; \\ (c)\,q^s, (d)\,q^{-s}; \end{bmatrix} q \end{bmatrix} ds$$

$$\sim \frac{1}{t} \sum_{\mu=1}^{D} \Pi \begin{bmatrix} (a)\,d_\mu, qd_\mu/z, (b)/d_\mu, z/d_\mu; \\ (c)\,d_\mu, (d)'/d_\mu, q; \end{bmatrix} q \end{bmatrix}$$

$$\times {}_{B+C}\Phi_{A+D-1} \begin{bmatrix} (c)\,d_\mu, qd_\mu/(b); \\ (a)\,d_\mu, qd_\mu/(d)'; \end{bmatrix} q, Q_\mu \end{bmatrix}, \quad (5.4.1)$$

where

$$Q_\mu = (-q^{\frac{1}{2}n+\frac{1}{2}}d_\mu)^{D-B} \frac{b_1 b_2 \dots b_B z}{d_1 d_2 \dots d_D},$$

and, when $C < A$,

$$\frac{1}{2\pi i} \int_{-i\pi/t}^{i\pi/t} \Pi \begin{bmatrix} (a)\,q^s, q^{1+s}/z, (b)\,q^{-s}, zq^{-s}; \\ (c)\,q^s, (d)\,q^{-s}; \end{bmatrix} q \end{bmatrix} ds$$

$$\sim \frac{1}{t} \sum_{\nu=1}^{C} \Pi \begin{bmatrix} (b)\,c_\nu, (a)/c_\nu, zc_\nu, q/zc_\nu; \\ (d)\,c_\nu, qc_\nu/(c)'; \end{bmatrix} q \end{bmatrix}$$

$$\times {}_{A+D}\Phi_{B+C-1} \begin{bmatrix} (d)\,c_\nu, qc_\nu/(a); \\ (b)\,c_\nu, qc_\nu/(c)'; \end{bmatrix} q, Q'_\nu \end{bmatrix}, \quad (5.4.2)$$

where

$$Q'_\nu = (-q^{\frac{1}{2}n+\frac{1}{2}}c_\nu)^{C-A} \frac{a_1 a_2 \dots a_A q}{c_1 c_2 \dots c_C z}.$$

As an example on the application of this theorem, let $A = B = 0$, $C = 1$ and $D = 2$, then we have

$$\frac{1}{2\pi i} \int_{-i\pi/t}^{i\pi/t} \Pi \begin{bmatrix} q^{1+s}/z, zq^{-s}; \\ aq^s, q^{1-s}/b, q^{-s}; \end{bmatrix} ds$$

$$= \Pi \begin{bmatrix} q/z, z; \\ a, q/b, q; \end{bmatrix} {}_1\Phi_1[a; b; q, -q^{\frac{1}{2}n-\frac{1}{2}}zb]$$

$$+ \Pi \begin{bmatrix} q^2/bz, bz/q; \\ qa/b, b/q, q; \end{bmatrix} {}_1\Phi_1[qa/b; q^2/b; q, -q^{\frac{1}{2}n+\frac{1}{2}}z]$$

$$\sim \Pi \begin{bmatrix} az, q/az; \\ qa/b, a, q; \end{bmatrix} {}_2\Phi_0[qa/b, a; ; q, q/az], \quad (5.4.3)$$

and if $A = C = D = 1$, $B = 0$, then

$$\frac{1}{2\pi i} \int_{-i\pi/t}^{i\pi/t} \Pi \begin{bmatrix} bq^s, q^{1-s}/x, xq^{-s}; \\ aq^s, q^{-s}; \end{bmatrix} ds = \Pi \begin{bmatrix} b, 1/x, x; \\ a; \end{bmatrix} {}_1\Phi_1[a; b; q, x]$$

$$\sim \Pi \begin{bmatrix} b/a, 1/ax, ax; \\ a, q; \end{bmatrix} {}_2\Phi_0[a, qa/b; ; q, -qbq^{-\frac{1}{2}n-\frac{1}{2}}/2ax]. \quad (5.4.4)$$

These two results provide basic analogues of the asymptotic forms for the confluent hypergeometric functions.†

## 5.5 Contour integrals of basic functions

A process of generalization, exactly similar to that for the ordinary hypergeometric functions, will now be carried out for the basic functions. The first step is to replace the integrand

$$\Pi \begin{bmatrix} (a)\, q^s, (b)\, q^{-s}; \\ (c)q^s, (d)\, q^{-s}; \end{bmatrix} q \quad (5.5.1)$$

in the main theorems of §5.2 and §5.4 by the general function

$$\sum_{m=0}^{\infty} \Pi \begin{bmatrix} (a)\, q^{m+s}, (b)\, q^{m-s}; \\ (c)q^{m+s}, (d)\, q^{m-s}; \end{bmatrix} q \frac{((e); q)_m\, q^{xm}}{((f); q)_m\, (q; q)_m}. \quad (5.5.2)$$

Then, under conditions which make the convergence of the above series absolute and uniform, we can state our main theorem thus;

$$\text{let}\quad I = \frac{t}{2\pi i} \int_{-i\pi/t}^{i\pi/t} \Pi \begin{bmatrix} (a)\, q^s, (b)\, q^{-s}; \\ (c)\, q^s, (d)\, q^{-s}; \end{bmatrix} q$$

$$\times {}_{C+D+E}\Phi_{A+B+F-1} \begin{bmatrix} (c)\, q^s, (d)\, q^{-s}, (e); \\ (a)\, q^s, (b)\, q^{-s}, (f); \end{bmatrix} q, q^x \end{bmatrix} ds, \quad (5.5.3)$$

$$\Sigma_D = \sum_{\mu=1}^{D} \Pi \begin{bmatrix} (a)\, d_\mu, (b)/d_\mu; \\ (c)\, d_\mu, (d)'/d_\mu; \end{bmatrix} q \sum_{m=0}^{\infty} \sum_{n=0}^{\infty} \frac{((c)\, d_\mu; q)_{2m+n}}{((a)\, d_\mu; q)_{2m+n}}$$

$$\times \frac{(qd_\mu/(b); q)_n\, ((e); q)_m\, q^{xm}Q_\mu^n}{(qd_\mu/(d)'; q)_n\, (q; q)_n\, ((f); q)_m\, (q; q)_m}, \quad (5.5.4)$$

† Slater (1960), (4.1.2) and (4.1.5).

$$\Sigma_C = \sum_{\nu=1}^{C} \Pi\begin{bmatrix}(b)\,c_\nu,(a)/c_\nu; \\ (d)\,c_\nu,(c)'/c_\nu;\end{bmatrix} q \Bigg] \sum_{m=0}^{\infty} \sum_{n=0}^{\infty} \frac{((d)\,c_\nu;\,q)_{2m+n}}{((b)\,c_\nu;\,q)_{2m+n}}$$

$$\times \frac{(qc_\nu/(a);\,q)_n\,((e);\,q)_m\,q^{xm}Q_\nu'^{\,n}}{(qc_\nu/(c)';\,q)_n\,(q;\,q)_n\,((f);\,q)_m\,(q;\,q)_m}, \qquad (5.5.5)$$

*where*
$$Q_\mu \equiv (-d_\mu\, q^{\frac12 n+\frac12})^{D-B}\frac{b_1 b_2 \ldots b_B}{d_1 d_2 \ldots d_D},$$

*and*
$$Q_\nu' \equiv (-c_\nu\, q^{\frac12 n+\frac12})^{C-A}\frac{a_1 a_2 \ldots a_A}{c_1 c_2 \ldots c_C}.$$

*Then, for* $|q| < 1, t > 0$,

(i) $I = \Sigma_D$ *if* $D > B$, *or if* $D = B$ *and* $\mathrm{Rl}\,(b_1 b_2 \ldots b_B/d_1 d_2 \ldots d_D) > 0$, *and* $I \sim \Sigma_D$ *if* $D < B$; *also*

(ii) $I = \Sigma_C$ *if* $C > A$, *or if* $C = A$ *and* $\mathrm{Rl}\,(a_1 a_2 \ldots a_A/c_1 c_2 \ldots c_C) > 0$, *and* $I \sim \Sigma_C$ *if* $C < A$.

**5.5.1 The general theorem.** Again, it is not necessary to assume, in the integral of (5.5.3), that all the parameters which occur in the products in the integrand must also occur in the $\Phi$ function. Also, we can introduce an independent variable $z$, and thus we can state the theorem in its most general form;

*let*
$$\Pi(q^s) = \Pi\begin{bmatrix}(a)\,q^s,(b)\,q^{-s},(g)\,q^s,(h)\,q^{-s}, zq^s, q^{1-s}/z; \\ (c)\,q^s,(d)\,q^{-s},(j)\,q^s,(k)\,q^{-s};\end{bmatrix} q \Bigg], \quad (5.5.1.1)$$

$$I = \frac{t}{2\pi i}\int_{-i\pi/t}^{i\pi/t} \Pi(q^s)\,_{C+D+E}\Phi_{A+B+F}\begin{bmatrix}(c)\,q^s,(d)\,q^{-s},(e); \\ (a)\,q^s,(b)\,q^{-s},(f);\end{bmatrix} q, x \Bigg]\,ds,$$
$$(5.5.1.2)$$

$$\Sigma_C = \sum_{\nu=1}^{C} \Pi(1/c_\nu) \sum_{m=0}^{\infty} \sum_{n=0}^{\infty}$$

$$\times \frac{((d)\,c_\nu;\,q)_{2m+n}\,(qc_\nu/(a);\,q)_n\,((k)\,c_\nu;\,q)_{m+n}}{\begin{array}{c}(qc_\nu/(g);\,q)_{m+n}\,((e);\,q)_m\,x^m z^{m+n}\alpha_1^n\,\alpha_2^{m+n}\end{array}}{\dfrac{}{((b)\,c_\nu;\,q)_{2m+n}\,(qc_\nu/(c)';\,q)_n\,((h)\,c_\nu;\,q)_{m+n}}{(qc_\nu/(j);\,q)_{m+n}\,((f);\,q)_m\,(q;\,q)_m\,(q;\,q)_n}}$$

$$+ \sum_{\nu=1}^{J} \Pi(1/j_\nu) \sum_{m=0}^{\infty} \sum_{n=0}^{\infty}$$

$$\times \frac{((c)/j_\nu;\,q)_{m-n}\,((d)\,j_\nu;\,q)_{m+n}\,(qj_\nu/(g);\,q)_n}{\begin{array}{c}((k)\,j_\nu;\,q)_n\,((e);\,q)_m\,x^m \beta^n\end{array}}{\dfrac{}{((a)/j_\nu;\,q)_{m-n}\,((b)\,j_\nu;\,q)_{m+n}\,(qj_\nu/(j)';\,q)_n\,((h)\,j_\nu;\,q)_n}{((f);\,q)_m\,(q;\,q)_m\,(q;\,q)_n}}$$

$$(5.5.1.3)$$

*where*
$$\alpha_1 \equiv (-c_\nu q^{\frac{1}{2}n+\frac{1}{2}})^{C-A} \frac{a_1 a_2 \dots a_A}{c_1 c_2 \dots c_C},$$

$$\alpha_2 \equiv (-c_\nu q^{\frac{1}{2}m+\frac{1}{2}n+\frac{1}{2}})^{J-G-1} \frac{g_1 g_2 \dots g_G}{j_1 j_2 \dots j_J},$$
$$\tag{5.5.1.4}$$

*and*
$$\beta \equiv (-j_\nu z q^{\frac{1}{2}n+\frac{1}{2}})^{J-G-1} \frac{g_1 g_2 \dots g_G}{j_1 j_2 \dots j_J}; \tag{5.5.1.5}$$

$$\Sigma_D = \sum_{\mu=1}^{D} \Pi(d_\mu) \sum_{m=0}^{\infty} \sum_{n=0}^{\infty}$$

$$\times \frac{((c)\,d_\mu;\,q)_{2m+n}\,(qd_\mu/(b);\,q)_n\,((j)\,d_\mu;\,q)_{m+n}\,(qd_\mu/(h);\,q)_{m+n}\;((e);\,q)_m\,x^m \gamma_1^n \gamma_2^{m+n}}{((a)\,d_\mu;\,q)_{2m+n}\,(qd_\mu/(d)';\,q)_n\,((g)\,d_\mu;\,q)_{m+n}\,(qd_\mu/(k);\,q)_{m+n}\;((f);\,q)_m\,(q;\,q)_m\,(q;\,q)_n}$$

$$+ \sum_{\mu=1}^{K} \Pi(k_\mu) \sum_{m=0}^{\infty} \sum_{n=0}^{\infty}$$

$$\times \frac{((c)\,k_\mu;\,q)_{m+n}\,((b)/k_\mu;\,q)_{m+n}\,((j)\,k_\mu;\,q)_n\,(qk_\mu/(h);\,q)_n\;((e);\,q)_m\,x^m \delta^n}{((a)\,k_\mu;\,q)_{m+n}\,((d)/k_\mu;\,q)_{m+n}\,((g)\,k_\mu;\,q)_n\,(qk_\mu/(k)';\,q)_n\;((f);\,q)_m\,(q;\,q)_m\,(q;\,q)_n},$$
$$\tag{5.5.1.6}$$

*where*
$$\gamma_1 \equiv (-d_\mu q^{\frac{1}{2}n+\frac{1}{2}})^{D-B} \frac{d_1 d_2 \dots d_D}{b_1 b_2 \dots b_B},$$

$$\gamma_2 \equiv (-d_\mu q^{\frac{1}{2}m+\frac{1}{2}n+\frac{1}{2}})^{K-H-1} \frac{h_1 h_2 \dots h_H}{k_1 k_2 \dots k_K},$$
$$\tag{5.5.1.7}$$

*and*
$$\delta \equiv (-k_\mu q^{\frac{1}{2}n+\frac{1}{2}})^{K-H-1} \frac{h_1 h_2 \dots h_H}{k_1 k_2 \dots k_K}. \tag{5.5.1.8}$$

*Then, when* $|q| < 1$, $t > 0$ *and* $\mathrm{Rl}\,x > 0$, *we have*

(i) $I = \Sigma_D$ *if (a)* $D > B$ *or* $D = B$ *and* $\mathrm{Rl}\,(b_1 b_2 \dots b_B/d_1 d_2 \dots d_D) \geqslant 0$, *and (b)* $K > H+1$ *or* $K = H+1$ *and* $\mathrm{Rl}\,(h_1 h_2 \dots h_H/k_1 k_2 \dots k_K) \geqslant 0$, *and* $I \sim \Sigma_D$ *if (c)* $D < B$ *or (d)* $K < H+1$,

(ii) $I = \Sigma_C$ *if (a)* $C > A$ *or* $C = A$ *and* $\mathrm{Rl}\,(a_1 a_2 \dots a_A/c_1 c_2 \dots c_C) \geqslant 0$, *and (b)* $J > G+1$ *or* $J = G+1$ *and* $\mathrm{Rl}\,(g_1 g_2 \dots g_G/j_1 j_2 \dots j_J) \geqslant 0$, *and* $I \sim \Sigma_C$ *if (c)* $C < A$ *or (d)* $J < G+1$.

We shall give a short outline of the proof of this theorem. It will be assumed in this proof, that $q$ is real, and $q = \exp(-t)$, $t > 0$, though this restriction can always be removed from the final results, by analytic continuation over the circle $|q| < 1$. Now

$$aq^s = \exp(-t \log a - ts),$$

and this expression has a period of $2\pi i/t$. Hence the product $1/\pi(aq^s; q)$ has poles at all the points $s = -\log a - n + 2\pi i/t$, one set in each strip of width $2\pi/t$. So let us consider the contour integral

$$\int \overset{M}{\underset{}{\Sigma}} \overset{P}{\underset{}{\Pi}} (q^s) \, ds$$
$$= \frac{t}{2\pi i} \int \overset{M}{\underset{m=0}{\Sigma}} \overset{P}{\underset{}{\Pi}} \left[ \begin{matrix} (a) \, q^{m+s}, (b) \, q^{m-s}, (g) \, q^s, (h) \, q^{-s}, zq^s, q^{1-s}/z; \\ (c) \, q^{m+s}, (d) \, q^{m-s}, (j) \, q^s, (k) \, q^{-s}; \end{matrix} \, q \right] ds$$

$$(5.5.1.9)$$

taken round the rectangular contours of Fig. 5.1,

$$A(-\pi i/t), \quad B(\pi i/t), \quad C(2N + \pi i/t), \quad D(2N - \pi i/t)$$
and
$$A(-\pi i/t), \quad B(\pi i/t), \quad E(-2N' + \pi i/t), \quad F(-2N' - \pi i/t),$$

where $N$ and $N'$ are integers, such that $P > \max(N, N')$. Both contours are indented so that the first $2N$ of each ascending sequence of poles of
$$\overset{M}{\underset{}{\Sigma}} \overset{P}{\underset{}{\Pi}} (q^s)$$

fall inside $ABCD$ and the first $2N'$ of each descending sequence of poles fall inside $ABEF$. This implies that none of the sequences of poles coincide or overlap. By the periodicity of the integrand in $2\pi i/t$, it follows that

$$\int_{BC} \overset{M}{\underset{}{\Sigma}} \overset{P}{\underset{}{\Pi}} (q^s) \, ds = \int_{AD} \overset{M}{\underset{}{\Sigma}} \overset{P}{\underset{}{\Pi}} (q^s) \, ds$$

and
$$\int_{FA} \overset{M}{\underset{}{\Sigma}} \overset{P}{\underset{}{\Pi}} (q^s) \, ds = \int_{EB} \overset{M}{\underset{}{\Sigma}} \overset{P}{\underset{}{\Pi}} (q^s) \, ds.$$

Hence
$$I_{P,N,M} = \int_{AB} \overset{M}{\underset{}{\Sigma}} \overset{P}{\underset{}{\Pi}} (q^s) \, ds + \int_{CD} \overset{M}{\underset{}{\Sigma}} \overset{P}{\underset{}{\Pi}} (q^s) \, ds$$

and
$$-I_{P,N',M} = \int_{AB} \overset{M}{\underset{}{\Sigma}} \overset{P}{\underset{}{\Pi}} (q^s) \, ds + \int_{EF} \overset{M}{\underset{}{\Sigma}} \overset{P}{\underset{}{\Pi}} (q^s) \, ds.$$

But
$$I_{P,N,M} = t \, \Sigma \text{ (residues of } \overset{M}{\underset{}{\Sigma}} \overset{P}{\underset{}{\Pi}} (q^s) \text{ within } ABCD),$$

and
$$-I_{P,N',M} = t \, \Sigma \text{ (residues of } \overset{M}{\underset{}{\Sigma}} \overset{P}{\underset{}{\Pi}} (q^{-s}) \text{ within } ABEF)$$

so that
$$I_{P,N,M} = \overset{D}{\underset{}{\Sigma}} \overset{M}{\underset{}{\Sigma}} \overset{N}{\underset{}{\Sigma}} \overset{P}{\underset{}{\Pi}} (d_\mu q^{m+n}) + \overset{K}{\underset{}{\Sigma}} \overset{M}{\underset{}{\Sigma}} \overset{N}{\underset{}{\Sigma}} \overset{P}{\underset{}{\Pi}} (k_\mu q^n),$$

and
$$-I_{P,N',M} = \overset{C}{\underset{}{\Sigma}} \overset{M}{\underset{}{\Sigma}} \overset{N'}{\underset{}{\Sigma}} \overset{P}{\underset{}{\Pi}} (q^{-m-n}/c_\nu) + \overset{J}{\underset{}{\Sigma}} \overset{M}{\underset{}{\Sigma}} \overset{N'}{\underset{}{\Sigma}} \overset{P}{\underset{}{\Pi}} (q^{-n}/j_\nu).$$

Hence

$$\int_{AB} \overset{M}{\Sigma}\overset{P}{\Pi}(q^s)\,ds = \overset{D}{\Sigma}\overset{M}{\Sigma}\overset{N}{\Sigma}\overset{P}{\Pi}(d_\mu q^{m+n})$$

$$+ \overset{K}{\Sigma}\overset{M}{\Sigma}\overset{N}{\Sigma}\overset{P}{\Pi}(k_\mu q^n) + \int_{DC} \overset{M}{\Sigma}\overset{P}{\Pi}(q^s)\,ds,$$

$$= \overset{C}{\Sigma}\overset{M}{\Sigma}\overset{N'}{\Sigma}\overset{P}{\Pi}(q^{-m-n}/c_\nu)$$

$$+ \overset{J}{\Sigma}\overset{M}{\Sigma}\overset{N'}{\Sigma}\overset{P}{\Pi}(q^{-n}/j_\nu) + \int_{FE} \overset{M}{\Sigma}\overset{P}{\Pi}(q^s)\,ds,$$

where

$$\left| \int_{DC} \overset{M}{\Sigma}\overset{P}{\Pi}(q^s)\,ds \right| \leqslant \frac{t}{2\pi}\int_{-\pi/t}^{\pi/t} \overset{M}{\Sigma} \,|\,\overset{P}{\Pi}(q^{N+ir})\,|\,dr$$

and

$$\left| \int_{FE} \overset{M}{\Sigma}\overset{P}{\Pi}(q^s)\,ds \right| \leqslant \frac{t}{2\pi}\int_{-\pi/t}^{\pi/t} \overset{M}{\Sigma} \,|\,\overset{P}{\Pi}(q^{N'+ir})\,|\,dr.$$

Now $\int_{DC}$ is bounded above as $P \to \infty$, and, for $N$ fixed, $\int_{DC}$ is of the same order in $N$ as

$$\overset{D}{\Sigma}\overset{M}{\Sigma}\overset{\infty}{\Pi}(d_\mu q^{m+N+1}) + \overset{K}{\Sigma}\overset{M}{\Sigma}\overset{\infty}{\Pi}(k_\mu q^{N+1}).$$

Similarly, $\int_{FE}$ is bounded above as $P \to \infty$, and, for $N'$ fixed, $\int_{FE}$ is of the same order in $N'$ as

$$\overset{C}{\Sigma}\overset{M}{\Sigma}\overset{\infty}{\Pi}(q^{-m-N'-1}/c_\nu) + \overset{J}{\Sigma}\overset{M}{\Sigma}\overset{\infty}{\Pi}(q^{-N'-1}/j_\nu).$$

Hence, as $N, N' \to \infty$, we have

$$\int_{-i\pi/t}^{i\pi/t} \overset{M}{\Sigma}\overset{\infty}{\Pi}(q^s)\,ds = \overset{D}{\Sigma}\overset{M}{\Sigma}\overset{\infty}{\Sigma}\overset{\infty}{\Pi}(d_\mu q^{m+n}) + \overset{K}{\Sigma}\overset{M}{\Sigma}\overset{\infty}{\Sigma}\overset{\infty}{\Pi}(k_\mu q^n),$$

$$= \overset{C}{\Sigma}\overset{M}{\Sigma}\overset{\infty}{\Sigma}\overset{\infty}{\Pi}(q^{-m-n}/c_\nu) + \overset{J}{\Sigma}\overset{M}{\Sigma}\overset{\infty}{\Sigma}\overset{\infty}{\Pi}(q^{-n}/j_\nu), \quad (5.5.1.10)$$

that is to say, $\qquad\qquad I = \Sigma_D = \Sigma_C, \qquad\qquad (5.5.1.11)$

when all these series are convergent. Also,

$$I \sim \Sigma_C \qquad\qquad (5.5.1.12)$$

if $C < A$ or if $J < G+1$, and $\quad I \sim \Sigma_D \qquad\qquad (5.5.1.13)$

if $D < B$ or if $K < H+1$.

## 5.5.2 Some special cases.

We shall now give a few examples on this fundamental theorem. Firstly, we have

$$\frac{t}{2\pi i}\int_{-i\pi/t}^{i\pi/t} \Pi[q;\,cq^s, q^{-s};\,q]\,{}_1\Phi_0[cq^s;\,;\,q,x]\,ds$$

$$= \Pi[cx;\,c,x;\,q]\,{}_1\Phi_1[c;\,cx;\,q, -q^{\frac12 n + \frac12}]. \quad (5.5.2.1)$$

We have summed here by Heine's theorem (3.2.2.11). The function on the right is one of the basic confluent hypergeometric functions. This is the same result that we should have obtained if we had applied the theorem of § 5.2 to the integral

$$\frac{t}{2\pi i}\int_{-i\pi/t}^{i\pi/t} \Pi[q, cxq^s; cq^s, q^{-s}, x; q]\,ds. \qquad (5.5.2.2)$$

If, in the theorem of § 5.5.1, we take

$$A = 1, \quad B = C = D = E = F = H = 0, \quad J = K = G = 1,$$

we have

$$\frac{t}{2\pi i}\int_{-i\pi/t}^{i\pi/t} \Pi[aq^s, q; kq^s, q^{-s}; q]\,_0\Phi_1[\;;aq^s;q,x]\,ds$$
$$= \Pi[a; k; q]\sum_{n=0}^{\infty} \frac{(k;q)_n\,(-1)^n\,q^{\frac{1}{2}n(n+1)}}{(a;q)_n\,(q;q)_n}\,_0\Phi_1[\;;aq^n;q,x], \quad (5.5.2.3)$$

where $\mathrm{Rl}\,(q/k) > 0$, and $\mathrm{Rl}\,(x) > 0$. In terms of the basic Bessel function† defined as

$$_qj_\nu(x) \equiv \frac{x^\nu}{2^\nu(q;q)_\nu}\,_0\Phi_1[\;;\nu q;q,-x^2/4], \qquad (5.5.2.4)$$

(5.5.2.3) becomes

$$\frac{t}{2\pi i}\int_{-i\pi/t}^{i\pi/t} \Pi[q, q; kq^s, q^{-s}; q]\,_qj_{aq^s-1}(x)\,2^s x^{-s}\,ds$$
$$= \Pi[a, a; k, a^2/q; q]\sum_{n=0}^{\infty} \frac{(k;q)_n\,(-1)^n\,q^{\frac{1}{2}n(n+1)}\,2^n}{(a;q)_n\,(q;q)_n\,x^n}\,_qj_{aq^n-1}(x). \quad (5.5.2.5)$$

For the series $_1\Phi_1$, we find, from the integral

$$I(x) \equiv \frac{t}{2\pi i}\int_{-i\pi/t}^{i\pi/t} \Pi[aq^s, q; cq^s, q^{-s}; q]\,_1\Phi_1[cq^s; aq^s; q, x]\,ds, \quad (5.5.2.6)$$

that $\qquad I(x) = \Pi[a/c, cx; c, x; q]\,_2\Phi_1[c, qc/a; cx; q, a/c], \qquad (5.5.2.7)$

from a consideration of the sequence of poles

$$s = -\log c - m - n \quad (n = 0, 1, 2, \ldots),$$

where $\mathrm{Rl}\,(a/c) > 0$. Also

$$I(x) = \Pi[a; c; q]\sum_{n=0}^{\infty} \frac{(c;q)_n\,(-1)^n\,q^{\frac{1}{2}n(n+1)}}{(a;q)_n\,(q;q)_n}\,_1\Phi_1[cq^n; aq^n; q, x], \quad (5.5.2.8)$$

from a consideration of the sequence of poles

$$s = n \quad (n = 0, 1, 2, \ldots).$$

† Jackson (1905c).

In particular, when $x = q$, we can sum the $_2\Phi_1$ series by the basic analogue of Gauss's theorem (3.3.2.7), or, alternatively, we can see that the series of $_1\Phi_1$ functions is now orthogonal, and so will reduce to unity. Thus, we find that

$$\frac{t}{2\pi i} \int_{-i\pi/t}^{i\pi/t} \Pi[aq^s, q; cq^s, q^{-s}; q]\,_1\Phi_1[cq^s; aq^s; q, q]\,\mathrm{d}s = \Pi[a; c; q].$$

$$(5.5.2.9)$$

# 6

## BILATERAL SERIES

### 6.1 The process of generalization

There are many ways of extending the definition of the Gauss function. We have already considered increasing the number of parameters, in order to produce the generalized functions of Chapter 2, and the introduction of a base $q$ to produce the basic series of Chapter 3. Indeed, the whole historical development of the subject has come from the inborn habit of the hypergeometric mathematician of trying to produce a general theory, when faced with some elementary series which does not fit into one of the forms already known. Such a situation faced Dougall in 1907,† when he discovered the formula

$$\sum_{n=-\infty}^{n=\infty} \frac{\Gamma(a+n)\,\Gamma(b+n)}{\Gamma(c+n)\,\Gamma(d+n)}$$

$$= \frac{\pi^2}{\sin(\pi a)\sin(\pi b)} \frac{\Gamma(c+d-a-b-1)}{\Gamma(c-a)\,\Gamma(d-a)\,\Gamma(c-b)\,\Gamma(d-b)}, \quad (6.1.1)$$

where $\mathrm{Rl}\,(c+d-a-b-1) > 0$.

When $d = 1$, this reduces to Gauss's theorem (1.7.6) but the series on the left of (6.1.1) is not an ordinary Gauss series, for it is infinite in both directions. Such a series is called a bilateral series.

**6.1.1 Notation.** Let us extend the definition of $(a)_n$ to have a meaning for negative integer values of $n$. We shall write

$$(a)_{-n} = \frac{\Gamma(a-n)}{\Gamma(a)} = \frac{(-1)^n}{(1-a)_n} = \frac{(-1)^n}{(1-a)\,(2-a)\,(3-a)\dots(n-a)}, \quad (6.1.1.1)$$

and we shall use the symbol $H$ for those series which are infinite in both directions, so that the general bilateral series is defined as

$$\sum_{n=-\infty}^{\infty} \frac{(a_1)_n\,(a_2)_n\,(a_3)_n\dots(a_A)_n}{(b_1)_n\,(b_2)_n\,(b_3)_n\dots(b_B)_n} z^n \equiv {}_A H_B \left[ \begin{matrix} a_1, a_2, a_3, \dots, a_A; \\ b_1, b_2, b_3, \dots, b_B; \end{matrix} \ z \right]. \quad (6.1.1.2)$$

When there is no danger of ambiguity, this notation can be contracted further to

$$\sum_{n=-\infty}^{\infty} \frac{((a))_n}{((b))_n} z^n = {}_A H_B[(a);\,(b);\,z], \quad (6.1.1.3)$$

† Dougall (1907), eq. 26.

where it is understood that there are always $A$ of the $a$ parameters and $B$ of the $b$ parameters, as usual.

Alternative names for such series are Dirichlet series, or Laurent series.

We can always write

$$_AH_B[(a); (b); z] \equiv {}_BH_A[(1-b); (1-a); 1/z]. \qquad (6.1.1.4)$$

This merely expresses the fact that the series has an unaltered sum, when the order of the terms is exactly reversed, provided that both series are finite, in both directions.

The function $_AH_B[z]$ is defined for all real and complex values of the parameters $\qquad a_1, a_2, a_3, ..., a_A, b_1, b_2, b_3, ..., b_B$ except zero or integers, and for all values of the variable $z$ such that $|z| = 1$. If $z = -1$, we must have

$$\text{Rl}\,(b_1 + b_2 + ... + b_B - a_1 - a_2 - ... - a_A) > 1 \qquad (6.1.1.5)$$

for convergence, and if $z = 1$

$$\text{Rl}\,(b_1 + b_2 + ... + b_B - a_1 - a_2 - ... - a_A) > 0. \qquad (6.1.1.6)$$

If any one of the $a$ parameters is a negative integer, the series terminates above, and, if any one of the $b$ parameters is a positive integer the series terminates below. If any one of the $a$ parameters is a positive integer, or if any one of the $b$ parameters is a negative integer, the series is not defined.

### 6.1.2 The generalized Gauss theorem.

In this notation, the formula (6.1.1) becomes

$$_2H_2[a, b; c, d; 1] = \Gamma\begin{bmatrix} c, d, 1-a, 1-b, c+d-a-b-1 \\ c-a, d-a, c-b, d-b \end{bmatrix}. \qquad (6.1.2.1)$$

We shall now give two ways of deducing this theorem.

The first method of proof depends on the fact that any $H$ series can always be expressed as two $F$ series, so that, for general values of $d$, we can rewrite the $H$ series as

$$_2H_2[a, b; c, d; 1]$$

$$= ... + \frac{(c-2)_2\,(d-2)_2}{(a-2)_2\,(b-2)_2} + \frac{(c-1)\,(d-1)}{(a-1)\,(b-1)} + 1 + \frac{ab}{cd} + \frac{(a)_2\,(b)_2}{(c)_2\,(d)_2} + ...$$

$$= \frac{(c-1)\,(d-1)}{(a-1)\,(b-1)}\,{}_3F_2\begin{bmatrix} 1, c-2, d-2; \\ a-2, b-2; \end{bmatrix} + {}_3F_2\begin{bmatrix} 1, a, b; \\ c, d; \end{bmatrix}. \qquad (6.1.2.2)$$

But, if we put $c = 1$, in the formula (4.3.4) connecting three $_3F_2(1)$ series, we find immediately, that the expression on the right of

(6.1.2.2) is equal to the expression on the right of (6.1.2.1) above, by a simple use of Gauss's theorem.

In general, when $A = B$, it should be noted that we can always write

$$_AH_A[(a); (b); z] = {}_{A+1}F_A[1, (a); (b); z]$$

$$+ \prod_{\nu=1}^{A} \frac{(1-b_\nu)}{(1-a_\nu)} {}_{A+1}F_A[1, (2-b); (2-a); 1/z]. \quad (6.1.2.3)$$

Alternatively, we can prove (6.1.2.1) by evaluating the residues at the poles within semi-circles to the right and to the left of the imaginary axis, of the integral

$$I = \frac{1}{2\pi i} \int_{-i\infty}^{i\infty} \Gamma \begin{bmatrix} -s, 1+s, a+s, b+s \\ c+s, d+s \end{bmatrix} ds. \quad (6.1.2.4)$$

Another formula,† due to Dougall, is

$$_5H_5 \begin{bmatrix} 1+\tfrac{1}{2}a, & b, & c, & d, & e; \\ \tfrac{1}{2}a, 1+a-b, 1+a-c, 1+a-d, 1+a-e; \end{bmatrix}$$

$$= \Gamma \begin{bmatrix} 1-b, 1-c, 1-d, 1-e, 1+a-b, 1+a-c, 1+a-d, \\ 1+a, 1-a, 1+a-b-c, 1+a-b-d, 1+a-b-e, \end{bmatrix}$$

$$\begin{matrix} 1+a-e, 1+2a-b-c-d-e \\ 1+a-c-d, 1+a-c-e, 1+a-d-e \end{matrix} \Bigg],$$

$$(6.1.2.5)$$

where, for convergence, we must have

$$\mathrm{Rl}\,(3+4a-2b-2c-2d-2e) > 0.$$

This is the bilateral analogue of Dougall's theorem (2.3.4.1). It can be deduced directly from the result (4.3.7.8) which is a relation between three well-poised $_7F_6(1)$ series. If we write $f = 1$, then the second $_7F_6(1)$ series on the right reduces to a well-poised $_5F_4(1)$ series, which we can sum by (2.3.4.5). The other two $_7F_6(1)$ series combine to form the $_5H_5(1)$ series, and (6.1.2.5) above follows after a little reduction.

If we put $e = a$, (6.1.2.5) itself reduces to the summation theorem (2.3.4.5) for the well-poised $_5F_4(1)$ series. But if $e = \tfrac{1}{2}a$, (6.1.2.5) reduces to

$$_3H_3 \begin{bmatrix} b, & c, & d; \\ 1+a-b, 1+a-c, 1+a-d; \end{bmatrix}$$

$$= \Gamma \begin{bmatrix} 1-b, 1-c, 1-d, 1+a-b, 1+a-c, 1+a-d, \\ 1+a-c-d, 1+a-b-d, 1+a-b-c, 1+\tfrac{1}{2}a-b, \end{bmatrix}$$

$$\begin{matrix} 1-\tfrac{1}{2}a, 1+\tfrac{1}{2}a, 1+\tfrac{3}{2}a-b-c-d \\ 1+\tfrac{1}{2}a-c, 1+\tfrac{1}{2}a-d, 1+a, 1-a \end{matrix} \Bigg].$$

$$(6.1.2.6)$$

† Dougall (1907), eq. 33.

When $d = a$, (6.1.2.6) reduces to Dixon's theorem for the sum of a well-poised $_3F_2(1)$ series (2.3.3.5). Alternatively, if we replace $d$ and $e$ in (6.1.2.5) by $1+a-d$, and $1+a-e$, and then let $a \to +\infty$, we can deduce the $_2H_2(1)$ summation theorem (6.1.2.1) again. We cannot, however, deduce the sum of an infinite $_7H_7(1)$ series in this way, for, if we use Dougall's theorem, the finite $_7H_7(1)$ series is only the normal $_7F_6(1)$ series which has been displaced from the origin by $n$ terms.

## 6.2 A method of obtaining bilateral transformations

An elementary method of obtaining transforms between bilateral series,† due to Bailey, depends on the fact that a terminating series

$$\sum_{r=0}^{2n} u_r \quad \text{can be rewritten as} \quad \sum_{r=-n}^{n} u_{r+n}.$$

Hence, any known transformation of terminating hypergeometric series can be re-expressed as a relation between finite bilateral series, and, under suitable circumstances, we can then let $n \to \infty$, to obtain relations between infinite bilateral series.

We shall start from the well-known result of Whipple (4.3.6.2) connecting a well-poised $_7F_6(1)$ series and a Saalschutzian $_4F_3(1)$ series, and in this, we replace $m, a, b, c, d, e$ by $2n, a-2n, b-n, c-n, d-n, e-n$ respectively. Then, after a little reduction, we find that

$$_7H_7\left[\begin{matrix} a-n, 1+\tfrac{1}{2}a, & b, & c, & d, & e, & -n; \\ n+1, & \tfrac{1}{2}a, 1+a-b, 1+a-c, 1+a-d, 1+a-e, 1+a+n; \end{matrix} \, 1\right]$$

$$= \frac{(1+a)_n (1-a)_n (1+a-d-e)_n (1+a-b-c)_n}{(1-b)_n (1-c)_n (1+a-d)_n (1+a-e)_n}$$

$$\times {}_4H_4\left[\begin{matrix} 1+a-b-c+n, & d, & e, & -n; \\ & 1+n, 1+a-b, 1+a-c, d+e-a-n; \end{matrix} \, 1\right]. \quad (6.2.1)$$

This $_4H_4(1)$ series can be called Saalschutzian, since the sum of the denominator parameters equals two plus the sum of the numerator parameters. We can now let $n \to \infty$, through positive integer values, and apply Tannery's theorem to the upper and lower ends of the series, to obtain

$$_5H_5\left[\begin{matrix} 1+\tfrac{1}{2}a, & b, & c, & d, & e; \\ \tfrac{1}{2}a, 1+a-b, 1+a-c, 1+a-d, 1+a-e; \end{matrix} \, 1\right]$$

$$= \Gamma\left[\begin{matrix} 1-b, 1-c, 1+a-d, & 1+a-e \\ 1+a, 1-a, 1+a-d-e, 1+a-b-c \end{matrix}\right]$$

$$\times {}_2H_2\left[\begin{matrix} d, & e; \\ 1+a-b, 1+a-c; \end{matrix} \, 1\right], \quad (6.2.2)$$

† Bailey (1936), § 3.

provided that $\quad \mathrm{Rl}\,(1+2a-b-c-d-e) > 0.$

This result can be used to provide an alternative proof of (6.1.2.5), if we use (6.1.2.1) to sum the $_2H_2(1)$ series.

In a similar way, we can start from a relation between two well-poised $_9F_8(1)$ series (2.3.4.11) and deduce that

$$_6H_6\left[\begin{matrix}1+\tfrac{1}{2}a, & b, & c, & d, & e, & f; \\ \tfrac{1}{2}a, & 1+a-b, & 1+a-c, & 1+a-d, & 1+a-e, & 1+a-f;\end{matrix}\,-1\right]$$

$$= \Gamma\left[\begin{matrix}1-b,1-c,1-d,1+a-e,1+a-f,1+2a-b-c-d \\ 1+a,1-a,1+a-e-f,1+a-c-d,1+a-b-d,1+a-b-c\end{matrix}\right]$$

$$\times\,_3H_3\left[\begin{matrix}1+2a-b-c-d, & e, & f; \\ & 1+a-b, & 1+a-c, & 1+a-d;\end{matrix}\,1\right] \quad (6.2.3)$$

provided that $\quad \mathrm{Rl}\,(2+\tfrac{5}{2}a-b-c-d-e-f) > 0.$

If $\qquad 1+2a = 2b+c+d, \quad c = e \quad \text{and} \quad d = f,$

then the $_3H_3(1)$ series† can be summed by (6.1.2.6) and we find that

$$_6H_6\left[\begin{matrix}1+\tfrac{1}{2}a, & b, & c, & d, & c, & d; \\ \tfrac{1}{2}a, & 1+a-b, & 1+a-c, & 1+a-d, & 1+a-c, & 1+a-d;\end{matrix}\,-1\right]$$

$$= \Gamma\left[\begin{matrix}1-b,1-c,1-d,1+a-c,1+a-d,1+2a-b-c-d, \\ 1+a,1-a,1+a-c-d,1+a-c-d,1+a-b-d, \\ b+c+d-2a,1-c,1-d,b+c+d-a,1+a-c, \\ 1+a-b-c,1+a-c-d,b+c-a,b+d-a, \\ 1+a-d,1-\tfrac{1}{2}a,1+\tfrac{1}{2}a,b-\tfrac{1}{2}a \\ c+d-\tfrac{3}{2}a,1+\tfrac{1}{2}a-c,1+\tfrac{1}{2}a-d,1+a,1-a\end{matrix}\right]$$

$$(6.2.4)$$

provided that $\qquad 1+2a = 2b+c+d,$

and that $\qquad \mathrm{Rl}\,(1+\tfrac{1}{2}a+b+c-d) > 0.$

We can generalize the result (6.2.1) immediately, to series which do not terminate above. The series

$$_7H_7\left[\begin{matrix}a-n,1+\tfrac{1}{2}a, & b, & c, & d, & e, & f; \\ 1+n, & \tfrac{1}{2}a, & 1+a-b, & 1+a-c, & 1+a-d, & 1+a-e, & 1+a-f;\end{matrix}\,1\right]$$

which only terminates below, can be expressed as a well-poised $_7F_6(1)$ series displaced $n$ places from the origin. This, in its turn, is equivalent

† Bailey (1936), § 3.

to two Saalschutzian $_4F_3(1)$ series, by (2.4.4.3). One of these $_4F_3(1)$ series can be replaced by a Saalschutzian $_4H_4(1)$ series, and we find that

$$_7H_7\begin{bmatrix} a-n, 1+\tfrac12 a, & b, & c, & d, & e, & f; \\ 1+n, & \tfrac12 a, 1+a-b, 1+a-c, 1+a-d, 1+a-e, 1+a-f; \end{bmatrix}$$

$$= \Gamma\begin{bmatrix} 1+a-d, 1+a-e, 1+a-f, 1-a, 1-b+n, 1-c+n \\ 1+a, 1-b, 1-c, 1-a+n \end{bmatrix}$$

$$\times \left\{ \Gamma\begin{bmatrix} 1+a-d-e-f, 1+a-b-c \\ 1+a-e-f, 1+a-d-f, 1+a-d-e, 1+a-b-c+n \end{bmatrix} \right.$$

$$\times {}_4H_4\begin{bmatrix} d, & e, & f, 1+a-b-c+n; \\ 1+a-b, 1+a-c, 1+n, d+e+f-a; \end{bmatrix} 1$$

$$+ \Gamma\begin{bmatrix} 1+a-b, 1+a-c, d+e+f-1-a, 2+2a-b-c-d-e-f \\ 1+a-b-c, e, f, 2+2a-b-d-e-f \end{bmatrix}$$

$$\times \frac{(1)_n (2+2a-b-c-d-e-f)_n}{(2+2a-c-d-e-f)_n (2+2a-d-e-f)_n}$$

$$\times {}_4F_3\begin{bmatrix} 2+2a-b-c-d-e-f+n, 1+a-e-f, \\ 2+a-d-e-f+n, 2+2a-b-d-e-f, \\ & 1+a-d-f, 1+a-d-e; \\ & 2+2a-c-d-e-f; \end{bmatrix} 1 \right\}.$$

$$(6.2.5)$$

Next, let $n \to \infty$, and we have

$$_6H_6\begin{bmatrix} 1+\tfrac12 a, & b, & c, & d, & e, & f; \\ \tfrac12 a, 1+a-b, 1+a-c, 1+a-d, 1+a-e, 1+a-f; \end{bmatrix} -1$$

$$= \Gamma\begin{bmatrix} 1+a-d, 1+a-e, 1+a-f, 1-b, 1-c \\ 1-a, 1+a, 1+a-b-c \end{bmatrix}$$

$$\times \left\{ \Gamma\begin{bmatrix} 1+a-d-e-f \\ 1+a-e-f, 1+a-d-f, 1+a-d-e \end{bmatrix} \right.$$

$$\times {}_3H_3\begin{bmatrix} d, & e, & f; \\ 1+a-b, 1+a-c, d+e+f-a; \end{bmatrix} 1$$

$$+ \Gamma\begin{bmatrix} 1+a-b, 1+a-c, 2+a-d-e-f, d+e+f-a-1 \\ d, e, f, 2+2a-b-d-e-f, 2+2a-c-d-e-f \end{bmatrix}$$

$$\times {}_3F_2\begin{bmatrix} 1+a-e-f, 1+a-d-f, 1+a-d-e; \\ 2+2a-b-d-e-f, 2+2a-c-d-e-f; \end{bmatrix} 1 \right\}. \quad (6.2.6)$$

This is the generalization of (6.2.3) when the restriction on the parameters is removed.

## 6.3 General bilateral transforms and integrals

The first result we shall consider is a very elegant general relation between $M$ series of the type ${}_M H_M(1)$. We shall outline two methods of proof.

First, let us consider the general theorem (4.5.1.2), and take $A = B$, and $C = D$. Then we have

$$\sum_{\mu=1}^{A} \Gamma\begin{bmatrix}(a)-a_\mu, (b)+a_\mu \\ (c)-a_\mu, (d)+a_\mu\end{bmatrix} {}_{A+C}F_{A+C-1}\begin{bmatrix}(b)+a_\mu, 1+a_\mu-(c); \\ 1+a_\mu-(a)', (d)+a_\mu;\end{bmatrix}(-1)^{A+C}\end{bmatrix}$$
$$= \sum_{\nu=1}^{A} \Gamma\begin{bmatrix}(a)+b_\nu, (b)-b_\nu \\ (c)+b_\nu, (d)-b_\nu\end{bmatrix} {}_{A+C}F_{A+C-1}\begin{bmatrix}(a)+b_\nu, 1+b_\nu-(d); \\ (c)+b_\nu, 1+b_\nu-(b)';\end{bmatrix}(-1)^{A+C}\end{bmatrix},$$

$$(6.3.1)$$

where $\qquad \mathrm{Rl}\{\Sigma(c+d-a-b)\} > 0 \quad$ and $\quad A \geqslant C$.

Now, let us combine each series on the left with a similar series on the right and choose values of the parameters so as to produce a bilateral series, as in (6.1.2.3) above. After some algebra, we find the result

$$\sum_{\mu=1}^{A} \Gamma\begin{bmatrix}1+(a)-a_\mu, a_\mu-(a)' \\ 1+(b)-a_\mu, a_\mu-(c)\end{bmatrix} {}_M H_M\begin{bmatrix}1+(c)-a_\mu; \\ 1+(b)-a_\mu;\end{bmatrix} 1\end{bmatrix} = 0. \quad (6.3.2)$$

In this result, there are $A$ of the $a$, $b$ and $c$ parameters respectively, and

$$\mathrm{Rl}\{\Sigma(b-c)\} > 0.$$

In particular, if $A = 2$, we have

$$\Gamma\begin{bmatrix}1+a_2-a_1, a_1-a_2 \\ 1+b_1-a_1, 1+b_2-a_1, a_1-c_1, a_1-c_2\end{bmatrix} {}_2 H_2\begin{bmatrix}1+c_1-a_1, 1+c_2-a_1; \\ 1+b_1-a_1, 1+b_2-a_1;\end{bmatrix} 1\end{bmatrix}$$
$$+ \Gamma\begin{bmatrix}1+a_1-a_2, a_2-a_1 \\ 1+b_1-a_2, 1+b_2-a_2, a_2-c_1, a_2-c_2\end{bmatrix}$$
$$\times {}_2 H_2\begin{bmatrix}1+c_1-a_2, 1+c_2-a_2; \\ 1+b_1-a_2, 1+b_2-a_2;\end{bmatrix} 1\end{bmatrix} = 0. \quad (6.3.3)$$

This is a relation between two ${}_2 H_2(1)$ series. If, further $a_2 = b_2$, the second ${}_2 H_2(1)$ series reduces to a ${}_2 F_1(1)$ series, which can be summed by Gauss's theorem, and we have another proof of the summation theorem (6.1.2.1).

A second method of proof depends on the Barnes-type contour integral

$$I = \frac{1}{2\pi i}\int_{-i\infty}^{i\infty} \Gamma\begin{bmatrix}(a)+s, 1-(a)-s \\ (b)+s, 1-(c)-s\end{bmatrix} ds, \qquad (6.3.4)$$

where there are $A$ of the $a$, $b$ and $c$ parameters respectively, and the integration is along a path which ensures that all poles in the increasing sequences
$$s = 1-(a)+n \quad (n=1,2,3,\dots),$$
lie to the right of the contour, and all the poles in the decreasing sequences
$$s = -(a)-m \quad (m=1,2,3,\dots),$$
lie to its left. Then we have

$$I = \sum_{m=1}^{A} \Gamma\begin{bmatrix} 1+(a)-a_m, a_m-(a)' \\ 1+(b)-a_m, a_m-(c) \end{bmatrix} {}_{A+1}F_A\begin{bmatrix} 1, 1+a_m-(b); \\ 1+(c)-a_m; \end{bmatrix} 1$$

$$= \sum_{m=1}^{A} \Gamma\begin{bmatrix} (a)'-a_m, 1+a_m-(a) \\ (b)-a_m, 1+a_m-(c) \end{bmatrix} {}_{A+1}F_A\begin{bmatrix} 1, 1+a_m-(b); \\ 1+a_m-(c); \end{bmatrix} 1, \quad (6.3.5)$$

provided that $\mathrm{Rl}\{\Sigma(b-c)\} > 0$. From this result $(6.3.2)$ follows immediately.

### 6.3.1 Well-poised bilateral transforms.

There are four well-poised transforms of bilateral series corresponding to the four transforms of ordinary well-poised series $(4.5.2.2, 4, 6$ and $8)$. Again, there are two possible methods of proof, firstly by direct manipulation of these four transforms and $(6.3.2)$ above, and secondly, by the use of contour integrals.

Thus, from the integral

$$I_1 = \int_{-i\infty}^{i\infty} \Gamma\begin{bmatrix} (a)+s, 1+a_0-(a)+s, 1-(a)-s, (a)-a_0-s \\ 1+a_0-(b)+s, 1-(b)-s \end{bmatrix} e^{\pi i s}\, ds,$$
$$(6.3.1.1)$$

where there are $N+1$ of the $a$ parameters including $a_0$, and $2N$ of the $b$ parameters, provided that

$$\mathrm{Rl}\,(N-\tfrac{1}{2}+Na_0-b_1-b_2\dots-b_{2N}) > 0,$$

we can deduce

$$\Gamma\begin{bmatrix} a_0, (a), 1+a_0-(a), 1-a_0, 1-(a), (a)-a_0 \\ 1+a_0-(b), 1-(b), \tfrac{1}{2}a_0, 1-\tfrac{1}{2}a_0 \end{bmatrix} {}_{2N}H_{2N}\begin{bmatrix} (b); \\ 1+a_0-(b); \end{bmatrix} -1$$

$$= \sum_{\nu=1}^{N} \Gamma\begin{bmatrix} a_0-a_\nu, 1-a_\nu, a_\nu, (a)'-a_\nu, 2a_\nu-a_0, 1+a_0-2a_\nu, 1+a_\nu-a_0, \\ 1+a_\nu-(b), 1+a_0-a_\nu-(b), \end{bmatrix}$$
$$\begin{matrix} a_\nu+(a)-a_0, 1+a_\nu-(a), 1+a_0-a_\nu-(a) \\ \tfrac{1}{2}a_0-a_\nu, 1+a_\nu-\tfrac{1}{2}a_0 \end{matrix}$$

$$\times {}_{2N}H_{2N}\begin{bmatrix} a_\nu+(b)-a_0; \\ 1+a_\nu-(b); \end{bmatrix} -1. \quad (6.3.1.2)$$

This result expresses a ${}_{2N}H_{2N}(-1)$ series, well-poised in $a_0$ in terms of $N$ other ${}_{2N}H_{2N}(-1)$ series each well-poised in $2a_\nu-a_0$.

Similarly, from the integral

$$I_2 = \int_{-i\infty}^{i\infty} \Gamma\begin{bmatrix} (a)+s, 1+a_0-(a)+s, 1-(a)-s, (a)-a_0-s \\ 1+a_0-(b)+s, 1-(b)-s \end{bmatrix} e^{2\pi i s}\, ds,$$

(6.3.1.3)

where there are $N+2$ of the $a$ parameters, including $a_0$, and $2N$ of the $b$ parameters, we deduce

$$\Gamma\begin{bmatrix} (a), 1+a_0-(a), 1-(a), (a)-a_0 \\ 1+a_0-(b), 1-(b) \end{bmatrix} {}_{2N}H_{2N}\begin{bmatrix} (b); \\ 1+a_0-(b); \end{bmatrix}$$

$$= \sum_{\nu=1}^{N+1} \Gamma\begin{bmatrix} 1-a_\nu, 1+a_0-a_\nu, 1+a_0-a_\nu-(a), (a)'-a_\nu, a_\nu, a_\nu-a_0, \\ 1+a_\nu-(b), \end{bmatrix}$$

$$\begin{matrix} a_\nu+(a)-a_0, 1+a_\nu-(a) \\ 1+a_0-a_\nu-(b) \end{matrix} \Bigg]$$

$$\times {}_{2N}H_{2N}\begin{bmatrix} a_\nu-a_0+(b); \\ 1+a_\nu-(b); \end{bmatrix}, \quad (6.3.1.4)$$

provided that

$$\mathrm{Rl}\,(N + Na_0 - 1 - b_1 - b_2 - \ldots - b_{2N}) > 0.$$

This expresses a ${}_{2N}H_{2N}(1)$ series well-poised in $a_0$, in terms of $N+1$ other ${}_{2N}H_{2N}(1)$ series, each well-poised in $2a_\nu - a_0$.

From the two integrals

$$I_3 = \int_{-i\infty}^{i\infty} \Gamma\begin{bmatrix} (a)+s, 1+a_0-(a)+s, 1-(a)-s, (a)-a_0-s \\ 1+a_0-(b)+s, 1-(b)-s \end{bmatrix} e^{\pi i s}\, ds$$

(6.3.1.5)

and

$$I_4 = \int_{-i\infty}^{i\infty} \Gamma\begin{bmatrix} (a)+s, 1+a_0-(a)+s, 1-(a)-s, (a)-a_0-s \\ 1+a_0-(b)+s, 1-(b)-s \end{bmatrix} e^{2\pi i s}\, ds,$$

(6.3.1.6)

where there are $N+1$ of the $a$ parameters, including $a_0$, and $2N-1$ of the $b$ parameters in both integrals, we can deduce the two results

$$\Gamma\begin{bmatrix} a_0, 1+a_0-(a), 1-a_0, 1-(a), (a)-a_0 \\ 1+a_0-(b), 1-(b), \tfrac{1}{2}a_0, 1-\tfrac{1}{2}a_0, \tfrac{1}{4}(1-a_0)\mp\tfrac{1}{4}(1-a_0), \\ \tfrac{1}{4}(1+a_0)\mp\tfrac{1}{4}(1+a_0) \end{bmatrix}$$

$$\times {}_{2N-1}H_{2N-1}\begin{bmatrix} (b); \\ 1+a_0-(b); \end{bmatrix} \pm 1$$

$$= \sum_{\nu=1}^{N} \Gamma\begin{bmatrix} a_0-a_\nu, 1-a_\nu, 1+a_0-a_\nu-(a), (a)'-a_\nu, a_\nu, 1+a_\nu-a_0, \\ 1+a_\nu-(b), 1+a_0-a_\nu-(b), \tfrac{1}{2}a_0-a_\nu, 1+a_\nu-\tfrac{1}{2}a_0, \end{bmatrix}$$

$$\begin{matrix} a_\nu+(a)-a_0, 1+a_\nu-(a) \\ \tfrac{1}{2}a_\nu+\tfrac{1}{4}(1-a_0)\mp\{\tfrac{1}{2}a_\nu+\tfrac{1}{4}(1-a_0)\}, -\tfrac{1}{2}a_\nu+\tfrac{1}{4}(1-a_0) \\ \mp\{\tfrac{1}{4}(1-a_0)-\tfrac{1}{2}a_\nu\} \end{matrix} \Bigg]$$

$$\times {}_{2N-1}H_{2N-1}\begin{bmatrix} a_\nu+(b)-a_0; \\ 1+a_\nu-(b); \end{bmatrix} \pm 1 .$$

(6.3.1.7, 8)

Here either all the upper signs are to be taken throughout, in which case we must have

$$\mathrm{Rl}\{2N-4+(2N-1)a_0-2(b_1+b_2+\ \ldots\ +b_{2N-1})\} > 0,$$

or all the lower signs have to be taken throughout, in which case we must have

$$\mathrm{Rl}\{2N-3+(2N-1)a_0-2(b_1+b_2+\ \ldots\ +b_{2N-1})\} > 0.$$

These last two results express a $_{2N-1}H_{2N-1}(\pm 1)$ series, well-poised in $a_0$ in terms of $N$ other $_{2N-1}H_{2N-1}(\pm 1)$ series, each well-poised in $2a_\nu - a_0$.

It should be noted that the results (6.3.1.2), (6.3.1.7) and (6.3.1.8) can all be deduced directly from (6.3.1.4). Thus, (6.3.1.8) follows by taking $b_{2N} = \frac{1}{2} + \frac{1}{2}a_0$, in (6.3.1.2), (6.3.1.2) follows from (6.3.1.7) by putting $N+1$ for $N$ in (6.3.1.7) and then letting $b_{2N+1} \to -\infty$ and $a_{N+1} \to \infty$, and finally (6.3.1.7) follows from (6.3.1.4) by taking $a_{N+1} = b_{2N} = \frac{1}{2} + \frac{1}{2}a_0$ in (6.3.1.4).

In particular, when $N = 2$, in (6.3.1.4), and $a_1 = b_1$, $a_2 = b_2$, the two series on the right reduce to two well-poised $_3F_2(1)$ series, which can be summed by Dixon's theorem (2.3.3.5) to give the sum of a well-poised $_3H_3(1)$ series (6.1.2.6).

When $N = 3$, in (6.3.1.4), and $a_1 = b_1$, $a_2 = b_2$, $a_3 = b_3 = 1 + \frac{1}{2}a_0$, one of the $_5H_5(1)$ series on the right vanishes, and the other two reduce to well-poised $_5F_4(1)$ series, which can be summed by (2.3.4.5) to give, on reduction, the sum of a well-poised $_5H_5(1)$ series (6.1.2.5). Similarly, when $N = 4$, we have a relation between five well-poised $_7H_7(1)$ series, and the special case of this result is a relation expressing a well-poised $_7H_7(1)$ series with the special form of the first parameters, in terms of three well-poised $_7F_6(1)$ series, all with the special forms of the second parameters.

# 7

## BASIC BILATERAL SERIES

### 7.1 Introduction

The next step is to generalize the bilateral series by the introduction of the base $q$ to produce basic bilateral series, just as we did in Chapter 3, in order to produce basic hypergeometric series.

The first thing to do is to extend the definition of $(a; q)_n$ to include negative integers. We shall write

$$(a; q)_{-n} = \frac{1}{(1-a/q)(1-a/q^2)\dots(1-a/q^n)}$$

$$= \frac{(-1)^n q^{\frac{1}{2}n(n+1)}}{(q/a; q)_n a^n}, \tag{7.1.1}$$

$$= \Pi \begin{bmatrix} aq^{-n-1}; \\ aq^{-1} \end{bmatrix}. \tag{7.1.2}$$

Then the general basic bilateral series is written as

$$_A\Psi_B \begin{bmatrix} a_1, a_2, \dots, a_A; \\ b_1, b_2, \dots, b_B; \end{bmatrix} q, z \end{bmatrix} \equiv \sum_{n=-\infty}^{\infty} \frac{(a_1; q)_n (a_2; q)_n \dots (a_A; q)_n}{(b_1; q)_n (b_2; q)_n \dots (b_B; q)_n} z^n. \tag{7.1.3}$$

This series is convergent for $|q| < 1$, for all values, real or complex, of the parameters

$$a_1, a_2, \dots, a_A, \quad b_1, b_2, \dots, b_B,$$

and for $|z| \leqslant 1$.

When there is no danger of confusion, this notation can be contracted further, as usual, to

$$_A\Psi_B[(a); (b); q, z] = \sum_{n=-\infty}^{\infty} \frac{((a); q)_n}{((b); q)_n} z^n, \tag{7.1.4}$$

where there are always $A$ of the $a$ parameters and $B$ of the $b$ parameters.

In some ways, these basic bilateral series are of more fundamental importance than their ordinary bilateral counterparts, as these basic series contain, as special cases, many interesting indentities connected with theta functions and Ramanujan identities.

Let us use (7.1.1) to transform $(a; q)_{-n}$, then we can reverse the order of any basic bilateral series, thus

$$_A\Psi_A \begin{bmatrix} (a); \\ (b); \end{bmatrix} q, z \end{bmatrix} = {}_A\Psi_A \begin{bmatrix} (q/b); \\ (q/a); \end{bmatrix} q, \frac{b_1 b_2 \dots b_A}{a_1 a_2 \dots a_A z} \end{bmatrix}. \tag{7.1.5}$$

### 7.1.1 The $_6\Psi_6$ summation theorem.

Let us seek a relation connecting three well-poised $_8\Phi_7(q)$ series (§ 3.4.2) and in it, let us take $c = q/a$. Then the second series on the right reduces to a well-poised summable $_6\Phi_5(q)$ series, and the two remaining series combine together to form a well-poised $_6\Psi_6(q)$ series.

After some reduction, we find the $_6\Psi_6$ summation theorem

$$_6\Psi_6 \left[ \begin{matrix} q\sqrt{a}, -q\sqrt{a}, & b, & c, & d, & e; \\ \sqrt{a}, & -\sqrt{a}, aq/b, aq/c, aq/d, aq/e; \end{matrix} \; q, a^2q/bcde \right]$$

$$= \Pi \left[ \begin{matrix} aq, aq/bc, aq/bd, aq/cd, aq/ce, aq/de, q, q/a; \\ q/b, q/c, q/d, q/e, aq/c, aq/d, aq/e, a^2q/bcde; \end{matrix} \; q \right]. \quad (7.1.1.1)$$

This is the basic analogue of (6.1.2.5), the $_5H_5(1)$ summation theorem, and the extension to bilateral series of the result (3.3.1.3), the $_6\Phi_5$ summation theorem.

There are many interesting special cases of this result. In particular,[†] if $b = c = d = e = a^{\frac{1}{2}}$, and $a^2$ is put for $a$, we have

$$\sum_{n=-\infty}^{\infty} \frac{(1+aq^{2n})q^n}{(1-aq^n)^3} = \frac{1}{1-a} \Pi \left[ \begin{matrix} a^2, q, q, q, q, q, q/a^2; \\ q/a, q/a, q/a, q/a, a, a, a; \end{matrix} \; q \right], \quad (7.1.1.2)$$

Two direct proofs of this fundamental result (7.1.1.1) are given in Slater & Lakin (1956). The first of these uses operators, and the second is a direct proof using contour integration, based on the integral

$$\frac{1}{2\pi i} \int_{-i\pi/t}^{i\pi/t} P_N(s) \, ds, \quad (7.1.1.3)$$

where

$$P_N(s) = \Pi \left[ \begin{matrix} q^{1+s}/d, q^{1-s}/d, q^{1+s}/e, q^{1-s}/e, q^{1+s}/f, q^{1-s}/f; \\ aq^s, aq^{-s}, bq^s, bq^{-s}, cq^s, cq^{-s}; \end{matrix} \; q \right] q^s,$$

and, as usual, $q = e^{-t}, t > 0$. The integral is taken round the rectangular contour of Fig. 5.1,

$$F(-2N - i\pi/t), \quad E(-2N + i\pi/t), \quad C(2N + i\pi/t), \quad D(2N - i\pi/t).$$

It is assumed that none of the members of the sequences of poles of the integrand, $\quad aq^{\pm n}, bq^{\pm n}, cq^{\pm n} \quad (n = 0, 1, 2, ...)$

coincide or fall on the contour, and that the contour is indented, if necessary, to avoid this event.

Then, by the periodicity of the integrand, we have

$$\int_{EC} + \int_{DF} = 0.$$

† Watson (1933b), eq. 4.7.

Also
$$\int_{CD} = \frac{1}{2\pi i} \int_0^{\pi/t} \{P_N(2N - ir) - P_N(2N + ir)\}\, dr$$

and
$$\int_{FE} = \frac{1}{2\pi i} \int_0^{\pi/t} \{P_N(-2N - ir) - P_N(-2N + ir)\}\, dr.$$

Both these integrals tend to zero as $N \to \infty$, provided that

$$\mathrm{Rl}\,\{q^s/(abcde)\} > 0.$$

Thus, we can equate to zero the sum of the residues at the poles of $P_N(s)$ in the $s$-plane. Now
$$1/(aq^s; q)$$
has poles within $FECD$ at

$$s = \log a - n + 2\pi i k/t$$

for some integer $k$. Hence $P_N(s)$ has increasing sequences of poles at

$$s = \log a + n, \quad \log b + n, \quad \log c + n$$

and decreasing sequences of poles at

$$s = -\log a - n, \quad -\log b - n, \quad -\log c - n,$$

for $n = 0, 1, 2, \ldots$.

If we combine the residues at $s = \log a + n$ and $s = -\log a - n$ and make use of the symmetry of the integrand, we find that

$$\sum_{n=0}^{\infty} \Pi \begin{bmatrix} aq^{1+n}/d, q^{1-n}/ad, q^{1+n}a/e, q^{1-n}/ae, q^{1+n}a/f, q^{1-n}/af; \\ a^2 q^n, q, abq^n, bq^{-n}/a, acq^n, cq^{-n}/a; \end{bmatrix} q$$

$$\times \frac{(aq^n - q^{-n}/a)}{(q^{-n}; q)_n} + \mathrm{idem}\,(a; b, c) = 0. \quad (7.1.1.4)$$

Here 'idem $(a; b, c)$' means that all the preceding expression has to be repeated with $b$ and $a$ interchanged, and then with $c$ and $a$ interchanged. This notation, due to Sears, is sometimes useful as a contracted notation for cyclic expressions, instead of making use of summation signs and suffixes.

If we rewrite these series in the usual notation, we find that we have

$$\frac{1}{a} \Pi \begin{bmatrix} aq/d, q/ad, aq/e, q/ae, aq/f, q/af; \\ a^2 q, ab, b/a, ac, c/a; \end{bmatrix} q$$

$$\times {}_8\Phi_7 \begin{bmatrix} a^2, aq, -aq, ab, ac, ad, ae, af; \\ a, -a, aq/b, aq/c, aq/d, aq/e, aq/f; \end{bmatrix} q, q^2/(abcdef)$$

$$+ \mathrm{idem}\,(a; b, c) = 0. \quad (7.1.1.5)$$

This relation between three $_8\Phi_7(q)$ series is the basic analogue of (4.3.7.8), the result due to Whipple. If now, we put $c = q/a$, the first and third series combine to give

$$\frac{1}{a}\,\Pi\begin{bmatrix} aq/d, aq/e, aq/f, q/ad, q/ae, q/af; \\ a^2q, ab, q, b/a, q/a^2; \end{bmatrix}\, q$$

$$\times\, _6\Psi_6\begin{bmatrix} aq, -aq, ab, ad, ae, af; \\ a, -a, aq/b, aq/d, aq/e, aq/f; \end{bmatrix}\, q,\, q/(bdef)$$

and the second series reduces to

$$_6\Phi_5\begin{bmatrix} b^2, bq, -bq, bd, be, bf; \\ b, -b, bq/d, bq/e, bq/f; \end{bmatrix}\, q, q/(bdef)$$

$$= \Pi\begin{bmatrix} b^2q, q/de, q/ef, q/df; \\ bq/d, bq/e, bq/f, q/bdef; \end{bmatrix}\, q,$$

when it is summed by the theorem (3.3.1.3). Then, after a little further reduction, we have the required result (7.1.1.1).

## 7.2 General transformations

In the general theorem on basic series (5.2.4), let us write

$$d_1 = a_1, \quad d_2 = a_2, \ldots, \quad d_D = a_A, \quad b_1 = q,$$

and let $A = B - 1 = C = D$. Then all the series of the general type

$$_{A+B+1}\Phi_{A+B}(q, z)$$

reduce to series of the type

$$_{A+1}\Phi_A(q, z)$$

and we can rewrite (5.2.4) in the form

$$P\Phi = \sum_{i=1}^{A} P(a_i)\,\Phi(a_i) + P(q)\,\Phi(q) - \sum_{i=1}^{A} P(b_i)\,\Phi(b_i), \qquad (7.2.1)$$

where $P, P(a_i), P(b_i), P(q)$, are the products preceding the series $\Phi$, $\Phi(a_i)$, $\Phi(b_i)$, $\Phi(q)$ respectively. Now, if we write $-n-1$ for $n$ in the series $\Phi(q)$, since

$$\sum_{n=0}^{\infty} f(n) = \sum_{n=-\infty}^{-1} f(-n-1), \qquad (7.2.2)$$

we can reverse the $\Phi(q)$ series as in (7.1.5) above. If we denote the reversed series by $\Phi'(q)$, we find that

$$\Phi'(q) = K\Phi(q), \qquad (7.2.3)$$

where $K$ is some constant independent of $n$. Also

$$P(q) = -P/K,$$

so that $\qquad P\Phi - P(q)\,\Phi(q) = P\{\Phi + \Phi'(q)\}, \qquad (7.2.4)$

and the series $\Phi$ and $\Phi'(q)$ can be combined together to form a basic bilateral series $\Psi'(q)$. In a similar way, we find that each of the series $P(a_i)\,\Phi(a_i)$ will combine with one of the series $P(b_i)\,\Phi(b_i)$ to form a basic bilateral series $\Psi'(a_i)$.

If we write

$$c_1 \text{ for } b_2, \quad c_2 \text{ for } b_3, \quad \ldots, \quad c_A \text{ for } b_{A+1},$$

we have the general theorem

$$\Pi\begin{bmatrix} z\alpha, q/z\alpha, (b), (q/c); \\ (a), (q/a); \end{bmatrix} q \end{bmatrix} {}_A\Psi_A\begin{bmatrix} (c); \\ (b); \end{bmatrix} q, z \end{bmatrix}$$

$$= \sum_{i=1}^{A} \frac{q}{a_i} \Pi\begin{bmatrix} a_i z\alpha/q, q^2/a_i z\alpha, a_i/(c), q(b)/a_i; \\ a_i, q/a_i, a_i/(a), q(a)/a_i; \end{bmatrix} q \end{bmatrix} {}_A\Psi_A\begin{bmatrix} q(c)/a_i; \\ q(b)/a_i; \end{bmatrix} q, z \end{bmatrix},$$

$$(7.2.5)$$

where $\qquad \alpha \equiv c_1 c_2 \ldots c_A/a_1 a_2 \ldots a_A, \quad |z| < 1 \quad \text{and} \quad |q| < 1.$

This expresses a general ${}_A\Psi_A(z)$ series in terms of $A$ other series of the same type.

If we let $q \to 1$, we obtain the ordinary bilateral theorem (6.3.3) again.

If we put $b_A = q$, the $\Psi'(z)$ series on the left of (7.2.5) becomes an ${}_A\Phi_{A-1}(z)$ series, and if we put

$$a_1 = b_1, \quad a_2 = b_2, \quad \ldots, \quad a_Q = b_Q, \quad Q < A,$$

the first $Q$ of the $\Psi'$ series on the right of (7.2.5) become ${}_A\Phi_{A-1}(z)$ series. Similarly, if we reverse the last $A - R$ bilateral series and put

$$qc_{R+1} = a_{R+1}, \quad qc_{R+2} = a_{R+2}, \quad \ldots, \quad qc_A = a_A,$$

these $R$ series will also become ${}_A\Phi_{A-1}(z)$ series, where $R < A$.

If we carry out all these processes, and put $R = Q$ we find that (7.2.5) reduces to a re-statement of (5.2.4). Thus, although we have deduced (7.2.5) from (5.2.4), it is also possible to deduce (5.2.4) from (7.2.5), and neither should be considered as a special case of the other result.

In particular, if $A = 2$, $d = q$, and $z = c/ab$, in (7.2.5), the first series is summable by Gauss's analogue (3.3.2.5) and we have

$$\Pi\begin{bmatrix} c/ef, qef/c, q, q/a, q/b, c/a, c/b; \\ e, f, q/e, q/f, c/ab; \end{bmatrix} q$$

$$= \frac{q}{e} \Pi\begin{bmatrix} e/ab, q^2f/e, e/a, e/b, qc/e, q^2/e; \\ e, q/e, e/f, qf/e; \end{bmatrix} q \, {}_2\Psi_2\begin{bmatrix} e/c, e/q; \\ e/a, e/b; \end{bmatrix} q, q \Big] + \text{idem } (e; f).$$

$$(7.2.6)$$

This is a relation between two ${}_2\Psi_2(q)$ series.

If $A = 3$, and $a_1 = d$, $a_2 = e$, $a_3 = f$, in (7.2.5), in a similar way, we can obtain a relation which expresses a general ${}_3\Psi_3(z)$ series in terms of three ${}_3\Phi_2(z)$ series.

### 7.2.1 Well-poised bilateral transforms.

We shall now deduce four transforms of well-poised bilateral series, from the four general results for well-poised basic series.

In the result (5.3.2) which expressed a well-poised basic ${}_{2M}\Phi_{2M-1}(q)$ series in terms of $M$ other similar series, let us suppose that $M$ is odd, so that

$$N = 2M+1, \quad a_1 = a, \quad a_2 = q, \quad a_3 = qa_1/a_4,$$
$$a_5 = qa_1/a_6, \quad \dots, \quad a_N = qa_1/a_{M+1}.$$

Then, if we write

$$a_1 \text{ for } a_4, \quad a_2 \text{ for } a_6, \quad \dots, \quad a_N \text{ for } a_{M+1},$$

we find that there are now $4M + 2$ parameters in all. We can combine the series in pairs as before, and we have after some reduction,

$$\Pi\begin{bmatrix} aq/(b), q/(b), \sqrt{a}, -\sqrt{a}, q/\sqrt{a}, -q/\sqrt{a}; \\ a, (a), aq/(a), q/a, q/(a), (a)/a; \end{bmatrix} q$$

$$\times {}_{2N}\Psi_{2N}\left[(b); qa/(b); q, \frac{-a^N q^N}{b_1 b_2 \dots b_{2N}}\right]$$

$$= \sum_{\nu=1}^{N} a_\nu \, \Pi\begin{bmatrix} a_\nu q/(b), aq/a_\nu(b), \sqrt{a}/a_\nu, -\sqrt{a}/a_\nu, qa_\nu/\sqrt{a}, -qa_\nu/\sqrt{a}; \\ a/a_\nu, q/a_\nu, a_\nu, (a)/a_\nu, (a) a_\nu/a, qa/(a) a_\nu, qa_\nu/a, qa_\nu/(a); \end{bmatrix} q$$

$$\times {}_{2N}\Psi_{2N}\left[\begin{matrix} a_\nu(b)/a; \\ qa_\nu/(b); \end{matrix} q, \frac{-a^N q^N}{b_1 b_2 \dots b_{2N}}\right], \qquad (7.2.1.1)$$

where there are $N$ of the $a$ parameters and $2N$ of the $b$ parameters. This expresses a well-poised ${}_{2N}\Psi_{2N}(q)$ series in terms of $N$ other well-poised ${}_{2N}\Psi_{2N}(q)$ series. Here the element $a$ can be thought of as the pivot element. If, in this result, we write

$$a_1 = b_{N+1}, \quad a_2 = b_{N+2}, \quad \dots, \quad a_Q = b_{N+Q}, \quad Q < N,$$

we get a result which expresses a well-poised $_{2N}\Psi_{2N}(q)$ series in terms of $Q$ ordinary $_{2N}\Phi_{2N-1}(q)$ series and $N-Q$ bilateral $_{2N}\Psi_{2N}(q)$ series.

If, further, we let $N = Q$, and take $b_N = a$, we find that (7.2.1.1) reduces again to (5.3.2). Thus, once again, we can say that (7.2.1.1) can be deduced from (5.3.2) and also that (5.3.2) can be deduced from (7.2.1.1), so that the two results are complementary.

If we suppose that $M$ is odd in (5.3.2), we only obtain the same result as if we had supposed that $M$ was even and then put $a_N = b_{2N}$. When

$$a_1 = b_1 = q\sqrt{a} \quad \text{and} \quad a_2 = b_2 = -q\sqrt{a},$$

two of the series on the right of (7.2.1.1) vanish, and then the theorem becomes one which expresses a well-poised $_{2N}\Psi_{2N}(q)$ series in terms of $N-2$ well-poised $_{2N}\Psi_{2N}(q)$ series, all with the special forms of the first and second parameters.

In particular, if $N = 3$, we find† that

$$_6\Psi_6\left[\begin{array}{c} q\sqrt{a}, \; -q\sqrt{a}, \quad b, \quad c, \quad d, \quad e; \\ \sqrt{a}, \quad -\sqrt{a}, aq/b, aq/c, aq/d, aq/e; \end{array} q, a^2q/bcde\right]$$

$$= \Pi\left[\begin{array}{c} aq, q/a, qf/b, qf/c, qf/d, qf/e, aq/bf, aq/cf, aq/df, aq/ef; \\ aq/f^2, qf^2/a, aq/b, aq/c, aq/d, aq/e, q/b, q/c, q/d, q/e; \end{array} q\right]$$

$$\times {}_6\Psi_6\left[\begin{array}{c} qf/\sqrt{a}, \; -qf/\sqrt{a}, bf/a, cf/a, df/a, ef/a; \\ f/\sqrt{a}, -f/\sqrt{a}, af/b, af/c, af/d, af/e; \end{array} q, \frac{qa^2}{bcde}\right]. \quad (7.2.1.2)$$

In a similar way, from Sears's other three basic well-poised theorems (5.3.4, 6 and 8), we can obtain three basic bilateral well-poised theorems. Thus, from (5.3.4) we have

$$\Pi\left[\begin{array}{c} aq/(b), q/(b); \\ q, (a), aq/(a), q/(a), (a)/a; \end{array} q\right] {}_{2N}\Psi_{2N}\left[\begin{array}{c} (b); \\ aq/(b); \end{array} q, \frac{q^N a^N}{b_1 b_2 \ldots b_{2N}}\right]$$

$$= \sum_{\nu=1}^{N+1} \Pi\left[\begin{array}{c} a_\nu q/(b), aq/a_\nu(b); \\ (a)/a_\nu, aq/a_\nu(a), a_\nu(a)/a, qa_\nu/(a), a_\nu, qa/a_\nu, a_\nu/a; \end{array} q\right]$$

$$\times {}_{2N}\Psi_{2N}\left[\begin{array}{c} a_\nu(b)/a; \\ qa_\nu/(b); \end{array} q, \frac{q^N a^N}{b_1 b_2 \ldots b_{2N}}\right], \quad (7.2.1.3)$$

where there are $N+1$ of the $a$ parameters and $2N$ of the $b$ parameters. This expresses a well-poised $_{2N}\Psi_{2N}(q)$ series in terms of $N+1$ other series of the same type.

<hr>

† M. Jackson (1950a), §2.1.

From (5.3.6 and 8) we deduce, in the same way,

$$\Pi\left[\begin{matrix} qa/(b), q/(b), \pm \sqrt{(qa)}, \pm \sqrt{(q/a)}, \sqrt{a}, -\sqrt{a}, q/\sqrt{a}, -q/\sqrt{a}; \\ a, q, (a), q/a, q/(a), (a)/a, aq/(a); \end{matrix} \quad q\right]$$

$$\times\,{}_{2N-1}\Psi_{2N-1}\left[\begin{matrix} (b); \\ aq/(b); \end{matrix} \quad q, \frac{\mp q^{N-\frac{1}{2}}a^{N-\frac{1}{2}}}{b_1 b_2 \ldots b_{2N-1}}\right]$$

$$= \sum_{\nu=1}^{N} a_\nu \,\Pi\left[\begin{matrix} qa_\nu/(b), qa/a_\nu(b), \pm a_\nu\sqrt{(q/a)}, \pm\sqrt{(qa)}/a_\nu, \sqrt{a}/a_\nu, \\ qa/a_\nu(a), (a)/a_\nu, a/a_\nu, q/a_\nu, a_\nu(a)/a, qa_\nu/a, a_\nu, q, \\ \qquad\qquad -\sqrt{a}/a_\nu, qa_\nu/\sqrt{a}, -qa_\nu/\sqrt{a}; \\ \qquad\qquad qa_\nu/(a); \end{matrix} \quad q\right]$$

$$\times\,{}_{2N-1}\Psi_{2N-1}\left[\begin{matrix} a_\nu(b)/a; \\ a_\nu q/(b); \end{matrix} \quad q, \frac{\mp q^{N-\frac{1}{2}}a^{N-\frac{1}{2}}}{b_1 b_2 \ldots b_{2N-1}}\right]. \qquad (7.2.1.4)$$

Here either all the upper or all the lower signs must be taken throughout, and there are $N$ of the $a$ parameters and $2N-1$ of the $b$ parameters. These results express a well-poised ${}_{2N-1}\Psi_{2N-1}(q)$ series in terms of $N$ other similar series. Again, we can reduce any of the ${}_{2N-1}\Psi_{2N-1}(q)$ series to ${}_{2N-1}\Phi_{2N-2}(q)$ series, and so obtain re-statements of (5.3.4, 6 and 8).

## 7.3 The theta functions

Some of the best known of the functions which can be classed as basic bilateral series, are the $\vartheta$ (theta) functions, of elliptic function theory. There are four of these functions, and the first one is usually defined as

$$\vartheta_1(z) = -i \sum_{n=-\infty}^{\infty} (-1)^n \exp\{\pi i \tau(n+\tfrac{1}{2})^2 + 2(n+1)\,iz\}, \qquad (7.3.1)$$

where $\operatorname{Im}\tau > 0$. Let us write

$$q = e^{\pi i \tau} \quad \text{and} \quad a = e^{iz},$$

then
$$\vartheta_1(z) = -i \sum_{n=-\infty}^{\infty} (-1)^n q^{(n+\frac{1}{2})^2} a^{2n+2}.$$

Now, if $|z| < A$, any positive constant, we have

$$|q^{(n+\frac{1}{2})^2} a^{\pm 2n}| \leqslant |q|^{(n+\frac{1}{2})^2} e^{2nA}.$$

But, by D'Alambert's test, this is a term of a convergent series, so that

$$\sum_{n=1}^{\infty} q^{(n+\frac{1}{2})^2} a^{\pm 2n}$$

are both absolutely convergent series, that is, the bilateral series of $\vartheta_1(z)$ is absolutely convergent, for all $|z|$ and so $\vartheta_1(z)$ is an integral function.

The other $\vartheta$ functions are defined as

$$\vartheta_2(z) = \sum_{n=-\infty}^{\infty} q^{(n+\frac{1}{2})^2} a^{2n+1}, \tag{7.3.2}$$

$$\vartheta_3(z) = \sum_{n=-\infty}^{\infty} q^{n^2} a^{2n} \tag{7.3.3}$$

and $$\vartheta_4(z) = \sum_{n=-\infty}^{\infty} (-1)^n q^{n^2} a^{2n}. \tag{7.3.4}$$

The absolute convergence of these three series can be established in a similar way.

We can see immediately, from these definitions, that

$$\vartheta_2(z) = \vartheta_1(z + \tfrac{1}{2}\pi) \tag{7.3.5}$$

and that $$\vartheta_3(z) = \vartheta_4(z + \tfrac{1}{2}\pi). \tag{7.3.6}$$

If we select the terms in $n$ and $-n$ and combine them, we can express all four functions as trigonometric series, thus

$$\vartheta_1(z) = 2 \sum_{n=1}^{\infty} (-1)^n q^{(n+\frac{1}{2})^2} \sin\{(n+1)z\}, \tag{7.3.7}$$

$$\vartheta_2(z) = 2 \sum_{n=1}^{\infty} q^{(n+\frac{1}{2})^2} \cos\{(n+1)z\}, \tag{7.3.8}$$

$$\vartheta_3(z) = 1 + 2 \sum_{n=1}^{\infty} q^{n^2} \cos(2nz) \tag{7.3.9}$$

and $$\vartheta_4(z) = 1 + 2 \sum_{n=1}^{\infty} (-1)^n q^{n^2} \cos(2nz). \tag{7.3.10}$$

Thus, $\vartheta_1(z)$ is an odd function of $z$ and the other theta functions are even functions of $z$.

When $z = 0$, then $a = 1$, and

$$\vartheta_1(0) = 0, \tag{7.3.11}$$

$$\vartheta_2(0) = \sum_{n=-\infty}^{\infty} q^{(n+\frac{1}{2})^2}, \tag{7.3.12}$$

$$\vartheta_3(0) = \sum_{n=-\infty}^{\infty} q^{n^2}, \tag{7.3.13}$$

$$\vartheta_4(0) = \sum_{n=-\infty}^{\infty} (-1)^n q^{n^2}. \tag{7.3.14}$$

We can express all four $\vartheta$ functions in terms of infinite products. The usual method of proof shows that if

$$f(z) = \prod_{n=1}^{\infty} \{1 - 2q^{2n-1}\cos(2z) + q^{4n-2}\}$$

then

$$\vartheta_4(z)/f(z)$$

is a doubly periodic function with periods $\pi$ and $\pi\tau$, with no poles. Hence it is a constant, independent of $z$. This constant can then be shown to be

$$\prod_{n=1}^{\infty}(1 - q^{2n}),$$

so that we have finally.

$$\vartheta_4(z) = \prod_{n=1}^{\infty}\{(1 - q^{2n})(1 - 2q^{2n-1}\cos(2z) + q^{4n-2})\}. \qquad (7.3.15)$$

A more direct proof follows immediately from Jacobi's formula (3.1.12).

We have finally

$$\vartheta_1(z) = -iaq^{\frac{1}{4}}\prod_{n=1}^{\infty}\{(1 - aq^{2n})(1 - q^{2n})(1 - q^{2n-2}/a)\}, \qquad (7.3.16)$$

$$\vartheta_2(z) = aq^{\frac{1}{4}}\prod_{n=1}^{\infty}\{(1 - iaq^{2n})(1 - q^{2n})(1 + iq^{2n-2}/a)\}, \qquad (7.3.17)$$

$$\vartheta_3(z) = \prod_{n=1}^{\infty}\{(1 + aq^{2n-1})(1 - q^{2n})(1 + q^{2n-1}/a)\} \qquad (7.3.18)$$

and

$$\vartheta_4(z) = \prod_{n=1}^{\infty}\{(1 - aq^{2n-1})(1 - q^{2n})(1 - q^{2n-1}/a)\}. \qquad (7.3.19)$$

Many further relations can be found in standard works on elliptic functions and, in particular, in a series of papers by Rogers (1894).

### 7.3.1 Further identities of the Rogers–Ramanujan type.
Most of the known identities of this type can be deduced from the summation theorems for basic bilateral series, and we shall now consider one such process of deduction in greater detail.

In the $_6\Psi_6$ summation theorem (7.1.1.1), let us write $b = q^{-n/3}$, $c = q^{(1-n)/3}$ and $d = q^{(2-n)/3}$, and put $q^2$ for $q$. Then we find that

$$\sum_{r=-N}^{[n/3]} \frac{(1 - aq^{6r})(q^{-n}; q)_{3r} a^{2r}q^{3nr}(e; q^3)_r}{(1 - a)(aq^{1+n}; q)_{3r} e^r(q^3 a/e; q^3)_r}$$

$$= \Pi\begin{bmatrix} a, q^3/a, q^2 a/e, qa/e; \\ q, q^2, q^3/e, a^2/e; \end{bmatrix} q^3 \left] \frac{(q; q)_n (aq; q)_n (a^2/e; q^3)_n}{(a; q)_{2n} (aq/e; q)_n}, \qquad (7.3.1.1)$$

where $a$ is some power of $q$, so that the series terminates below. For example, let $a = q$, then (7.3.1.1) becomes

$$\sum_{r=-[n/3]}^{[n/3]} \frac{(1-q^{6r+1})(-1)^r (e; q^3)_r q^{\frac{1}{2}r(3r+1)}}{(q; q)_{n+3r+1}(q; q)_{n-3r}(q^4/e; q^3)_r e^r} = \frac{(q^2/e; q^3)_n}{(q; q)_{2n}(q^2/e; q)_n}. \quad (7.3.1.2)$$

Now let $e \to \infty$, and we get

$$\sum_{r=-[n/3]}^{[n/3]} \frac{(1-q^{6r+1})q^{r(6r-1)}}{(q; q)_{n+3r+1}(q; q)_{n-3r}} = \frac{1}{(q; q)_{2n}}. \quad (7.3.1.3)$$

But

$$q^{r(6r-1)}(1-q^{6r+1}) = q^{r(6r-1)}(1-q^{n+3r+1}) - q^{(2r+1)(3r+1)}(1-q^{n-3r}), \quad (7.3.1.4)$$

so that

$$\sum_{r=-[n/3]}^{[n/3]} \frac{q^{r(6r-1)}}{(q; q)_{n+3r}(q; q)_{n-3r}} - \sum_{r=-[(n+1)/3]}^{[(n-1)/3]} \frac{q^{(2r+1)(3r+1)}}{(q; q)_{n+3r+1}(q; q)_{n-3r-1}} = \frac{1}{(q; q)_{2n}},$$
$$(7.3.1.5)$$

that is

$$\frac{1-q}{(q; q)_n (q; q)_{n+1}} + \sum_{r=1}^{[n/3]} \frac{q^{r(6r-1)}+q^{r(6r+1)}}{(q; q)_{n-3r}(q; q)_{n+3r}} - \sum_{r=1}^{[(n+1)/3]} \frac{q^{(2r+1)(3r+1)}}{(q; q)_{n+3r+1}(q; q)_{n-3r-1}}$$

$$- \sum_{r=1}^{[(n-1)/3]} \frac{q^{(2r-1)(3r-1)}}{(q; q)_{n-3r+1}(q; q)_{n+3r-1}} = \frac{1}{(q; q)_{2n}}. \quad (7.3.1.6)$$

If, in (3.4.9), which is Bailey's transform with $\gamma_n$ summed by the basic Gauss theorem, we put

$$\beta_n = \frac{1}{(q; q)_{2n}},$$

and

$$\left. \begin{aligned} \alpha_{3n-1} &= -q^{(2n-1)(3n-1)}, \\ \alpha_{3n} &= q^{n(6n-1)}+q^{n(6n+1)}, \\ \alpha_{3n+1} &= -q^{(3n+1)(2n+1)}, \end{aligned} \right\} \quad (7.3.1.7)$$

and make use of the above identity (7.3.1.6), we can deduce the result

$$\sum_{n=0}^{\infty} \frac{(y; q)_n (z; q)_n x^n}{(q; q)_{2n} y^n z^n} = \Pi \begin{bmatrix} x/y, x/z; \\ x, x/yz; \end{bmatrix} \sum_{n=0}^{\infty} \frac{(y; q)_n (z; q)_n x^n \alpha_n}{(x/y; q)_n (x/z; q)_n y^n z^n}.$$

$$(7.3.1.8)$$

In particular, when $x = q$, $y = q^{\frac{1}{2}}/u$ and $z = q^{\frac{1}{2}}/v$, we have

$$\sum_{n=0}^{\infty} \frac{(q^{\frac{1}{2}}/u; q)_n (q^{\frac{1}{2}}/v; q)_n}{(q; q)_{2n}} u^n v^{3n} = \Pi\begin{bmatrix} uq^{\frac{1}{2}}, vq^{\frac{1}{2}}; \\ q, uv; \end{bmatrix}$$

$$\times \sum_{n=0}^{\infty} \frac{(q^{\frac{1}{2}}/u; q)_{3n} (q^{\frac{1}{2}}/v; q)_{3n}}{(uq^{\frac{1}{2}}; q)_{3n} (vq^{\frac{1}{2}}; q)_{3n}} u^{3n} v^{3n}$$

$$\times \left\{ q^{n(6n-1)} - \frac{(1 - uq^{3n-\frac{1}{2}})(1 - vq^{3n-\frac{1}{2}}) q^{(2n-1)(3n-1)}}{(u - q^{3n-\frac{1}{2}})(v - q^{3n-\frac{1}{2}})} \right.$$

$$\left. + q^{n(6n+1)} - \frac{(u - q^{3n+\frac{1}{2}})(v - q^{3n+\frac{1}{2}}) q^{(2n+1)(3n+1)}}{(1 - uq^{3n+\frac{1}{2}})(1 - vq^{3n+\frac{1}{2}})} \right\}. \quad (7.3.1.9)$$

This is one of the classical results proved by L. J. Rogers (1894 and 1916).

He wrote it in the form

$$a_0 + a_2 + a_4 + \ldots = b_0 - b_2 q - b_4 q^2 + b_6(q^5 + q^7) - b_8 q^{12} - b_{10} q^{15} + \ldots, \quad (7.3.1.10)$$

where
$$a_{2n} = \frac{(q^{\frac{1}{2}}/u; q)_n (q^{\frac{1}{2}}/v; q)_n u^n v^n}{(q; q)_{2n}}$$

and
$$b_{2n} = \Pi\begin{bmatrix} uq^{\frac{1}{2}}, vq^{\frac{1}{2}}; \\ q, uv; \end{bmatrix} \frac{(q^{\frac{1}{2}}/u; q)_n (q^{\frac{1}{2}}/v; q)_n u^n v^n}{(uq^{\frac{1}{2}}; q)_n (vq^{\frac{1}{2}}; q)_n}.$$

Many sets of values of $\alpha_n$ and $\beta_n$ can be deduced, similar to those in (7.3.1.7) by application of the various basic summation theorems, and a little algebra. In every case, a result corresponding to the transform (7.3.1.9) can be found.

Now, if special values are given to $u$ and $v$ in (7.3.1.9), one side of the result reduces to series which can be summed by Jacobi's theorem (3.1.12). For example, let $u = v = 0$. Then

$$\prod_{n=1}^{\infty} (1 - q^n) \sum_{n=0}^{\infty} \frac{q^{n^2}}{(q; q)_{2n}} = \prod_{n=1}^{\infty} (1 + q^{30n-16})(1 + q^{30n-14})(1 - q^{30n})$$

$$- q^2 \prod_{n=1}^{\infty} (1 + q^{30n-26})(1 + q^{30n-4})(1 - q^{30n}), \quad (7.3.1.11)$$

and, if $u = 1, v = 0$, on writing $q^2$ in place of $q$, we get

$$\prod_{n=1}^{\infty} \frac{(1 - q^{2n})}{(1 - q^{2n-1})} \sum_{n=0}^{\infty} \frac{(-1)^n q^{n^2}}{(-q; q^2)_n (q^4; q^4)_n}$$

$$= \prod_{n=1}^{\infty} (1 - q^{42n-19})(1 - q^{42n-23})(1 - q^{42n})$$

$$+ q^3 \prod_{n=1}^{\infty} (1 - q^{42n-5})(1 - q^{42n-37})(1 - q^{42n}). \quad (7.3.1.12)$$

Since at least eighty sets of values of $\alpha_n$ and $\beta_n$ have already been discovered, and there are at least twenty sets of values of $u$ and $v$, which have been investigated in the literature, the total possible number of such identities is very large indeed. We give below a few of the most interesting. For further lists, see Slater (1951) and (1952$a$).

$$\prod_{n=1}^{\infty} (1-q^{7n-1})(1-q^{7n-6})(1-q^{7n})$$

$$= \prod_{n=1}^{\infty} (1-q^{2n}) \sum_{n=0}^{\infty} \frac{q^{2n(n+1)}}{(q^2; q^2)_n (-q; q)_{2n+1}}, \tag{7.3.1.13}$$

$$\prod_{n=1}^{\infty} (1-q^{7n-2})(1-q^{7n-5})(1-q^{7n})$$

$$= \prod_{n=1}^{\infty} (1-q^{2n}) \sum_{n=0}^{\infty} \frac{q^{2n(n+1)}}{(q^2; q^2)_n (-q; q)_{2n}}, \tag{7.3.1.14}$$

$$\prod_{n=1}^{\infty} (1-q^{7n-3})(1-q^{7n-4})(1-q^{7n})$$

$$= \prod_{n=1}^{\infty} (1-q^{2n}) \sum_{n=0}^{\infty} \frac{q^{2n^2}}{(q^2; q^2)_n (-q; q)_{2n}}, \tag{7.3.1.15}$$

$$\prod_{n=1}^{\infty} (1-q^{20n-8})(1-q^{20n-12})(1-q^{20n})$$

$$= \prod_{n=1}^{\infty} \frac{(1-q^{2n})}{(1+q^{2n-1})} \sum_{n=0}^{\infty} \frac{q^{2n^2}}{(q; q)_{2n}}, \tag{7.3.1.16}$$

$$\prod_{n=1}^{\infty} (1-q^{27n-3})(1-q^{27n-24})(1-q^{27n})$$

$$= \prod_{n=1}^{\infty} (1-q^n) \sum_{n=0}^{\infty} \frac{(q^3; q^3)_n q^{n(n+3)}}{(q; q)_n (q; q)_{2n+2}}, \tag{7.3.1.17}$$

$$\prod_{n=1}^{\infty} (1-q^{27n-6})(1-q^{27n-21})(1-q^{27n})$$

$$= \prod_{n=1}^{\infty} (1-q^n) \sum_{n=0}^{\infty} \frac{(q^3; q^3)_n q^{n(n+2)}}{(q; q)_n (q; q)_{2n+2}}, \tag{7.3.1.18}$$

$$\prod_{n=1}^{\infty} (1-q^{27n-9})(1-q^{27n-18})(1-q^{27n})$$

$$= \prod_{n=1}^{\infty} (1-q^n) \sum_{n=0}^{\infty} \frac{(q^3; q^3)_n q^{n(n+1)}}{(q; q)_n (q; q)_{2n+1}}, \tag{7.3.1.19}$$

$$\prod_{n=1}^{\infty} (1-q^{27n-12})(1-q^{27n-15})(1-q^{27n})$$

$$= \prod_{n=1}^{\infty} (1-q^n) \sum_{n=0}^{\infty} \frac{(q^3; q^3)_{n-1} q^{n^2}}{(q; q)_n (q; q)_{2n-1}}, \tag{7.3.1.20}$$

$$\prod_{n=1}^{\infty} (1-q^{36n-3})(1-q^{36n-33})(1-q^{36n})$$

$$= \prod_{n=1}^{\infty} \frac{(1-q^{2n})}{(1+q^{2n-1})} \sum_{n=0}^{\infty} \frac{(-q;q^2)_{n+1}(q^6;q^6)_n q^{n(n+4)}}{(q^2;q^2)_{2n+2}(q^2;q^2)_n}, \qquad (7.3.1.21)$$

$$\prod_{n=1}^{\infty} (1-q^{36n-9})(1-q^{36n-27})(1-q^{36n})$$

$$= \prod_{n=1}^{\infty} \frac{(1-q^{2n})}{(1+q^{2n-1})} \sum_{n=0}^{\infty} \frac{(-q;q^2)_{n+1}(q^6;q^6)_n q^{n(n+2)}}{(q^2;q^2)_{2n+2}(q^2;q^2)_n} \qquad (7.3.1.22)$$

and

$$\prod_{n=1}^{\infty} (1-q^{36n-15})(1-q^{36n-21})(1-q^{36n})$$

$$= \prod_{n=1}^{\infty} \frac{(1-q^{2n})}{(1+q^{2n-1})} \sum_{n=0}^{\infty} \frac{(-q;q^2)_n(q^6;q^6)_{n-1} q^{n^2}}{(q^2;q^2)_{2n-1}(q^2;q^2)_n}. \qquad (7.3.1.23)$$

There are similar results,† involving products of powers of $q^{42}$, $q^{48}$, $q^{54}$ and $q^{64}$, and it is probable that many other similar results remain to be discovered. Thus, in place of (3.4.9), which depends on the basic analogue of Gauss's theorem, we might try to use (3.4.7) as our fundamental equation, and sum the series for $\gamma_n$ by the basic analogue of Saalschutz's theorem as a $_3\Phi_2$ series, or by Jackson's theorem (3.3.1.1) as a well-poised $_8\Phi_7$ series, or even by (7.1.1.1) as a basic $_6\Psi_6$ series. All these summation theorems would provide more general results than (3.4.9) which was used above. So they would probably lead to even more general identities of the Rogers–Ramanujan type.

## 7.4 Equivalent products

The simple question 'When is a set of products of the general type

$$\Pi[(a);(b);q^k]$$

equivalent to other sets of similar products?' quickly leads us deep into some of the more recondite branches of the theories of elliptic functions, generalized basic hypergeometric functions, modular functions and partition functions. Thus, since a sigma function can be written

$$\sigma(z) \equiv 2\omega_1 \exp\left[(\eta_1 z^2 + \pi i z)/2\omega_1\right]$$

$$\times \prod_{n=1}^{\infty} \left\{ \frac{[1-\exp(\pi i z/2\omega_1)q^{2n}]}{(1-q^{2n})^2}[1-\exp(-\pi i z/2\omega_1)q^{2n-2}] \right\}, \qquad (7.4.1)$$

† Slater (1952a).

we shall now show that the general theorem on sigma functions† which is usually stated as

$$\sum_{r=1}^{n} \frac{\sigma(a_r - b_1)\, \sigma(a_r - b_2) \dots \sigma(a_r - b_n)}{\sigma(a_r - a_1)\, \sigma(a_r - a_2) \dots \sigma(a_r - a_{r-1})\, \sigma(a_r - a_{r+1}) \dots \sigma(a_r - a_n)} = 0,$$
$$(7.4.2)$$

where $\qquad\qquad a_1 + a_2 + \dots + a_n = b_1 + b_2 + \dots + b_n,$$

can be rewritten in terms of ordinary products.

Let us put
$$a_r \quad \text{for} \quad \exp(\pi i a_r / 2\omega_1)$$

and
$$b_r \quad \text{for} \quad \exp(\pi i b_r / 2\omega_1)$$

in each sigma function, and replace $q$ by $q^{\frac12}$. Then (7.4.2) becomes

$$\sum_{r=1}^{n} \Pi \begin{bmatrix} a_r/(b),\, q(b)/a_r,\, q; \\ a_r/(a)',\, q(a)/a_r; \end{bmatrix} = 0, \qquad (7.4.3)$$

where $a_1 a_2 \dots a_n = b_1 b_2 \dots b_n$. This relation involves $n$ products each of $\frac12 n(n-1)$ theta functions and there are $2n - 2$ independent parameters.

In particular, if $n = 3$, we have

$$\Pi[aq/b, b/a, aq/ef, ef/a, aq/df, df/a, aq/bde, bde/a; q]$$

$$= \Pi[aq/f, f/a, aq/be, be/a, aq/bd, bd/a, aq/def, def/a; q]$$

$$- \frac{b}{a} \Pi[d, q/d, e, q/e, bq/f, f/b, a^2 q/bdef, bdef/a^2; q]. \quad (7.4.4)$$

This result is due to Bailey (1936), §5.2. He deduced it indirectly from the relation connecting three basic $_8\Phi_7$ series referred to in §3.4.2. In the notation for theta functions, (7.4.4) can also be rewritten as

$$\vartheta_3(a)\, \vartheta_3(b+c)\, \vartheta_3(b+d)\, \vartheta_3(a+c+d)$$

$$= \vartheta_3(b)\, \vartheta_3(a+c)\, \vartheta_3(a+d)\, \vartheta_3(b+c+d)$$

$$+ \vartheta_1(c)\, \vartheta_1(d)\, \vartheta_1(b-a)\, \vartheta_1(a+b+c+d). \quad (7.4.5)$$

If, in (7.4.4), we put $b = a\sqrt{q} = -f$, $d = e = -1$, and replace $q$ by $q^2$, we find that

$$\left\{ \prod_{n=1}^{\infty} (1 - q^{2n-1}) \right\}^{8} + 16q \left\{ \prod_{n=1}^{\infty} (1 + q^{2n}) \right\}^{8} = \left\{ \prod_{n=1}^{\infty} (1 + q^{2n-1}) \right\}^{8}. \quad (7.4.6)$$

This is another of Jacobi's classical results.

--------

† Whittaker & Watson (1947), p. 451, ex. 3.

Again, if in (7.4.4) we replace $q$ by $q^3$, write $q$ for $a$, $qz$ for $b$, $z$ for $d$, $qz$ for $e$, and $q^2$ for $f$, and multiply across by

$$\prod_{n=1}^{\infty} (1-q^{3n})/\{(1-q^n/z)(1-q^{n-1}z)\},$$

we find, after some algebra, that

$$\prod_{n=1}^{\infty} (1+q^n/z)(1+q^{n-1}z)(1-q^n)(1-q^{2n-1}/z^2)(1-q^{2n-1}z^2)$$

$$= \prod_{n=1}^{\infty} (1-q^{3n-2}z^3)(1-q^{3n})(1-q^{3n-1}/z^3)$$

$$+ z \prod_{n=1}^{\infty} (1-q^{3n-2}/z^3)(1-q^{3n})(1-q^{3n-1}z^3). \quad (7.4.7)$$

If $n = 4$, in (7.4.3), we have

$ab\Pi[bc/a, aq/bc, bd/a, aq/bd, be/a, aq/be, bf/a, aq/bf, g/a, aq/g, h/a,$

$aq/h, g/h, hq/g; q] - ab\Pi[ch/a, aq/ch, dh/a, aq/dh, eh/a, aq/eh,$

$fh/a, aq/fh, b/a, aq/b, g/a, aq/g, g/b, bq/g; q] - ag\Pi[cg/a, aq/cg,$

$dg/a, aq/dg, eg/a, aq/eg, fg/a, aq/fg, b/a, aq/b, h/a, aq/h, b/h, hq/b; q]$

$+ bh\Pi[c, q/c, d, q/d, e, q/e, f, q/f, b/h, hq/b, g/h, hq/g, g/b,$

$bq/g; q] = 0,$ $\qquad\qquad (7.4.8)$

provided that $\qquad\qquad a^3q^2 = bcdefgh.$

This is equivalent to a relation between four products, each containing seven theta functions, and involving six independent parameters. It can be deduced from the relation between $_{10}\Phi_9$ series (3.4.2.4), just as (7.4.3) can be deduced from a relation between $_8\Phi_7$ series.

These facts suggests that (7.4.3) can be deduced directly from the general theorem (7.2.5) on basic bilateral series, and this is in fact the case. For, in (7.2.5), let us take

$$b_1 = c_1, b_2 = c_2, \dots, b_{M-1} = c_{M-1}, b_M = qzc_M.$$

Then we have (7.4.3) above immediately.

Even this result (7.4.3) is not the most general known theorem on infinite products. A more general result is given by taking

$$b_2 = b_1 q^{1/M}, \dots, b_M = b_1 q^{1-1/M},$$

$$c_2 = c_1 q^{1/M}, \dots, c_M = c_1 q^{1-1/M},$$

in (7.2.5), if we put $q^2$ for $q$, $-zq/c$ for $z$, and then let $c_1 \to \infty$ and $b_1 \to 0$. We find the theorem

$$\frac{\Pi[zq^{1+M}/\alpha, \alpha/zq^{M-1}; \; q^2] \, \Pi[q^{-M}z, q^{-M}/z; \; q^{2M}]}{\Pi[(a), q^2/(a); \; q^2]}$$

$$= \sum_{s=1}^{M} \frac{q^2}{a_s} \frac{\Pi[a_s q^{M-1}z/\alpha, \alpha/a_s zq^{M-3}; \; q^2] \, \Pi[z/a_s^M, a_s^M/zq^{2M}; \; q^{2M}]}{\Pi[a_s, q^2/a_s, (a)'/a_s, q^2(a)'/a_s; \; q^2]}, \quad (7.4.9)$$

where $\alpha = a_1 a_2 a_3 \ldots a_M$, and there are $M$ of the $a$ parameters. This is a modular equation connecting products in $q^{2M}$, and products in $q^2$. Similar results follow from the theorems on transformations of well-poised bilateral series.

Thus, in (7.2.1.1), since

$$\lim_{b \to \infty} \frac{(b; q)_n \, (-1)^n a^n q^n}{(q/b; q)_n \, b^n} = q^{\frac{1}{2}n(n+1)} a^n,$$

if we let $b_1, b_2, b_3, \ldots, b_{2N} \to \infty$, we have, on summing by Jacobi's theorem (3.1.12),

$$\frac{\Pi[a, q^2/a; \; q^2] \, \Pi[a^N q^N, a^{-N} q^N; \; q^{2N}]}{\Pi[a, (a), qa/(a), q/a, q/(a), (a)/a; \; q]}$$

$$= \sum_{\nu=1}^{N} a_\nu \frac{\Pi[a/a_\nu^2, q^2 a_\nu^2; \; q^2] \, \Pi[a_\nu^{2N} q^N/a^N, a^N q^N/a_\nu^{2N}; \; q^{2N}]}{\Pi[a/a_\nu, (a)'/a_\nu, qa_\nu/(a), a_\nu, a_\nu(a)/a, q/a_\nu, qa/(a) \, a_\nu; \; q]}. \quad (7.4.10)$$

This is a general relation between products in $q^{2N}$ and products in $q$, with $N+1$ free parameters $a, a_1, a_2, \ldots, a_N$.

Three further similar results can be deduced, from (7.2.1.3) and (7.2.1.4) in the same way, though, since (7.2.1.1) to (7.2.1.4) are all special cases of (5.3.2), so (7.4.10) above, and these three further results can all be thought of as special cases of (7.2.5). When seeking to prove any given modular equation, it is probably best to start from (7.2.5) which is the general theorem, and make the necessary transformations and passages to the limit as required to prove each special case. We shall now give some examples of this process.

In (7.2.5), put $a = b_N$, $a_1 = b_{N+1}$, $a_2 = b_{N+2}$, ..., and $a_N = b_{2N}$ in order to deduce Sears's general theorem for well-poised series (5.3.2). Further, let

$$b_1, b_2, \ldots, b_N \to \infty, \quad a_2 = a_1 q^{1/N}, a_3 = a_1 q^{2/N}, \ldots, a_{N+1} = q^{1-1/N},$$

and write $q^N$ for $q$, $q^N$ for $a_0$ and $q^{\frac{1}{2}(N+1)}$ for $a_1$. Then, since

$$(a; \; q^{1/N})_{Nn} = (a; q)_n \, (aq^{1/N}; q)_n \, (aq^{2/N}; q)_n \ldots (aq^{1-1/N}; q)_n, \quad (7.4.11)$$

after considerable reduction, we find, for $N$ even and positive, that

$$\frac{\Pi[q^N, q^N; q^{2N}]}{\Pi[q^{2N}; q^{2N}]\,\Pi[q^{\frac{1}{2}(N+1)}, q^{\frac{1}{2}(1-N)}; q]}\sum_{n=0}^{\infty}(-1)^{Nn}q^{\frac{1}{2}N(N-1)n\,(n+1)}$$

$$=\sum_{s=0}^{N-1}\frac{q^{\frac{1}{2}(1+N)+s}\,\Pi[q^{-2s-1}, q^{2N+2s+1}; q^{2N}]}{\Pi[q^{-s-\frac{1}{2}+\frac{1}{2}N}, q^{s+\frac{1}{2}+\frac{1}{2}N}, q^{-s}, q^{1-s}, \ldots, q^{-1}, q, q^2 \ldots, q^{N-s-1}; q^N]}$$

$$\times\frac{1}{\Pi[q^{s+1}; q]}\sum_{n=0}^{\infty}(-1)^{Nn}q^{s(N-1)\,n+\frac{1}{2}(N-1)\,(Nn+1)\,n}. \quad (7.4.12)$$

Let us now put $q^2$ for $q$, and $2N$ for $N$ throughout, so that we can reduce the resulting products to their simplest forms. Then we find that we can reverse the order of the second half of the series in $n$, that is put $2N-s-1$ for $s$, and then put $-n-1$ for $n$ in the second half of each series. We see that the power of $q$ is now

$$2N(2N-1)\,n^2+(2s+1)\,(2N-1)\,n+s(s-1)+2N-1,$$

in each series. Thus the first half of each of the $n$ series will combine with the second half of each of the $n$ series, to give series summable by Jacobi's theorem. When we have carried out this summation we have the result

$$(-1)^N q^{2N-2}\prod_{n=1}^{\infty}\frac{(1-q^{8Nn-4})^2\,(1-q^{2n})^2\,(1+q^{4N(2N-1)\,n})^2}{(1-q^{4Nn})^2\,(1-q^{2n-1})^2}$$

$$=\sum_{s=0}^{N-1}(-1)^s q^{s(s-1)}\prod_{n=1}^{\infty}\frac{(1-q^{8Nn-4s-2})\,(1-q^{8Nn-8N+4s+2})}{(1-q^{4N-2N-2s-1})\,(1-q^{4Nn-2N+2s+1})}$$

$$\times\prod_{n=1}^{\infty}(1+q^{(2N-1)\,(4Nn-2N+2s+1)})\,(1+q^{(2N-1)\,(4Nn-2N-2s-1)}). \quad (7.4.13)$$

This is a general modular equation connecting $q^{4N}$ and $q^{4N(2N-1)}$. For example, if $N=2$, this result gives

$$q\prod_{n=1}^{\infty}\frac{(1-q^{16n-4})^2\,(1-q^{2n})^2\,(1+q^{24n})^2}{(1-q^{8n})^2\,(1-q^{2n-1})^2}$$

$$+\prod_{n=1}^{\infty}\frac{(1-q^{16n-2})\,(1-q^{16n-14})\,(1+q^{24n-15})\,(1+q^{24n-9})}{(1-q^{8n-3})\,(1-q^{8n-5})}$$

$$=\prod_{n=1}^{\infty}\frac{(1-q^{16n-6})\,(1-q^{16n-10})\,(1+q^{24n-3})\,(1+q^{24n-21})}{(1-q^{8n-1})\,(1-q^{8n-7})}, \quad (7.4.14)$$

and, if $N = 3$,

$$q^4 \prod_{n=1}^{\infty} \frac{(1 - q^{24n-4})^2 (1 - q^{2n})^2 (1 + q^{60n})^2}{(1 - q^{12n})^2 (1 - q^{2n-1})^2}$$

$$= \prod_{n=1}^{\infty} \frac{(1 - q^{24n-2})(1 - q^{24n-22})(1 + q^{60n-25})(1 + q^{60n-35})}{(1 - q^{12n-7})(1 - q^{12n-5})}$$

$$- \prod_{n=1}^{\infty} \frac{(1 - q^{24n-6})(1 - q^{24n-18})(1 + q^{60n-5})(1 + q^{60n-45})}{(1 - q^{12n-9})(1 - q^{12n-3})}$$

$$+ q^2 \prod_{n=1}^{\infty} \frac{(1 - q^{24n-10})(1 - q^{24n-14})(1 + q^{60n-5})(1 + q^{60n-55})}{(1 - q^{12n-11})(1 - q^{12n-1})}. \quad (7.4.15)$$

As a second example, let us put

$$a = b_N, a_1 = b_{N+1}, a_2 = b_{N+2}, \dots, a_N = b_{2N},$$

in (7.2.5) again, and suppose now that there are $2M + 4$ parameters in all. We let the last $M + 1$ of these parameters tend to infinity, and then write

$$a_{M+2} = q\sqrt{a_1}, \quad a_{M+3} = -q\sqrt{a_1}, \quad a_3 = q^{1/M} a_2, \quad a_{M+1} = q^{1-1/M} a_2.$$

We can now put $s - 2$ for $s$, $q^{2M}$ for $q$, $q^{2M}$ for $a_1$, and $q^{M+1}$ for $a_2$. Then, after all these substitutions, we have, for $M$ even and positive,

$$\sum_{n=0}^{\infty} (1 - q^{2M+4Mn})(-1)^n q^{nM\{(M+1)\, n + M - 1\}}$$

$$= \sum_{s=0}^{M-1} \frac{-q^{M-1-2s} \Pi[q^{2M};\, q^{2M}] \, \Pi[q^{1-M}, q^{1+M};\, q^2]}{\Pi[q^{2s+2};\, q^2] \, \Pi[q^{M-2s-1}, q^{M+2s+1}, q^{-2s}, q^{2-2s}, \dots, q^{-2};\, q^{2M}]}$$

$$\times \frac{\displaystyle\sum_{n=0}^{\infty} (1 - q^{4Mn+4s+2})(-1)^n q^{n\, M(M+1)\, n - M + 1 + 2s(M+1)}}{\Pi[q^2, q^4, q^6, \dots, q^{2M-2s-2};\, q^{2M}]}. \quad (7.4.16)$$

Again, we can put $2N$ for $M$ in this result and remove the negative powers of $q$ from the products. We find that the first series is of the Jacobi type, so again we can put $2N - 1 - s$ for $s$ in the second half of each series, and combine the resulting series together in pairs. These series are again summable by Jacobi's theorem, and we find the general result

$$(-1)^{N-1} q^{2N-2} \prod_{n=1}^{\infty} \frac{(1 - q^{4N(2N+1)\, n - 8N^2})(1 - q^{4N(2N+1)\, n - 4N})(1 - q^{2n})^2}{(1 - q^{4Nn})^2 (1 - q^{2n-1})^2}$$

$$= \sum_{s=0}^{2N-1} (-1)^s q^{s(s-1)} \prod_{n=1}^{\infty} \frac{(1 - q^{4N(2N+1)\, n + 2s(2N+1) - 4N^2 - 4N + 1})}{(1 - q^{4Nn-2N-2s-1})(1 - q^{4Nn-2N+2s+1})}$$

$$\times \prod_{n=1}^{\infty} (1 - q^{4N(2N+1)\, n - 2s(2N+1) - 4N^2 - 1}). \quad (7.4.17)$$

Again this is a type of general modular equation, connecting powers of $q^{4N(2N+1)}$ and $q^{4N}$.

If $N = 1$, we have

$$\prod_{n=1}^{\infty} \frac{(1 - q^{12n-8})(1 - q^{12n-4})(1 - q^{2n})^2}{(1 - q^{4n})^2(1 - q^{2n-1})^2}$$

$$= \prod_{n=1}^{\infty} \frac{(1 - q^{12n-7})(1 - q^{12n-5})}{(1 - q^{4n-3})(1 - q^{4n-1})} - \prod_{n=1}^{\infty} \frac{(1 - q^{12n-1})(1 - q^{12n-11})}{(1 - q^{4n+1})(1 - q^{4n-5})}, \quad (7.4.18)$$

and, if $N = 2$,

$$-q \prod_{n=1}^{\infty} \frac{(1 - q^{40n-32})(1 - q^{40n-8})(1 - q^{2n})^2}{(1 - q^{8n})^2(1 - q^{2n-1})^2}$$

$$= \prod_{n=1}^{\infty} \frac{(1 - q^{40n-23})(1 - q^{40n-17})}{(1 - q^{8n-5})(1 - q^{8n-3})} - \prod_{n=1}^{\infty} \frac{(1 - q^{40n-13})(1 - q^{40n-27})}{(1 - q^{8n-7})(1 - q^{8n-1})}$$

$$+ q^2 \prod_{n=1}^{\infty} \frac{(1 - q^{40n-3})(1 - q^{40n-37})}{(1 - q^{8n-9})(1 - q^{8n+1})} - q^6 \prod_{n=1}^{\infty} \frac{(1 - q^{40n+7})(1 - q^{40n-47})}{(1 - q^{8n-11})(1 - q^{8n+3})}.$$

$$(7.4.19)$$

## 7.5 Basic bilateral integrals

The main integral is

$$I = \int_{-i\pi/t}^{i\pi/t} \Pi \left[ \begin{matrix} q^{1-s}/(c), (b) q^s, q^{1+s}/(x\alpha), q^{-s}x\alpha; \\ (a) q^s, q^{1-s}/(a), q^{1+s}, q^{-s}; \end{matrix} q \right] ds, \quad (7.5.1)$$

where
$$\alpha \equiv \frac{c_1 c_2 \dots c_A}{a_1 a_2 \dots a_A},$$

and there are $A$ of the $a$ parameters, and $A$ of the $b$ and $c$ parameters also. By a consideration of the residues of this integral, in strip contours, similar to those of Fig. (5.1), we can give a direct proof of the general transformation of bilateral series (7.2.5), which was deduced above from the general transformation of ordinary basic series.

In a similar way, from the integral

$$I_1 = \int_{-i\pi/t}^{i\pi/t} \Pi \left[ \begin{matrix} q^{1-s}/(b), q^{1+s}a_0/(b), \\ a_0 q^s, (a) q^s, q^{1-s}/a_0, q^{1-s}/(a), \\ q^{1-s}a_0^{\frac{1}{2}}, q^{1+s}a_0^{-\frac{1}{2}}, q^{1+s+\pi i/t}a_0^{\frac{1}{2}}, q^{1-s+\pi i/t}a_0^{-\frac{1}{2}}; \\ q^{1+s}a_0/(a), (a) q^{-s}/a_0; \end{matrix} q \right] q^s ds,$$

$$(7.5.2)$$

and three similar integrals, we can give direct proofs of the four general transformation theorems for well-poised basic bilateral series of any order.

# 8

## APPELL SERIES

### 8.1 Notation

We have already generalized the Gauss series by increasing the number of parameters, and by making the series infinite in both directions. Another way in which the series can be generalized is by increasing the number of variables. Thus we are led to the study of double series in two variables. Such series are called Appell series.

The simplest case is the product of two Gauss functions

$$
{}_2F_1[a,b;\,c;\,x]\,{}_2F_1[a',b';\,c';\,y] = \sum_{m=0}^{\infty} \sum_{n=0}^{\infty} \frac{(a)_m\,(a')_n\,(b)_m\,(b')_n\,x^m y^n}{(c)_m\,(c')_n\,m!\,n!}.
$$

$$(8.1.1)$$

This series, in itself, gives us nothing new, but if one or more of the three pairs of products

$$
(a)_m\,(a')_n, \quad (b)_m\,(b')_n, \quad (c)_m\,(c')_n
$$

is replaced by a composite product of the general type

$$
(a)_{m+n}
$$

we are led to some entirely new functions.

Five possibilities arise,

$$
\sum_{m=0}^{\infty} \sum_{n=0}^{\infty} \frac{(a)_{m+n}\,(b)_{m+n}\,x^m y^n}{(c)_{m+n}\,m!\,n!}
$$

$$
= \sum_{N=0}^{\infty} \sum_{m=0}^{N} \frac{(a)_N\,(b)_N\,x^m y^{N-m}}{(c)_N\,(N-m)!\,m!}
$$

$$
= \sum_{N=0}^{\infty} \frac{(a)_N\,(b)_N\,(x+y)^N}{(c)_N\,N!}
$$

$$
= {}_2F_1[a,b;\,c;\,x+y]. \qquad (8.1.2)
$$

This is just an ordinary Gauss series. The summation is by the use of the binomial theorem. Next

$$
F_1[a;\,b,b';\,c;\,x,y] \equiv \sum_{m=0}^{\infty} \sum_{n=0}^{\infty} \frac{(a)_{m+n}\,(b)_m\,(b')_n\,x^m y^n}{(c)_{m+n}\,m!\,n!}. \qquad (8.1.3)
$$

This is the Appell function of the first kind. It exists for all real or complex values of $a, b, b', c, x$ and $y$, except $c$ a negative integer.

$$F_2[a; b, b'; c, c'; x, y] \equiv \sum_{m=0}^{\infty} \sum_{n=0}^{\infty} \frac{(a)_{m+n} (b)_m (b')_n x^m y^n}{(c)_m (c')_n m! n!}, \quad (8.1.4)$$

which exists for all real or complex values of $a, b, b', c, c', x$ and $y$ except $c, c'$ negative integers.

$$F_3[a, a'; b, b'; c; x, y] \equiv \sum_{m=0}^{\infty} \sum_{n=0}^{\infty} \frac{(a)_m (a')_n (b)_m (b')_n x^m y^n}{(c)_{m+n} m! n!}, \quad (8.1.5)$$

which exists for all values except $c$ a negative integer.

$$F_4[a; b; c, c'; x, y] \equiv \sum_{m=0}^{\infty} \sum_{n=0}^{\infty} \frac{(a)_{m+n} (b)_{m+n} x^m y^n}{(c)_m (c')_n m! n!}, \quad (8.1.6)$$

which exists for all values except $c, c'$ negative integers.

The standard work on these functions is Appell & Kampé de Fériét (1926). A great amount of work was also done by Jakob Horn, which he published in a long series of papers extending over fifty years.

Horn defined thirty types of double series, including series with suffixes of the type $m - n$ as well as the normal $m + n$ suffixes, and he investigated the relationships between them, {see Erdélyi and others (1953), Vol. I, Chap. 5}.

We can rewrite the series for $F_1$ in terms of a Gauss function, so that

$$F_1[a; b, b'; c; x, y] = \sum_{m=0}^{\infty} \frac{(a)_m (b)_m x^m}{(c)_m m!} {}_2F_1[a+m, b; c+m; y]. \quad (8.1.7)$$

All four Appell functions reduce to ordinary Gauss series ${}_2F_1[x]$ when $y = 0$. The first three functions also reduce to ordinary ${}_2F_1[x]$ series when $b'$ is zero.

## 8.1.1 The convergence of the double series.
We shall now prove that the four double series for $F_1, F_2, F_3$ and $F_4$ are in fact convergent under certain conditions. The general term of $F_1$ is

$$A_{mn} x^m y^n \equiv \Gamma \begin{bmatrix} c, a+m+n, b+m, b'+n \\ a, b, b', c+m+n, m+1, n+1 \end{bmatrix} x^m y^n.$$

By Stirling's formula, for large values of $m$ and $n$,

$$A_{mn} \sim \Gamma \begin{bmatrix} c \\ a, b, b' \end{bmatrix} m^{b-1} n^{b'-1} (m+n)^{a-c}.$$

Let $N$ be an integer, such that

$$N > \left| \Gamma \left[ \begin{matrix} c \\ a, b, b' \end{matrix} \right] \right|,$$

then     $|A_{mn} x^m y^n| < N |x|^m |y|^n (m+n)^{-\text{Rl}(c-a)} m^{\text{Rl}(b)-1} n^{\text{Rl}(b')-1},$

so that $F_1$ is convergent when $|x| < 1$ and $|y| < 1$. Conversely, this series is divergent if either $|x| > 1$, or if $|y| > 1$.

Similarly, if the general term for $F_2$ is $A_{mn} x^m y^n$, then

$$|A_{mn} x^m y^n| < N \frac{(1)_{m+n}}{(1)_m (1)_n} (m+n)^{\text{Rl}(a)-1} m^{\text{Rl}(b-c)} n^{\text{Rl}(b'-c')} |x|^m |y|^n.$$

If $k$ is a positive number, such that

$$k > \max[\text{Rl}(b-c), \text{Rl}(b'-c')],$$

then     $m^{\text{Rl}(b-c)} n^{\text{Rl}(b'-c')} < m^k n^k \leqslant [\tfrac{1}{2}(m+n)]^{2k},$

so that

$$\sum_{m=0}^{\infty} \sum_{n=0}^{\infty} |A_{mn} x^m y^n| < \frac{N}{4^k} \sum_{m=0}^{\infty} \sum_{n=0}^{\infty} \frac{(1)_{m+n}}{(1)_m (1)_n} (m+n)^{2k+\text{Rl}(a)-1} |x|^m |y|^n,$$

$$< \frac{N}{4^k} \sum_{r=0}^{\infty} r^{2k+\text{Rl}(a)-1} (|x| + |y|)^r,$$

where $r = m+n$, and this series is convergent when $|x| + |y| < 1$. Hence $F_2$ is convergent when $|x| + |y| < 1$.

For the series $F_3$, we find that

$$|A_{mn} x^m y^n| < N m^{\text{Rl}(a+b)-2} n^{\text{Rl}(a'+b')-2} (m+n)^{1-\text{Rl}(c)} \frac{(1)_m (1)_n}{(1)_{m+n}} |x|^m |y|^n,$$

that is

$$|A_{mn} x^m y^n| < N m^{\text{Rl}(a+b)-2} n^{\text{Rl}(a'+b')-2} (m+n)^{1-\text{Rl}(c)} |x|^m |y|^n,$$

and so the series $F_3$ is convergent when $|x| < 1$ and $|y| < 1$.

Finally, for $F_4$, we find that

$$|A_{mn} x^m y^n| < N(m+n)^{\text{Rl}(a+b)-2} m^{1-\text{Rl}(c)} n^{1-\text{Rl}(c')} \left[ \frac{(1)_{m+n}}{(1)_m (1)_n} \right]^2 |x|^m |y|^n.$$

Again, if $k$ is a positive number such that

$$k > \max[1-\text{Rl}(c), 1-\text{Rl}(c')],$$

then

$$\sum_{m=0}^{\infty} \sum_{n=0}^{\infty} |A_{mn} x^m y^n| < \frac{N}{4^k} \sum_{m=0}^{\infty} \sum_{n=0}^{\infty} (m+n)^{2k+\text{Rl}(a+b)-2} \left[ \frac{(1)_{m+n}}{(1)_m (1)_n} \right]^2 |x|^m |y|^n,$$

$$< \sum_{r=0}^{\infty} r^{2k+\text{Rl}(a+b)-2} \sum_{s=0}^{r} \binom{r}{s}^s |x|^{r-s} |y|^s,$$

(where $r = m+n$)

$$< \sum_{r=0}^{\infty} r^{2k+\mathrm{Rl}\,(a+b)-2} \left(\sqrt{|x|} + \sqrt{|y|}\right)^{2r}$$

and this series is convergent if $\sqrt{|x|} + \sqrt{|y|} < 1$. Hence $F_4$ is convergent under the same condition.

## 8.1.2 Partial differential equations satisfied by the Appell functions. If

$$F_1 = \sum_{m=0}^{\infty} \sum_{n=0}^{\infty} A_{mn}\, x^m y^n, \qquad (8.1.2.1)$$

then

$$A_{m+1,\, n} = \frac{(a+m+n)\,(b+m)}{(1+m)\,(c+m+n)}\, A_{m,\,n} \qquad (8.1.2.2)$$

and

$$A_{m,\, n+1} = \frac{(a+m+n)\,(b'+n)}{(1+n)\,(c+m+n)}\, A_{m,\,n}. \qquad (8.1.2.3)$$

Let

$$\theta \equiv x\frac{\partial}{\partial x} \quad \text{and} \quad \phi \equiv y\frac{\partial}{\partial y}.$$

Then by direct substitution in the equations, we see that $F_1$ satisfies the differential equations

$$\left.\begin{aligned}\left\{(\theta+\phi+a)\,(\theta+b) - \frac{1}{x}\theta(\theta+\phi+c-1)\right\} z &= 0, \\[2mm] \left\{(\theta+\phi+a)\,(\phi+b') - \frac{1}{x}\theta(\theta+\phi+c-1)\right\} z &= 0.\end{aligned}\right\} \qquad (8.1.2.4)$$

Now let $p \equiv \dfrac{\partial z}{\partial x},\quad q \equiv \dfrac{\partial z}{\partial y},\quad r \equiv \dfrac{\partial z}{\partial x\,\partial y},\quad s \equiv \dfrac{\partial z^2}{\partial x^2}\quad \text{and} \quad t \equiv \dfrac{\partial z^2}{\partial y^2}.$

Then $F_1$ satisfies the equations

$$\left.\begin{aligned} x(1-x)\,r + y(1-x)\,s + \{c-(a+b+1)\,x\}\,p - byq - abz &= 0, \\ y(1-y)\,t + x(1-y)\,s + \{c-(a+b'+1)\,y\}\,q - b'xp - ab'z &= 0. \end{aligned}\right\} \qquad (8.1.2.5)$$

Similarly, we can show that $F_2$ satisfies the equations

$$\left.\begin{aligned} x(1-x)\,r - xys + \{c-(a+b+1)\,x\}\,p - byq - abz &= 0, \\ y(1-y)\,t - xys + \{c'-(a+b'+1)\,y\}\,q - b'xp - ab'z &= 0, \end{aligned}\right\} \qquad (8.1.2.6)$$

$F_3$ satisfies the equations

$$\left.\begin{aligned} x(1-x)\,r + ys + \{c-(a+b+1)\,x\}\,p - abz &= 0, \\ y(1-y)\,t + xs + \{c-(a'+b'+1)\,y\}\,q - a'b'z &= 0, \end{aligned}\right\} \qquad (8.1.2.7)$$

and $F_4$ satisfies the equations

$$x(1-x)\,r - y^2t - 2xys + cp - (a+b+1)\,(xp+yq) - abz = 0,$$
$$y(1-y)\,t - x^2r - 2xys + c'q - (a+b+1)\,(xp+yq) - abz = 0.$$

$$(8.1.2.8)$$

## 8.2 Integrals representing Appell functions

Let us consider the integral

$$I = \iint u^{b-1}v^{b'-1}(1-u-v)^{c-b-b'-1}(1-ux-vy)^{-a}\,du\,dv, \quad (8.2.1)$$

taken over the triangular region $0 \leqslant u$, $0 \leqslant v$, $u+v \leqslant 1$. The parameters $a, b, b', c$ are assumed to be such that the double integral has a meaning and is convergent.

Now, provided that

$$\left|\frac{vy}{1-ux}\right| < 1,$$

$$(1-ux-vy)^{-a} = (1-ux)^{-a} \sum_{m=0}^{\infty} \frac{(a)_m}{(1)_m} \left(\frac{vy}{1-ux}\right)^m$$

$$= \sum_{m=0}^{\infty} \frac{(a)_m}{(1)_m} v^m y^m (1-ux)^{-a-m}$$

$$= \sum_{m=0}^{\infty} \frac{(a)_m}{(1)_m} v^m y^m \sum_{n=0}^{\infty} \frac{(a+m)_n}{(1)_n} u^n x^n,$$

so that

$$I = \sum_{m=0}^{\infty} \sum_{n=0}^{\infty} \frac{(a)_{m+n}}{(1)_m (1)_n} x^n y^m \iint u^{b-1+n}v^{b'-1+n}(1-u-v)^{c-b-b'-1}\,du\,dv$$

$$= \sum_{m=0}^{\infty} \sum_{n=0}^{\infty} \frac{(a)_{m+n}}{(1)_m (1)_n} x^n y^m \, \Gamma\!\left[\begin{matrix} b+n, b'+n, c-b-b' \\ c+m+n \end{matrix}\right].$$

Finally, we find that

$$I = \Gamma\!\left[\begin{matrix} c \\ b, b', c-b-b' \end{matrix}\right] F_1[a;\, b, b';\, c;\, x, y]. \quad (8.2.2)$$

In a similar way, by expanding the integrand in powers of $x$ and $y$ and integrating term by term, we can show that

$$\int_0^1\!\!\int_0^1 u^{b-1}v^{b'-1}(1-u)^{c-b-1}(1-v)^{c'-b'-1}(1-ux-vy)^{-a}\,du\,dv$$

$$= \Gamma\!\left[\begin{matrix} b, b', c-b, c'-b' \\ c, c' \end{matrix}\right] F_2[a;\, b, b';\, c, c';\, x, y], \quad (8.2.3)$$

and $\displaystyle\iint u^{b-1}v^{b'-1}(1-u-v)^{c-b-b'-1}(1-ux)^{-a}(1-vy)^{-a'}\,du\,dv$

$$= \Gamma\begin{bmatrix} b,b',c-b-b' \\ c \end{bmatrix} F_3[a,a';\,b,b';\,c;\,x,y], \quad (8.2.4)$$

where the integral is taken over the triangle

$$u \geqslant 0, \quad v \geqslant 0, \quad u+v \leqslant 1.$$

No similar integral for $F_4$ has been found.

We can also deduce a single integral for the function $F_1$. Let

$$I' = \int_0^1 u^{a-1}(1-u)^{c-a-1}(1-ux)^{-b}(1-vy)^{-b'}\,du.$$

Then $\displaystyle I' = \sum_{m=0}^{\infty}\sum_{n=0}^{\infty}\int_0^1 u^{a-1}(1-u)^{c-a-1}\frac{(b)_m}{(1)_m}u^m x^m \frac{(b')_n}{(1)_n}u^n y^n\,du,$

$$= \sum_{m=0}^{\infty}\sum_{n=0}^{\infty}\frac{(b)_m(b')_n}{(1)_m(1)_n}x^m y^n \int_0^1 u^{a+m+n-1}(1-u)^{c-a-1}\,du,$$

$$= \sum_{m=0}^{\infty}\sum_{n=0}^{\infty}\frac{(b)_m(b')_n}{(1)_m(1)_n}x^m y^n \Gamma\begin{bmatrix} a+m+n,c-a \\ c+m+n \end{bmatrix},$$

so that $\displaystyle I' = \Gamma\begin{bmatrix} a,c-a \\ c \end{bmatrix} F_1[a;\,b,b';\,c;\,x,y]. \quad (8.2.5)$

In the four integrals given above, powers of various linear functions of $u$ and $v$ occur. Thus, in (8.2.1), we have powers of $u,v,1-u-v$, and $1-ux-vy$. In (8.2.3) we have powers of $u,v,1-u,1-v$ and $1-ux-vy$, in (8.2.4) we have powers of $u,v,1-u-v,1-ux$ and $1-vy$, and in (8.2.5) we have powers of $u,1-u,1-ux$ and $1-vy$.

These suggest that more general integrals of the type

$$I = \iint u^{a-1}v^{b-1}(1-u)^{c-1}(1-v)^{d-1}(1-u-v)^{e-1}(1-ux)^{f-1}$$

$$\times (1-vy)^{g-1}(1-ux-vy)^{h-1}\,du\,dv \quad (8.2.6)$$

might be studied in order to discover more general relations between similar functions of two variables.

## 8.2.1 Single Barnes-type integrals for Appell functions.

From the general theorem of §4.7.2, we can deduce immediately single contour integrals of ordinary hypergeometric functions which represent the various Appell functions, and lead to their asymptotic expansions when one of the variables is large. In particular, for the four functions $F_1$, $F_2$, $F_3$ and $F_4$, we find that, if $|z| < 1$, $|x| < 1$ and $|\arg z| < \pi$, then

$$\frac{1}{2\pi i}\int_{-i\infty}^{i\infty} \Gamma\begin{bmatrix} b-s, g+s, h-s \\ d-s \end{bmatrix} {}_2F_1\begin{bmatrix} b-s, e; \\ d-s; \end{bmatrix} z^{-s}\,ds$$

$$= z^g\, \Gamma\begin{bmatrix} b+g, h+g \\ d+g \end{bmatrix} F_1[b+g;\, e, h+g;\, d+g;\, x, -z], \quad (8.2.1.1)$$

$$\sim z^{-b}\, \Gamma\begin{bmatrix} b+g, h-b \\ d-b \end{bmatrix} \sum_{m=0}^{\infty}\sum_{n=0}^{\infty} \frac{(b+g)_{m+n}(1+b-d)_n(e)_m}{(1+b-h)_{m+n}\,m!\,n!} x^m(-1)^n z^{-m-n}$$

$$+ z^{-h}\, \Gamma\begin{bmatrix} b-h, g+h \\ d-h \end{bmatrix} \sum_{m=0}^{\infty}\sum_{n=0}^{\infty} \frac{(b-h)_{m-n}(g+h)_n(e)_m}{(d-h)_{m-n}\,m!\,n!} x^m(-1)^n z^{-n}.$$

$$(8.2.1.2)$$

$$\frac{1}{2\pi i}\int_{-i\infty}^{i\infty} \Gamma\begin{bmatrix} b-s, g+s \\ j+s, k-s \end{bmatrix} {}_2F_1[b-s, e;\, f;\, x]\, z^{-s}\,ds$$

$$= z^g\, \Gamma\begin{bmatrix} b+g \\ j-g, k+g \end{bmatrix} F_2[b+g;\, e, 1+g-j;\, k+g, f;\, x, z], \quad (8.2.1.3)$$

$$\sim z^{-b}\, \Gamma\begin{bmatrix} b+g \\ j+b, k-b \end{bmatrix} \sum_{m=0}^{\infty}\sum_{n=0}^{\infty} \frac{(g+b)_{m+n}(1+b-k)_{m+n}(e)_m}{(j+b)_{m+n}(f)_m\,m!\,n!} (-x)^m z^{-m-n}.$$

$$(8.2.1.4)$$

$$\frac{1}{2\pi i}\int_{-i\infty}^{i\infty} \Gamma\begin{bmatrix} g+s, h-s, k-s \\ d-s \end{bmatrix} {}_2F_1[e, f;\, d-s;\, x]\, z^{-s}\,ds$$

$$= z^g\, \Gamma\begin{bmatrix} g+h, g+k \\ g+d \end{bmatrix} F_3[g+h, e;\, g+k, f;\, g+d;\, x, -z], \quad (8.2.1.5)$$

$$\sim z^{-h}\, \Gamma\begin{bmatrix} g+h, k-h \\ d-h \end{bmatrix} \sum_{m=0}^{\infty}\sum_{n=0}^{\infty} \frac{(g+h)_n(e)_m(f)_m\, x^m z^{-n}}{(1+h-k)_n(d-h)_{m-n}\,m!\,n!}$$

$$+ z^{-k}\, \Gamma\begin{bmatrix} g+k, h-k \\ d-k \end{bmatrix} \sum_{m=0}^{\infty}\sum_{n=0}^{\infty} \frac{(g+k)_n(e)_m(f)_m\, x^m z^{-n}}{(1+k-h)_n(d-k)_{m-n}\,m!\,n!} \quad (8.2.1.6)$$

and

$$\frac{1}{2\pi i}\int_{-i\infty}^{i\infty} \Gamma\begin{bmatrix} b-s, c-s, g+s \\ k-s \end{bmatrix} {}_2F_1[b-s, c-s;\, f;\, x]\, z^{-s}\,ds$$

$$= z^g\, \Gamma\begin{bmatrix} b+g, c+g \\ k+g \end{bmatrix} F_4[b+g, c+g;\, k+g, f;\, x, -z], \quad (8.2.1.7)$$

$$\sim z^{-b}\, \Gamma\begin{bmatrix} c-b, g+b \\ k-b \end{bmatrix} \sum_{m=0}^{\infty}\sum_{n=0}^{\infty} \frac{(g+b)_{m+n}(1+b-k)_{m+n}\, x^m}{(1+b-c)_n(f)_m\,m!\,n!} (-1)^{m+n} z^{-m-n}$$

$$+ z^{-c}\, \Gamma\begin{bmatrix} b-c, g+c \\ k-c \end{bmatrix} \sum_{m=0}^{\infty}\sum_{n=0}^{\infty} \frac{(g+c)_{m+n}(1+c-k)_{m+n}\, x^m}{(1+c-b)_n(f)_m\,m!\,n!} (-1)^{m+n} z^{-m-n}.$$

$$(8.2.1.8)$$

These integrals are due to Appell (1926) and the asymptotic expansions are due to Slater (1955 b, § 5). In a similar way, contour integrals and asymptotic expansions can be deduced for the thirty other series defined by Horn.

## 8.3 Linear transformations

Several linear transformations of Appell functions are known. In the single integral (8.2.5) for the $F_1$ series given above, let us put

$$u = 1 - v.$$

Then

$$F_1 = \Gamma \begin{bmatrix} c \\ a, c-a \end{bmatrix} \int_0^1 v^{c-a-1}(1-v)^{a-1}(1-x+vx)^{-b}(1-y+vy)^{-b'} \, dv$$

$$= \Gamma \begin{bmatrix} c \\ a, c-a \end{bmatrix} (1-x)^{-b}(1-y)^{-b'} \int_0^1 v^{c-a-1}(1-v)^{a-1}\{1+xv/(1-x)\}^{-b}$$

$$\times \{1+vy/(1-y)\}^{-b'} \, dv, \quad (8.3.1)$$

so that

$$F_1[a; b, b'; c; x, y]$$

$$= (1-x)^{-b}(1-y)^{-b'} F_1[c-a; b, b'; c; x/(x-1), y/(y-1)]. \quad (8.3.2)$$

If $b' = 0$, this result reduces to (1.7.1.3), a relation between two Gauss functions. There exist five such changes of variable which leave unaltered the fundamental form of the integral.

Next let us try

$$u = v/(1-x+vx).$$

This transformation is chosen so that $u = 0$ when $v = 0$, and $u = 1$, when $v = 1$. Then

$$1 - ux = (1-x)/(1-x+vx),$$

$$1 - uy = (1-x+vx-vy)/(1-x+vx),$$

$$1 - u = (1-x)(1-v)/(1-x+vx)$$

and

$$du = \{(1-x)/(1-x+vx)^2\} \, dv.$$

Hence

$$F_1[a; b, b'; c; x, y] = \Gamma \begin{bmatrix} c, \\ a, c-a \end{bmatrix} (1-x)^{c-a-b}(1-x)^{-b'}(1-x)^{b+b'-c}$$

$$\times \int_0^1 v^{a-1}(1-v)^{c-a-1}\{1-(x-vx)/(1-x)\}^{b+b'-c}\{1-v(y-x)/(1-x)\}^{-b'} \, dv,$$

so that

$$F_1[a; b, b'; c; x, y]$$
$$= (1-x)^{-a} F_1[a; -b-b'+c, b'; c; -x(1-x), (y-x)/(1-x)]. \quad (8.3.3)$$

Again, if $b' = 0$, this result reduces to (1.7.1.3).

In particular, if $c = b+b'$,

$$F_1[a; b, b'; b+b'; x, y] = (1-x)^{-a} {}_2F_1[a, b'; b+b'; (y-x)/(1-x)].$$
$$(8.3.4)$$

Thus, in this particular case, $F_1$ reduces to an ordinary Gauss function. Appell (1926) gives several further cases of such reductions.

In a similar way, we can prove that

$$F_1[a; b, b'; c; x, y]$$
$$= (1-y)^{-a} F_1[a; b, c-b-b'; c; (x-y)/(1-y), y/(y-1)]. \quad (8.3.5)$$

If we write
$$u = v/(1-x+vx) \quad \text{and} \quad v = 1-V,$$

so that $u = (1-V)/(1-xV)$, we find that

$$F_1[a; b, b'; c; x, y]$$
$$= (1-x)^{c-a-b} (1-y)^{-b'} F_1[c-a; c-b-b', b'; c; x, (x-y)/(1-y)]$$
$$(8.3.6)$$

and, in a similar way,

$$F_1[a; b, b'; c; x, y]$$
$$= (1-x)^{-b} (1-y)^{c-a-b'} F_1[c-a; b, c-b-b'; c; (y-x)/(1-x), y].$$
$$(8.3.7)$$

From these five transformations, we can see that there are at least six equivalent solutions of the differential equations (8.1.2.5) satisfied by the $F_1$ function.

It has been shown† that there are sixty integrals of these equations. Each integral involves an $F_1$ function and there exists a linear relation connecting any four of these integrals. These sixty solutions thus correspond to Kummer's twenty-four solutions for the ordinary Gauss equation.

We can deduce three transformations of the $F_2$ function in a similar way. First put $u = 1-u'$ in the double integral (8.2.3). Then

$$F_2[a; b, b'; c, c'; x, y]$$
$$= (1-x)^{-a} F_2[a; c-b, b'; c, c'; x/(x-1), y/(1-x)]. \quad (8.3.8)$$

† Appell (1926), pp. 62–4.

Next put $v = 1 - v'$, then

$$F_2[a; b,b'; c,c'; x,y]$$
$$= (1-y)^{-a} F_2[a; b,c'-b'; c,c'; x/(1-y),y/(y-1)], \quad (8.3.9)$$

and finally put $u = 1 - u'$ and $v = 1 - v'$, so that

$$F_2[a; b,b'; c,c'; x,y]$$
$$= (1-x-y)^{-a} F_2[a; c-b,c'-b'; c,c'; x/(x+y-1),y/(x+y-1)].$$
$$(8.3.10)$$

Various more complicated substitutions have been tried, but these all break down and lead to no useful results. In transformation of the integral for the $F_3$ function, another trouble arises, as the integration in this case should be carried out over a triangular region. Similarly, for the $F_4$ function there are no known transformations.

### 8.3.1 Cases of reducibility of $F_1$, $F_2$ and $F_3$.

If $x = y$, the result (8.2.6) gives us the special case

$$F_1[a; b,b'; c; x,x] = (1-x)^{c-a-b-b'} {}_2F_1[c-a,c-b-b'; c; x]$$
$$= {}_2F_1[a,b+b'; c; x]. \quad (8.3.1.1)$$

If $c = b+b'$, (8.2.5) gives us the special case

$$F_1[a; b,b'; b+b'; x,y] = (1-y)^{-a} {}_2F_1[a,b; b+b'; (x-y)/(1-y)].$$
$$(8.3.1.2)$$

If $c = b$, (8.2.8) gives us the special case

$$F_2[a; b,b'; b,c'; x,y] = (1-x)^{-a} {}_2F_1[a,b'; c'; y/(1-x)]. \quad (8.3.1.3)$$

Of these formulae, the second one shows that the function $F_1$ reduces to an ordinary hypergeometric function when $c = b+b'$, and the third one shows that the function $F_2$ reduces in the same way, when $c = b$, and, by symmetry, when $c' = b'$.

We can write

$$F_1[a; b,b'; c; x,y] = \sum_{m=0}^{\infty} \frac{(a)_m (b)_m x^m}{(1)_m (c)_m} {}_2F_1[a+m,b; c+m; y]. \quad (8.3.1.4)$$

But $\qquad {}_2F_1[a,b; c; y] = (1-y)^{-b} {}_2F_1[a,b; c; y/(y-1)]$,

so that

$$F_1[a; b,b'; c; x,y]$$
$$= (1-y)^{-b} \sum_{m=0}^{\infty} \frac{(a)_m (b)_m}{(1)_m (c)_m} x^m (1-y)^{-b'} {}_2F_1[a,b; c; y/(y-1)]$$
$$= (1-y)^{-b'} F_3[a,c-a; b,b'; c; x,y/(y-1)]. \quad (8.3.1.5)$$

Thus, any $F_1$ function can always be expressed in terms of an $F_3$ function, but the converse is not true, except in the special case when $c = a + a'$. Since the $F_1$ function reduces to an ordinary hypergeometric series when $c = b + b'$, we can also expect that the $F_3$ function will reduce to an ordinary hypergeometric series when $c = a + a' = b + b'$. In fact we have

$$F_3[a, c-a; b, c-b; c; x, y/(y-1)]$$
$$= (1-y)^{c-b}(1-y)^{-a}{}_2F_1\left[a, b; c-b; \frac{(x-y)}{(1-y)}\right]. \quad (8.3.1.6)$$

In a similar way, we can always express any general $F_1$ function in terms of a special $F_2$ function, in which $c' = a$. In fact,

$$(1-y)^{-b'} F_2[a; b, b'; c, a; x, y/(y-1)]$$
$$= (1-y)^{-b'} \sum_{m=0}^{\infty} \frac{(a)_m (b)_m x^m}{(c)_m (1)_m} {}_2F_1[a+m, b'; a; y/(y-1)]$$
$$= \sum_{m=0}^{\infty} \frac{(a)_m (b)_m x^m}{(c)_m (1)_m} {}_2F_1[b', -m; a; y]$$
$$= \sum_{m=0}^{\infty} \frac{(a)_m (b)_m x^m (a-b')_m}{(c)_m (1)_m (a)_m} {}_2F_1[b', -m; 1+b'-a-m; 1-y]$$
$$\hspace{5cm} \{\text{from } (1.8.1.12)\},$$
$$= \sum_{m=0}^{\infty} \sum_{n=0}^{m} \frac{(b)_m (a-b')_m (b')_n (-m)_n x^m (1-y)^n}{(1)_m (1)_n (c)_m (1+b'-a-m)_n}$$
$$= \sum_{n=0}^{\infty} \sum_{s=0}^{\infty} \frac{(b)_{n+s} (a-b')_{n+s} (b')_n (-1)^n x^{n+s} (1-y)^n}{(1)_s (1)_n (c)_{n+s} (1+b'-a-n-s)_n}$$
$$= \sum_{n=0}^{\infty} \sum_{s=0}^{\infty} \frac{(b)_{n+s} (a-b')_s (b')_n x^{n+s} (1-y)^n}{(1)_s (1)_n (c)_{n+s}}$$
$$= F_1[b; a-b', b'; c; x, x(1-y)]. \quad (8.3.1.7)$$

Thus, any $F_2$ function always reduces to an $F_1$ function when $c' = a$. If, further, $c - a = a'$, we can express a special $F_2$ series in terms of a Gauss function. The formula is

$$F_2[a; b, b'; a, a; x, y]$$
$$= (1-x)^{-b}(1-y)^{-b'}{}_2F_1[b, b'; a; xy/\{(1-x)(1-y)\}]. \quad (8.3.1.8)$$

This can be proved in a similar way, by expanding the right-hand side in powers of $x$ and $y$.

All these results hold for infinite series, provided that $|x|$ and $|y|$ are small enough; that is, they hold for values of $x$ and $y$ in some simply connected domain about the origin.

In some particular cases, transformations exist between an $F_2$ function and an $F_3$ function, but no general transformations of this type have been found.

## 8.4 The expansion of the $F_4$ function in terms of Gauss functions

In 1920, G. N. Watson† proved that

$$_2F_1[a,b;\ c;\ z]\,_2F_1[a,b;\ c;\ Z]$$

$$= \Gamma\begin{bmatrix} c, c-a-b \\ c-a, c-b \end{bmatrix} F_4[a,b;\ c,a+b-c+1;\ zZ,(1-z)(1-Z)]$$

$$+ \Gamma\begin{bmatrix} c, a+b-c \\ a, b \end{bmatrix} (1-z)^{c-a-b}(1-Z)^{c-a-b}$$

$$\times F_4[c-a,c-b;\ c,c-a-b+1;\ zZ,(1-z)(1-Z)]. \quad (8.4.1)$$

This expresses the sum of two $F_4$ series, in terms of the product of two Gauss functions.‡ The second part of the expression on the right-hand side vanishes when $a$ or $b$ is a negative integer, and we have

$$_2F_1[-n,b;\ c;\ z]\,_2F_1[-n,b;\ c;\ Z]$$

$$= \frac{(c-b)_n}{(c)_n} F_4[-n,b;\ c,1-n+b-c;\ zZ,(1-z)(1-Z)]. \quad (8.4.2)$$

But

$$_2F_1[-n,b;\ c;\ Z] = \frac{(c-b)_n}{(c)_n}\,_2F_1[-n,b;\ 1-n+b-c;\ 1-Z]. \quad (8.4.3)$$

Hence

$$_2F_1[-n,b;\ c;\ z]\,_2F_1[-n,b;\ 1-n+b-c;\ 1-Z]$$

$$= F_4[-n,b;\ c,1-n+b-c;\ zZ,(1-z)(1-Z)]. \quad (8.4.4)$$

We shall now prove that in fact this result holds when $a$ is not a negative integer. Let us write $x/(x-1)$ for $z$ and $1/(1-y)$ for $Z$, and consider $U_{mn}$, the coefficient of $x^m y^n$ in the expansion in series of the function

$$\Phi(x,y) \equiv (1-x)^{-a}(1-y)^{-b}$$

$$\times F_4\left[ a,b;\ c,a+b-c+1;\ \frac{x}{(1-x)(y-1)},\ \frac{y}{(1-x)(y-1)} \right]. \quad (8.4.5)$$

† Watson (1920).        ‡ See also Watson (1948a), §11.6.

We find that

$$
\begin{aligned}
U_{mn} &= \sum_{r=0}^{m} \sum_{s=0}^{n} (-1)^{r+s} \frac{(a)_{r+s}(b)_{r+s}(a+r+s)_{m-r}(b+r+s)_{n-s}}{(1)_r(1)_s(c)_r(1+a+b-c)_s(1)_{m-r}(1)_{n-s}} \\
&= \frac{1}{m!\,n!} \sum_{r=0}^{m} \sum_{s=0}^{n} \frac{(a)_{m+s}(-m)_s(b)_{n+r}(-m)_r}{r!\,(1+a+b-c)_s s!\,(c)_r} \\
&= \frac{(a)_m (b)_n}{m!\,n!}\, {}_2F_1[a+m,\,-n;\,1+a+b-c;\,1]\,{}_2F_1[b+n,\,-m;\,c;\,1] \\
&= \frac{(a)_m (b)_n (c-b-n)_m (1+b-c-m)_n}{m!\,n!\,(c)_m (1+a+b-c)_n},
\end{aligned}
$$

provided that $|x|$ and $|y|$ are small enough. The ${}_2F_1$ series have been summed by Vandermonde's theorem. Also

$$
(c-b-n)_m \equiv (c-b)_{n-m}(-1)^n (1+b-c)_n,
$$

so that
$$
U_{mn} = \frac{(a)_m (b)_n (1+b-c)_n (c-b)_m}{m!\,n!\,(c)_m (1+a+b-c)_n}, \tag{8.4.6}
$$

after some reduction. Hence

$$
\Phi(x,y) = {}_2F_1[a, c-b;\, c;\, x]\,{}_2F_1[b, 1+b-c;\, 1+a+b-c;\, y], \tag{8.4.7}
$$

$$
= (1-x)^{-a}\,{}_2F_1[a,b;\,c;\,x/(x-1)]\,(1-y)^{-b}
$$
$$
\times\,{}_2F_1[a,b;\,1+a+b-c;\,y/(y-1)], \tag{8.4.8}
$$

provided that

$$
|x/\{(x-1)(1-y)\}|^{\frac{1}{2}} + |y/\{(x-1)(1-y)\}|^{\frac{1}{2}} < 1.
$$

Hence finally we have the result that

$$
F_4[a,b;\,c,1+a+b-c;\,x/\{(x-1)(1-y)\}, y/\{(x-1)(1-y)\}]
$$
$$
= {}_2F_1[a,b;\,c;\,x/(x-1)]\,{}_2F_1[a,b;\,1+a+b-c;\,y/(y-1)]. \tag{8.4.9}
$$

This result gives the complete expression of the $F_4$ function in terms of ordinary Gauss functions when

$$
c + c' = 1 + a + b,
$$

but this result is only a special case of a still more general expansion, due to Burchnall & Chaundy (1940) and (1941$a$), which we shall now prove.

Let us consider the coefficient $U_{mn}$ of $x^m y^n$ in the expansion in series of the function

$$
\Phi(x,y) \equiv (1-x)^{-a}(1-y)^{-b}
$$
$$
\times F_4[a,b;\,c,c';\,x/\{(x-1)(1-y)\}, y/\{(x-1)(1-y)\}]. \tag{8.4.10}
$$

We find that

$$U_{mn} = \sum_{r=0}^{m} \sum_{s=0}^{n} \frac{(-1)^{r+s}(a)_{r+s}(b)_{r+s}(a+r+s)_{m-r}(b+r+s)_{n-s}}{r!\,s!\,(c)_r\,(c')_s\,(m-r)!\,(n-s)!},$$

$$= \frac{(a)_m\,(b)_n}{m!\,n!}\,{}_2F_1[a+m,\,-m;\,c;\,1]\,{}_2F_1[b+n,\,-n;\,c';\,1],$$

$$= \frac{(a)_m\,(b)_n\,(c-b-n)_m\,(c'-a-m)_n}{m!\,n!\,(c)_m\,(c')_n},$$

by Vandermonde's theorem. But, by Saalschutz's theorem, we can re-expand these last two products to give

$$U_{mn} = \frac{(a)_m\,(b)_n\,(c'-a)_m\,(c-b)_n}{m!\,n!\,(c)_m\,(c')_n}\,{}_3F_2\!\left[\begin{array}{c} 1+a+b-c-c',\,-m,\,-n; \\ 1+b-c-m,\,1+a-c'-n; \end{array}\,1\right].$$

$$\text{(8.4.11)}$$

Now let $r$ be the subscript in the ${}_3F_2(1)$ series and let us replace $m$ and $n$ by $r+s$ and $r+t$ respectively, in $U_{mn}$. Then

$$U_{r+s,\,r+t} = \sum_{r=0}^{\min(m,n)} \frac{(a)_{r+s}(b)_{r+t}(c'-a)_{r+s}(c-b)_{r+t}(1+a+b-c-c')_r}{r!\,s!\,t!\,(c)_{r+s}(c')_{r+t}(c-b+s)_r(c'-a+t)_r},$$

and so

$$\Phi(x,y)$$

$$= \sum_{r=0}^{\infty} \sum_{s=0}^{\infty} \sum_{t=0}^{\infty} \frac{(a)_{r+s}(b)_{r+t}(c'-a)_t(c-b)_s(1+a+b-c-c')_r\,x^{r+s}y^{r+t}}{r!\,s!\,t!\,(c)_{r+s}(c')_{r+t}}$$

$$= \sum_{r=0}^{\infty} \frac{x^r y^r (a)_r\,(b)_r\,(1+a+b-c-c')_r}{r!\,(c)_r\,(c')_r}$$

$$\times\,{}_2F_1\!\left[\begin{array}{c} a+r,\,c-b; \\ c+r; \end{array}\,x\right]{}_2F_1\!\left[\begin{array}{c} b+r,\,c'-a; \\ c'+r; \end{array}\,y\right] \qquad \text{(8.4.12)}$$

$$= (1-x)^{-a}(1-y)^{-b}\sum_{r=0}^{\infty} \frac{x^r(1-x)^r y^r(1-y)^r (a)_r\,(b)_r\,(1+a+b-c-c')_r}{r!\,(c)_r\,(c')_r}$$

$$\times\,{}_2F_1[a+r,\,b+r;\,c+r;\,x/(x-1)]\,{}_2F_1[a+b,\,b+r;\,c'+r;\,y/(y-1)].$$

$$\text{(8.4.13)}$$

Now put $x$ for $x/(x-1)$, and $y$ for $y/(y-1)$, and we find the final result that

$$F_4[a,b;\,c,c';\,x(1-y),\,y(1-x)]$$

$$= \sum_{r=0}^{\infty} \frac{(a)_r\,(b)_r\,(1+a+b-c-c')_r}{r!\,(c)_r\,(c')_r}\,x^r y^r$$

$$\times\,{}_2F_1\!\left[\begin{array}{c} a+r,\,b+r; \\ c+r; \end{array}\,x\right]{}_2F_1\!\left[\begin{array}{c} a+r,\,b+r; \\ c'+r; \end{array}\,y\right]. \qquad \text{(8.4.14)}$$

If we put $1+a+b-c = c'$ in this result we can deduce the previous result (8.4.9) as a special case.

**8.4.1 An integral for $F_4$.** Now let us replace the two Gauss functions in (8.4.14) above by their integral representations, using (1.6.6). We find that

$$F_4 \equiv F_4[a,b;\,c,c';\,x(1-y),y(1-x)]$$
$$= \sum_{r=0}^{\infty} \frac{(a)_r\,(b)_r\,(1+a+b-c-c')_r\,x^r y^r}{(1)_r\,(c)_r\,(c')_r}$$
$$\times \Gamma\begin{bmatrix} c+r \\ a+r, c-a \end{bmatrix} \int_0^1 u^{a+r-1}(1-u)^{c-a-1}(1-ux)^{-b-r}\,du$$
$$\times \Gamma\begin{bmatrix} c'+r \\ b+r, c'-b \end{bmatrix} \int_0^1 v^{b+r-1}(1-v)^{c'-b-1}(1-vy)^{-a-r}\,dv, \quad (8.4.1.1)$$

provided that $\mathrm{Rl}\,(c') > \mathrm{Rl}\,(b) > 0$, and $\mathrm{Rl}\,(c) > \mathrm{Rl}\,(a) > 0$.

Now let us interchange the order of summation and integration, so that

$$F_4 = \Gamma\begin{bmatrix} c,c' \\ a,b,c-a,c'-b \end{bmatrix}$$
$$\times \int_0^1 \int_0^1 u^{a-b}v^{b-1}(1-u)^{c-a-1}(1-v)^{c'-b-1}(1-ux)^{-b}(1-vy)^{-a}$$
$$\times \sum_{r=0}^{\infty} \frac{(1+a+b-c-c')_r\,(uvxy)^r}{(1)_r\,(1-ux)^r\,(1-vy)^r}\,du\,dv.$$

Then, by using the binomial theorem to perform the summation, we deduce that

$$F_4[a,b;\,c,c';\,x(1-y),y(1-x)]$$
$$= \Gamma\begin{bmatrix} c,c' \\ a,b,c-a,c'-b \end{bmatrix}$$
$$\times \int_0^1 \int_0^1 u^{a-1}v^{b-1}(1-u)^{c-a-1}(1-v)^{c'-b-1}(1-ux)^{-b}(1-vy)^{-a}$$
$$\times [1-(uvxy)/\{(1-ux)(1-vy)\}]^{c+c'-a-b-1}\,du\,dv, \quad (8.4.1.2)$$

provided that $|x|$ and $|y|$ are small enough to make the double integral convergent. This proof is due to Bailey (1941a). An alternative proof is that of Burchnall & Chaundy (1941a).

## 8.5 Double Barnes-type integrals

Just as it is possible to generalize the ordinary hypergeometric function into a double series, or Appell function, so it is possible to generalize the ordinary Barnes-type integral into a double contour integral. We shall let

$$\Gamma(s,t) \equiv \Gamma\begin{bmatrix} (a)+s, (b)-s, (c)+t, (d)-t \\ (e)+s, (f)-s, (g)+t, (h)-t, (j)+s+t, (k)-s-t \end{bmatrix}, \quad (8.5.1)$$

where, as usual, there are $A$ of the $a$ parameters, $B$ of the $b$ parameters and so on.

Now consider the integral

$$I_1 = \iint \Gamma(s,t)\, ds\, dt, \quad (8.5.2)$$

taken round a semi-circular contour $ABCD$ to the right of the imaginary axis in the $s$-plane, as in Fig. 4.1, and a similar contour $A'B'C'D'$ in the $t$-plane. Then

$$I_1 = \int_{-\frac{1}{2}\pi}^{\frac{1}{2}\pi} \int_{-\frac{1}{2}\pi}^{\frac{1}{2}\pi} \Gamma(Re^{i\theta}, Re^{i\phi})\, R^2 e^{i(\theta+\phi)}\, d\theta\, d\phi$$

$$+ \int_{-\frac{1}{2}\pi}^{\frac{1}{2}\pi} \int_{-iR}^{iR} \Gamma(s, Re^{i\phi})\, Ri\, e^{i\phi}\, ds\, d\phi$$

$$+ \int_{-iR}^{iR} \int_{-\frac{1}{2}\pi}^{\frac{1}{2}\pi} \Gamma(Re^{i\theta}, t)\, Ri\, e^{i\theta}\, d\theta\, dt$$

$$- \int_{-iR}^{iR} \int_{-iR}^{iR} \Gamma(s,t)\, ds\, dt,$$

$$= J_1 + J_2 + J_3 + J_4 \quad \text{say.} \quad (8.5.3)$$

Also $I_1$ is equal to the sum of the residues with respect to $t$ at those poles within $A'B'C'D'$ and the residues with respect to $s$ at those poles within $ABCD$, of $\Gamma(s,t)$.

Now, under suitable convergence conditions,

$$J_1, J_2 \text{ and } J_3 \to 0 \quad \text{as} \quad R \to \infty.$$

Hence $\quad -I_1 \to - \lim_{R\to\infty} J_4 = I \equiv \dfrac{1}{4\pi^2} \displaystyle\int_{-i\infty}^{i\infty} \int_{-i\infty}^{i\infty} \Gamma(s,t)\, ds\, dt. \quad (8.5.4)$

Also $\quad -I_1 \to \displaystyle\sum_{\mu=1}^{B} \sum_{\nu=1}^{D} \sum_{m=0}^{\infty} \sum_{n=0}^{\infty} \Gamma(b_\mu+m, d_\nu+n), \quad (8.5.5)$

provided that these double series are all absolutely convergent.

In a similar way, it follows from the integral

$$\int_{A'C'D'A'}\int_{ACDA}\Gamma(s,t)\,ds\,dt,$$

where $ACDA$ is a semi-circular contour in the $s$-plane to the left of $AC$, and $A'C'D'A'$ is a similar contour in the $t$-plane to the left of $A'C'$, that

$$I = \sum_{\mu=1}^{A}\sum_{\nu=1}^{C}\sum_{m=0}^{\infty}\sum_{n=0}^{\infty}\Gamma(-a_{\mu}-m,\,-c_{\nu}-n). \qquad (8.5.6)$$

Hence, under the convergence conditions

(i) $\text{Rl}\{\Sigma(a+b+c+d-e-f-g-h-j-k)$
$\qquad\qquad +2+F+H+K-B-D\} < 0,$

(ii) $\text{Rl}\{\Sigma(a+b-e-f-j-k)+1-B+F+K\} < 0,$

(iii) $\text{Rl}\{\Sigma(c+d-g-h-j-k)+1-D+H+K\} < 0,$

(iv) $A-B-E+F-J+K = 0$

and (v) $C-D-G+H-J+K = 0,$

we find that

$$\frac{-1}{4\pi^2}\int_{-i\infty}^{i\infty}\int_{-i\infty}^{i\infty}\Gamma\begin{bmatrix}(a)+s,\,(b)-s,\,(c)+t,\,(d)-t\\ (e)+s,\,(f)-s,\,(g)+t,\,(h)-t,\,(j)+s+t,\,(k)-s-t\end{bmatrix}ds\,dt$$

$$= \sum_{\mu=1}^{A}\sum_{\nu=1}^{C}\left\{\Gamma\begin{bmatrix}(a)-a_{\mu},\,(b)+a_{\mu},\,(c)-c_{\nu},\,(d)+c_{\nu}\\ (e)-a_{\mu},\,(f)+a_{\mu},\,(g)-c_{\nu},\,(h)+c_{\nu},\,(j)-a_{\mu}-c_{\nu},\,(k)+a_{\mu}+c_{\nu}\end{bmatrix}\right.$$

$$\times\sum_{m=0}^{\infty}\sum_{n=0}^{\infty}\frac{((b)+a_{\mu})_m\,((d)+c_{\nu})_n\,(1-(e)+a_{\mu})_m\,(1-(g)+c_{\nu})_n}{(1+a_{\mu}-(a))_m\,(1+c_{\nu}-(c))_n\,((f)+a_{\mu})_m\,((h)+c_{\nu})_n}$$

$$\times\frac{(1-(j)+a_{\mu}+c_{\nu})_{m+n}}{((h)+a_{\mu}+c_{\nu})_{m+n}}\,(-1)^{(A+E+J)\,m+(C+G+J)\,n}\Bigg\},$$

$$= \sum_{\mu=1}^{B}\sum_{\nu=1}^{C}\left\{\Gamma\begin{bmatrix}(a)+b_{\mu},\,(b)-b_{\mu},\,(c)+d_{\nu},\,(d)-d_{\nu}\\ (e)+b_{\mu},\,(f)-b_{\mu},\,(g)+d_{\nu},\,(h)-d_{\nu},\,(j)+b_{\mu}+d_{\nu},\,(k)-b_{\mu}-d_{\nu}\end{bmatrix}\right.$$

$$\times\sum_{m=0}^{\infty}\sum_{n=0}^{\infty}\frac{((a)+b_{\mu})_m\,((c)+d_{\nu})_n\,(1-(f)+b_{\mu})_m\,(1-(h)+d_{\nu})_n}{(1+b_{\mu}-(b))_m\,(1+d_{\nu}-(d))_n\,((e)+b_{\mu})_m\,((g)+d_{\nu})_n}$$

$$\times\frac{(1-(k)+b_{\mu}+d_{\nu})_{m+n}}{((j)+b_{\mu}+d_{\nu})_{m+n}}\,(-1)^{(B+F+K)\,m+(D+H+K)\,n}\Bigg\}. \qquad (8.5.7)$$

If there are no terms in $j$ and $k$ in the integrand, that is if $J = K = 0$, the theorem reduces to one about series of products of ordinary hypergeometric functions. We have already had an example of this type of relation in (8.4.14). This result can be deduced directly from (8.5.7), if we take $J = K = 0$ in that relation.

## 8.5.1 Integrals of Appell functions.

The general theorem of §4.7.2 can be amended to provide a Barnes-type single contour integral of each of the four Appell functions. These integrals can be summed up in the formula

$$I(z) = \frac{1}{2\pi i} \int_{-i\infty}^{i\infty} \Gamma\begin{bmatrix} (a) + s, (b) - s \\ (c) + s, (d) - s \end{bmatrix} F^{(\nu)}(x, y) \, z^s \, ds, \quad (8.5.1.1)$$

where $F^{(\nu)}(x, y)$, $(\nu = 1, 2, 3, 4)$ are the four Appell functions.

If we substitute the parameters of each Appell function in turn in the expansions of $\Sigma_{A,\infty}(z)$ and $\Sigma_{B,\infty}(z)$, given in §4.7.2, and amend the convergence conditions accordingly, we can find, for each Appell function, three complicated expansions of the general forms

$$I(z) = \Sigma_{A,\infty}(z) \sim \Sigma_{B,\infty}(z), \quad (8.5.1.2)$$

$$I(z) = \Sigma_{B,\infty}(z) \sim \Sigma_{A,\infty}(z) \quad (8.5.1.3)$$

and

$$I(1) = \Sigma_{A,\infty}(1) = \Sigma_{B,\infty}(z). \quad (8.5.1.4)$$

The exact forms of these expansions have not been published nor investigated in detail.

## 8.6 Lauricella functions

The concept of a double hypergeometric series can be extended to triple, quadruple or multiple sums, in general, though the results become progressively more complicated. Such series were first studied by Lauricella (1893), whose name they carry. The theory of general multiple series was investigated more fully by Appell (1926, Chapter VII).

Lauricella defined the four functions

$$F_A[a, b_1, b_2, \ldots, b_n, c_1, c_2, \ldots, c_n, x_1, x_2, \ldots, x_n]$$

$$= \sum_{m_1=0}^{\infty} \cdots \sum_{m_n=0}^{\infty} \frac{(a)_{m_1+m_2+\ldots+m_n}(b_1)_{m_1}(b_2)_{m_2}\cdots(b_n)_{m_n}}{(c_1)_{m_1}(c_2)_{m_2}\cdots(c_n)_{m_n}(1)_{m_1}(1)_{m_2}\cdots(1)_{m_n}} x_1^{m_1} x_2^{m_2} \ldots x_n^{m_n}, \quad (8.6.1)$$

where, for convergence,

$$|x_1| + |x_2| + \ldots + |x_n| < 1.$$

$$F_B[a_1, a_2, \ldots, a_n, b_1, b_2, \ldots, b_n, c, x_1, x_2, \ldots, x_n]$$

$$= \sum_{m_1=0}^{\infty} \cdots \sum_{m_n=0}^{\infty} \frac{(a_1)_{m_1}(a_2)_{m_2}\cdots(a_n)_{m_n}(b_1)_{m_1}(b_2)_{m_2}\cdots(b_n)_{m_n}}{(c)_{m_1+m_2+\ldots+m_n}(1)_{m_1}(1)_{m_2}\cdots(1)_{m_n}} x_1^{m_1} x_2^{m_2} \ldots x_n^{m_n}, \quad (8.6.2)$$

where, for convergence

$$|x_1| < 1, |x_2| < 1, ..., |x_n| < 1.$$

$$F_C[a, b, c_1, c_2, ..., c_n, x_1, x_2, ..., x_n]$$

$$= \sum_{m_1=0}^{\infty} \cdots \sum_{m_n=0}^{\infty} \frac{(a)_{m_1+m_2+...+m_n} (b)_{m_1+m_2+...+m_n} x_1^{m_1} x_2^{m_2} \cdots x_n^{m_n}}{(c_1)_{m_1} (c_2)_{m_2} \cdots (c_n)_{m_n} (1)_{m_1} (1)_{m_2} \cdots (1)_{m_n}}, \quad (8.6.3)$$

where, for convergence

$$|x_1^{\frac{1}{2}}| + |x_2^{\frac{1}{2}}| + ... + |x_n^{\frac{1}{2}}| < 1$$

and $F_D[a, b_1, b_2, ..., b_n, c, x_1, x_2, ..., x_n]$

$$= \sum_{m_1=0}^{\infty} \cdots \sum_{m_2=0}^{\infty} \frac{(a)_{m_1+m_2+...+m_n} (b_1)_{m_1} (b_2)_{m_2} \cdots (b_n)_{m_n}}{(c)_{m_1+m_2+...+m_n} (1)_{m_1} (1)_{m_2} \cdots (1)_{m_n}} x_1^{m_1} x_2^{m_2} \cdots x_n^{m_n},$$
$$(8.6.4)$$

where, for convergence,

$$|x_1| < 1, |x_2| < 1, ..., |x_n| < 1.$$

Appell gives multiple Euler-type integrals for these functions and he states the general forms of the systems of partial differential equations which are satisfied by them.

As we shall see below, there exist single contour integrals representing such functions, but the method of getting at these series by the use of multiple contour integrals has not been worked out systematically.

### 8.6.1 Integrals of products of hypergeometric functions.

The general theorem of § 4.7.2 can be extended further to provide general transformations and asymptotic expansions of products of hypergeometric functions, by considering integrals of the general type

$$I(z) = \frac{1}{2\pi i} \int_{-i\infty}^{i\infty} \Gamma\begin{bmatrix} (a)+s, (b)-s \\ (c)+s, (d)-s \end{bmatrix}$$

$$\times {}_{A'+B'+E}F_{C'+D'+F}\begin{bmatrix} (a')+s, (b')-s, (e); \\ (c')+s, (d')-s, (f); \end{bmatrix} x$$

$$\times {}_{A''+B''+G}F_{C''+D''+H}\begin{bmatrix} (a'')+s, (b'')-s, (g); \\ (c'')+s, (d'')-s, (h); \end{bmatrix} y \bigg] z^s \, ds. \quad (8.6.1.1)$$

Among the simplest integrals of this type are a group of integrals of $_0F_1(x)$ functions, that is Bessel functions. Thus

$$\frac{1}{2\pi i}\int_{-i\infty}^{i\infty}\Gamma[1-b+s,1-a-s]\,_0F_1[\;;a+s;x]\,_0F_1[\;;b-s;y]z^s\,ds$$

$$=z^{1-a}(1+z)^{a+b-2}\,\Gamma[2-a-b]$$

$$\times\sum_{m=0}^{\infty}\frac{x^m(1+z)^m z^{-m}}{(a+b-1)_m(1)_m}\,_0F_1[\;;a+b+m-1;(1+z)\,y],\quad(8.6.1.2)$$

where $|x|<|z|<1$, $|y|<1$, and $|\arg z|<\pi$.

In particular, if $z=1$, from the third part of our theorem, we shall have

$$\frac{1}{2\pi i}\int_{-i\infty}^{i\infty}\Gamma[1-b+s,1-a-s]\,_0F_1[\;;a+s;x]\,_0F_1[\;;b-s;y]\,ds$$

$$=2^{a+b-2}\Gamma[2-a-b]\sum_{m=0}^{\infty}\frac{2^m x^m}{(a+b-1)_m(1)_m}\,_0F_1[\;;a+b+m-1;2y],$$

$$(8.6.1.3)$$

or, since $\quad J_a(x)=x^a 2^{-a}\{\Gamma[1+a]\}^{-1}\,_0F_1[\;;1+a;-\tfrac{1}{4}x^2],$

$$\frac{1}{2\pi i}\int_{-i\infty}^{i\infty}J_{a+s}(x/2^{\frac{1}{2}})\operatorname{cosec}\{\pi(a+s)\}J_{b-s}(y/2^{\frac{1}{2}})\operatorname{cosec}\{\pi(b-s)\}y^s x^{-s}\,ds$$

$$=x^a y^b\,\Gamma[-a-b]\sum_{m=0}^{\infty}\frac{(-\tfrac{1}{4}x^2)^m}{(1+a+b)_m(1)_m}J_{a+b+m}(y).\quad(8.6.1.4)$$

Also, since

$$I_a(x)=(\tfrac{1}{2}x)^a\{\Gamma[1+a]\}^{-1}\,_0F_1[\;;1+a;\tfrac{1}{4}x^2],$$

$$\frac{1}{2\pi i}\int_{-i\infty}^{i\infty}I_{a+s}(x/2^{\frac{1}{2}})\operatorname{cosec}\{\pi(a+s)\}I_{b-s}(y/2^{\frac{1}{2}})\operatorname{cosec}\{\pi(b-s)\}y^s x^{-s}\,ds$$

$$=x^a y^b\,\Gamma[-a-b]\sum_{m=0}^{\infty}\frac{(\tfrac{1}{2}x)^{2m}}{(1+a+b)_m(1)_m}I_{a+b+m}(y).\quad(8.6.1.5)$$

Similar results hold for the other types of Bessel function.

Another example of this type of integral of products of series is Meixner's integral

$$\frac{1}{2\pi i}\int_{-i\infty}^{i\infty}\Gamma[\tfrac{1}{2}a+b+s,\tfrac{1}{2}a+b-s,\tfrac{1}{2}a-b-s,\tfrac{1}{2}a-b+s]$$

$$\times\,_1F_1[\tfrac{1}{2}a-b-s;a;x]\,_1F_1[\tfrac{1}{2}a-b+s;a;y]\,ds$$

$$=\Gamma[a+2b,a-2b]\,_1F_1[a-2b;2a;x+y].\quad(8.6.1.6)$$

There is a group of similar results, involving other confluent hypergeometric functions and Whittaker functions, but they do not all reduce so elegantly.

Integrals of the general type (8.6.1.1) can be used to deduce contour integrals representing the Lauricella functions. Thus, for the Gauss function, an integral corresponding to Meixner's integral above is

$$\frac{1}{2\pi i}\int_{-i\infty}^{i\infty} \Gamma[a+s, a'-s]\,_2F_1[a+s, b;\ c;\ x]\,_2F_1[a'-s, b';\ c';\ y]\,ds$$

$$= 2^{-a-a'}\,\Gamma[a+a']\sum_{m=0}^{\infty}\sum_{n=0}^{\infty}\frac{(a+a')_{m+n}\,(b)_m\,(b')_n}{(c)_m\,(c')_n\,(1)_m\,(1)_n}\,(\tfrac{1}{2}x)^m\,(\tfrac{1}{2}y)^n, \quad (8.6.1.7)$$

where $|x| < 1$ and $|y| < 1$. This Lauricella function on the right-hand side, is an example of the simplest kind of generalized Appell function, with three denominator parameters and two numerator parameters, and two variables.

A second example is

$$\frac{1}{2\pi i}\int_{-i\infty}^{i\infty} \Gamma\begin{bmatrix} a+s, a'-s \\ c+s, c'-s \end{bmatrix}\,_2F_1\begin{bmatrix} a+s, b; \\ c+s; \end{bmatrix} x\end{bmatrix}\,_2F_1\begin{bmatrix} a'-s, b'; \\ c'-s; \end{bmatrix} y\end{bmatrix}\,ds$$

$$= \Gamma\begin{bmatrix} a+a', c+c'-a-a'-1 \\ c-a, c'-a', c+c'-1 \end{bmatrix}\sum_{m=0}^{\infty}\sum_{n=0}^{\infty}\frac{(a+a')_{m+n}\,(b)_m\,(b')_n}{(c+c'-1)_{m+n}\,(1)_m\,(1)_n}x^m y^n,$$

$$(8.6.1.8)$$

where $|x| < 1$ and $|y| < 1$. Many similar examples can be constructed, and isolated examples of integrals of this type are scattered throughout the literature.

## 8.7 The general contour integral of Lauricella functions

We have already given several applications and extensions of the theorem of §4.7.2. We shall now state the extension of this general theorem to integrals of Lauricella functions. Let

$$I = \frac{1}{2\pi i}\int_{-i\infty}^{i\infty}\sum_{m=0}^{\infty}\sum_{n=0}^{\infty}$$

$$\times \Gamma\begin{bmatrix} (a)+u_a m+v_a n+s,\ (b)+u_b m+v_b n-s \\ (c)+u_c m+v_c n+s,\ (d)+u_d m+v_d n-s \end{bmatrix}z^s A_m B_n\,ds, \quad (8.7.1)$$

$$\Sigma_A = \sum_{\mu=1}^{A}\sum_{m=0}^{\infty}\sum_{n=0}^{\infty}\sum_{p=0}^{\infty}\Gamma\begin{bmatrix} (a)'+u_a m+v_a n-a_\mu-u_\mu m-v_\mu n-p, \\ (c)+u_c m+v_c n-a_\mu-u_\mu m-v_\mu n-p, \end{bmatrix}$$

$$(b)+u_b m+v_b n+a_\mu+u_\mu m+v_\mu n+p \\ (d)+u_d m+v_d n+a_\mu+u_\mu m+v_\mu n+p \end{bmatrix}$$

$$\times A_m B_n z^{-a_\mu-u_\mu m-v_\mu n-p}, \quad (8.7.2)$$

and

$$\Sigma_B = \sum_{\nu=1}^{B} \sum_{m=1}^{\infty} \sum_{n=0}^{\infty} \sum_{p=0}^{\infty} \Gamma \begin{bmatrix} (a) + u_a m + v_a n + b_\nu + u_\nu m + v_\nu n + p, \\ (c) + u_c m + v_c n + b_\nu + u_\nu m + v_\nu n + p, \end{bmatrix}$$

$$\begin{aligned} (b)' + u_b m + v_b n - b_\nu - u_\nu m - v_\nu n - p \\ (d) + u_d m + v_d n - b_\nu - u_\nu m - v_\nu n - p \end{bmatrix}\end{aligned}$$

$$\times A_m B_n z^{b_\nu + u_\nu m + v_\nu n + p}. \quad (8.7.3)$$

In these expressions, the $u$'s and $v$'s are zero or positive or negative integers, and the $A_m$ and $B_n$ are expressions independent of $s$, such that the double series under the integral is absolutely and uniformly convergent in $x$ and $y$. Then, provided that

$$\tfrac{1}{2}\pi(A+B-C-D) > |\arg z|,$$

we have

(i)  $I = \Sigma_A \sim \Sigma_B$, either $(a)$ when $A+D > B+C$, or $(b)$ when $A+D = B+C$ and $|z| > 1$,

(ii)  $I = \Sigma_B \sim \Sigma_A$, either $(a)$ when $A+D < B+C$, or $(b)$ when $A+D = B+C$ and $|z| < 1$,

(iii)  $I = \Sigma_A = \Sigma_B$, when $A-C = B-D \geqslant 0$, provided that $z = 1$, and $\mathrm{Rl}\,\Sigma(c+d-a-b) > 0$.

Similar theorems will hold for integrals of all the other types of function, involving double, triple, or multiple summation of products of Gamma functions of the general type

$$\Gamma[a + um + vn + wp + \ldots + s],$$

where $a$ is complex, and $u, v, w, \ldots$ are integers or zero.

From such integrals, we can deduce contour integrals representing the known types of functions which can properly be called hypergeometric. In every case, each integral will lead to an asymptotic expansion, an analytic continuation, or a simple equation connecting finite sums of such functions. It is clear also that a theory of the multiple solutions of the differential equations which underlie each type of integrand, can be developed on lines similar to the investigations of Meijer (1946 $a$–56), which are based on the first theorem above, that is (4.6.1). However, such results for these general functions become progressively more complicated and numerous at every stage. Also, the problem of devising reasonably short notations, which express clearly the results found, quickly becomes acute.

# 9

# BASIC APPELL SERIES

## 9.1 Notation

Just as we can extend the concept of a hypergeometric function to a basic function, so we can extend the concept of a double hypergeometric series to a basic double series. The four basic Appell series, analogous to the four ordinary Appell functions, can be defined as

$$\Phi^{(1)}[a;\,b,b';\,c;\,x,y] = \sum_{m=0}^{\infty}\sum_{n=0}^{\infty} \frac{(a;\,q)_{m+n}\,(b;\,q)_m\,(b';\,q)_n\,x^m y^n}{(q;\,q)_m\,(q;\,q)_n\,(c;\,q)_{m+n}}, \quad (9.1.1)$$

$$\Phi^{(2)}[a;\,b,b';\,c,c';\,x,y] = \sum_{m=0}^{\infty}\sum_{n=0}^{\infty} \frac{(a;\,q)_{m+n}\,(b;\,q)_m\,(b';\,q)_n\,x^m y^n}{(q;\,q)_m\,(q;\,q)_n\,(c;\,q)_m\,(c';\,q)_n}, \quad (9.1.2)$$

$$\Phi^{(3)}[a,a';\,b,b';\,c;\,x,y] = \sum_{m=0}^{\infty}\sum_{n=0}^{\infty} \frac{(a;\,q)_m\,(a';\,q)_n\,(b;\,q)_m\,(b';\,q)_n\,x^m y^n}{(q;\,q)_m\,(q;\,q)_n\,(c;\,q)_{m+n}} \quad (9.1.3)$$

and

$$\Phi^{(4)}[a;\,b;\,c,c';\,x,y] = \sum_{m=0}^{\infty}\sum_{n=0}^{\infty} \frac{(a;\,q)_{m+n}\,(b;\,q)_{m+n}\,x^m y^n}{(q;\,q)_m\,(q;\,q)_n\,(c;\,q)_m\,(c';\,q)_n}. \quad (9.1.4)$$

These basic Appell functions were first discussed by F. H. Jackson (1942, 1944). Alternative forms of all four functions can occur, in which, in the general term, $y^n q^{\frac{1}{2}n(n-1)}$ replaces $y^n$.

## 9.2 Integrals representing these functions

We can form basic contour integrals representing the four basic Appell functions, by the use of the general theorem of §5.5.1. Thus

$$\frac{t}{2\pi i}\int_{-i\pi/t}^{i\pi/t} \Pi \begin{bmatrix} dq^{-s},zq^s,q^{1-s}/z; \\ bq^{-s},q^s,kq^{-s}; \end{bmatrix} q \Big]\,{}_2\Phi_1\begin{bmatrix} bq^{-s},e; \\ dq^{-s}; \end{bmatrix} q,x \Big]\,ds$$

$$= \Pi \begin{bmatrix} d,z,q/z; \\ b,k,q; \end{bmatrix} q \Big]\,\Phi^{(1)}[b;\,e,k;\,d;\,x,z]$$

$$\sim \Pi \begin{bmatrix} d/b,zb,q/zb; \\ b,k/b,q; \end{bmatrix} q \Big]\,\Phi^{(1)}[b;\,e,qb/d,qb/k;\,qx/(bz),q/(bz)]$$

$$+ \Pi \begin{bmatrix} d/k,zk,q/(zk); \\ b/k,k,q; \end{bmatrix} q \Big]\,\sum_{m=0}^{\infty}\sum_{n=0}^{\infty} \frac{(b/k;\,q)_{m-n}\,(e;\,q)_m\,(k;\,q)_n\,x^m q^n}{(d/k;\,q)_{m-n}\,(q;\,q)_m\,(q;\,q)_n\,z^n k^n}, \quad (9.2.1)$$

where $\quad \mathrm{Rl}\,x > 0, \quad \mathrm{Rl}\,(b/d) > 0, \quad \mathrm{Rl}\,z > 0, \quad \mathrm{Rl}\,(q/kz) > 0.$

The three other integrals for the other basic Appell functions are:

$$\frac{t}{2\pi i}\int_{-i\pi/t}^{i\pi/t} \Pi\begin{bmatrix} zq^s, q^{1-s}/z, hq^{-s}; \\ dq^{-s}, kq^{-s}, q^s; \end{bmatrix} q \Bigg]_2\Phi_1[dq^{-s}, e; f; q, x]\,ds$$

$$= \Pi\begin{bmatrix} z, q/z, h; \\ d, k, q; \end{bmatrix} q \Bigg] \Phi^{(2)}[d; k, e; h, f; z, x]$$

$$\sim \Pi\begin{bmatrix} zk, q/zk, h/k; \\ d/k, k, q; \end{bmatrix} q \Bigg] \sum_{m=0}^{\infty}\sum_{n=0}^{\infty} \frac{(qk/h; q)_n\, (d/k; q)_{m-n}\, (k; q)_n}{(f; q)_m\, (q; q)_m\, (q; q)_n}$$

$$\times (e; q)_m\, x^m q^{-\frac{1}{2}n(n-1)}(-1)^n\, h^n k^{-2n} z^{-n} + \Pi\begin{bmatrix} zd, q/zd, h/d; \\ k/d, d, q; \end{bmatrix} q \Bigg]$$

$$\times \sum_{m=0}^{\infty}\sum_{n=0}^{\infty} \frac{(qd/h; q)_{m+n}\, (d; q)_{m+n}\, (e; q)_m\, x^m}{(qd/k; q)_{m+n}\, (f; q)_m\, (q; q)_m\, (q; q)_n}$$

$$\times h^{m+n}(dzk)^{-m-n}\, q^{m+n-\frac{1}{2}m(m+2n+1)}, \qquad (9.2.2)$$

where $\mathrm{Rl}\, z > 0$, and $\mathrm{Rl}\, x > 0$.

$$\frac{t}{2\pi i}\int_{-i\pi/t}^{i\pi/t} \Pi\begin{bmatrix} bq^{-s}, zq^s, q^{1-s}/z; \\ jq^s, kq^{-s}, q^{-s}; \end{bmatrix} q \Bigg]_2\Phi_1[e, f; bq^{-s}; q, x]\,ds$$

$$= \Pi\begin{bmatrix} bj, z/j, qj/z; \\ kj, j, q; \end{bmatrix} q \Bigg] \Phi^{(3)}[kj, e; j, f; bj; z/j, x]$$

$$\sim \Pi\begin{bmatrix} b/k, zk, q/zk; \\ jk, 1/k, q; \end{bmatrix} q \Bigg] \sum_{m=0}^{\infty}\sum_{n=0}^{\infty} \frac{(jk; q)_n\, (e; q)_m\, (f; q)_m\, x^m(-1)^n\, q^{\frac{1}{2}n(n+3)}}{(b/k; q)_{m-n}\, (q; q)_m\, (q; q)_n\, (qk; q)_n\, z^n}$$

$$+ \Pi\begin{bmatrix} b, z, q/z; \\ j, k, q; \end{bmatrix} q \Bigg] \sum_{m=0}^{\infty}\sum_{n=0}^{\infty} \frac{(j; q)_n\, (e; q)_m\, (f; q)_m\, x^m q^{\frac{1}{2}n(n+3)}(-1)^n}{(b; q)_{m-n}\, (q; q)_m\, (q/k; q)_n\, (q; q)_n\, k^n z^n},$$

$$(9.2.3)$$

where $\mathrm{Rl}\, z > 0$ and $\mathrm{Rl}\, (z/j) > 0$.

Finally,

$$\frac{t}{2\pi i}\int_{-i\pi/t}^{i\pi/t} \Pi\begin{bmatrix} hq^{-s}, zq^s, q^{1-s}/z; \\ dq^{-s}, q^{-s}, jq^s; \end{bmatrix} q \Bigg]_2\Phi_1[dq^{-s}, q^{-s}; f; q, x]\,ds$$

$$= \Pi\begin{bmatrix} hj, z/j, qj/z; \\ dj, j, q; \end{bmatrix} q \Bigg] \Phi^{(4)}[dj; j; hj, f; z/j, x]$$

$$\sim \Pi\begin{bmatrix} h/d, z/j, qj/z; \\ qj, 1/d, q; \end{bmatrix} q \Bigg] \sum_{m=0}^{\infty}\sum_{n=0}^{\infty} \frac{(dj; q)_{m+n}\, (qd/h; q)_{m+n}\, (-x)^m}{(dq; q)_{m+n}\, (b; q)_m\, (q; q)_m\, (q; q)_n}$$

$$\times q^{\frac{1}{2}m+n-\frac{1}{2}m(m+2n)}\, h^{m+n}(dz)^{-m-n}$$

$$+ \Pi\begin{bmatrix} h, z, q/z; \\ d, j, q; \end{bmatrix} q \Bigg] \sum_{m=0}^{\infty}\sum_{n=0}^{\infty} \frac{(j; q)_{m+n}\, (qd/h; q)_{m+n}\, (-x)^m}{(q/d; q)_{m+n}\, (q; q)_m\, (f; q)_m\, (q; q)_n}$$

$$\times q^{\frac{1}{2}m+n-\frac{1}{2}m(m+2n)}\, h^{m+n}(dz)^{-m-n}, \qquad (9.2.4)$$

where $\mathrm{Rl}\, (z/j) > 0$, and $\mathrm{Rl}\, x > 0$.

In particular, from (9.2.1), when $z = qb^2/d^2$, $e = d/k$ and $x = qkb^2/d^3$, we can sum the series on the left-hand side and we find that

$$\Phi^{(1)}[b; d/k, k; d; qkb^2/d^3, qb^2/d^2]$$

$$= \Pi \left[ \begin{matrix} k, q/k, d/b, qb/d, qb^3/d^2, d^2/b^3, d^3/(b^2k); \\ d, q/d, k/b, qb/k, qb^2/d^2, d^2/b^2, d^3/(b^3k); \end{matrix} \; q \right]$$

$$+ \Pi \left[ \begin{matrix} b, q/b, d/k, d/b, qb^2k/d^2, d^2/(b^2k), qb^3/d^3; \\ d, b/k, qk/b, qb^2/d^2, d^2/b^2, d/(bk), qb^2/d^3; \end{matrix} \; q \right]. \quad (9.2.5)$$

This is an example of a summation theorem for one of the basic Appell functions. There does not seem to have been any systematic attempt to find any other such summation theorems. Agarwal (1954) has given several other integrals, but he has not investigated any possible special cases which might be summable.

## 9.3 Basic double integrals

We seek next the basic analogue of theorem (8.5.6) for Appell series, so we are led to consider the integral

$$I_1 = \frac{-1}{4\pi^2} \iint \Pi(s, t) \, ds \, dt, \quad (9.3.1)$$

where

$$\Pi(s, t) \equiv \Pi \left[ \begin{matrix} (e) \, q^s, (f) \, q^{-s}, (g) \, q^t, (h) \, q^{-t}, (j) \, q^{s+t}, (k) \, q^{-s-t}; \\ (a) \, q^s, (b) \, q^{-s}, (c) \, q^t, (d) \, q^{-t}; \end{matrix} \; q \right]. \quad (9.3.2)$$

This integral is taken first round the contour $ABCD$ of Fig. (5.1) in the $s$-plane, and then round a similar contour $A'B'C'D'$ in the $t$-plane, This process leads us to one double sequence of residues. Next the integral is taken round the contour $ADEF$ of Fig. (5.1) to the left of the $Oy$ axis in the $s$-plane, and then round a similar contour $A'D'E'F'$ in the $t$-plane. This process leads us to a second double sequence of residues. Both the contours in the $s$-plane are indented so that the first $N + 1$ poles of each increasing sequence of poles fall within $ABCD$ and the first $N + 1$ poles of each decreasing sequence of poles fall within $ADEF$. The two contours in the $t$-plane are indented in a similar way.

Now

$$I_1 = \frac{-1}{4\pi^2} \int_{-i\pi/t}^{i\pi/t} \int_{-i\pi/t}^{i\pi/t} \{\Pi(R+s, R+t) - \Pi(s, R+t)$$
$$- \Pi(R+s, t) + \Pi(s, t)\} \, ds \, dt$$
$$= J_1 - J_2 - J_3 + I, \text{ say.} \quad (9.3.3)$$

But, as $I_1$ is equal to the double sum of the residues to the right of the imaginary axis, we find that

$$I_1 \to \sum_{\mu=1}^{A} \sum_{\nu=1}^{C} \sum_{m=0}^{\infty} \sum_{n=0}^{\infty} \Pi(q^{-m}/a_\mu, q^{-n}/c_\nu) \quad \text{as} \quad N \to \infty.$$

Also, under suitable convergence conditions, $J_1, J_2$, and $J_3 \to 0$, as $N \to \infty$. Hence $I_1 \to I$ as $N \to \infty$. Similarly, from the integration round contours to the left of the imaginary axis, we get

$$I = \sum_{\mu=1}^{B} \sum_{\nu=1}^{D} \sum_{m=0}^{\infty} \sum_{n=0}^{\infty} \Pi(b_\mu q^m, d_\nu q^n). \tag{9.3.4}$$

The complete theorem is then

$$-\frac{1}{4\pi^2} \int_{-i\pi/t}^{i\pi/t} \int_{-i\pi/t}^{i\pi/t} \Pi(s,t)\, ds\, dt$$

$$= \sum_{\mu=1}^{A} \sum_{\nu=1}^{C} \Pi(1/a_\mu, 1/c_\nu) \sum_{m=0}^{\infty} \sum_{n=0}^{\infty} \frac{((b)\,a_\mu;\,q)_m\,((d)\,c_\nu;\,q)_n}{((f)\,a_\mu;\,q)_m\,((h)\,c_\nu;\,q)_n}$$

$$\times \frac{(qa_\mu/(e);\,q)_m\,(qc_\nu/(g);\,q)_n\,(qa_\mu c_\nu/(j);\,q)_{m+n}}{(qa_\mu/(a);\,q)_m\,(qc_\nu/(c);\,q)_n} Q_{\mu\nu}$$

$$= \sum_{\mu=1}^{\infty} \sum_{\nu=1}^{\infty} \Pi(b_\mu, d_\nu) \sum_{m=0}^{\infty} \sum_{n=0}^{\infty} \frac{((a)\,b_\mu;\,q)_m\,((c)\,d_\nu;\,q)_n\,(qb_\mu/(f);\,q)_m}{((e)\,b_\mu;\,q)_m\,((g)\,d_\nu;\,q)_n\,((j)\,b_\mu d_\nu;\,q)_{m+n}}$$

$$\times \frac{(qd_\nu/(h);\,q)_n\,(q/((b_\mu d_\nu(k));\,q)_{m+n}}{(qb_\mu/(b);\,q)_m\,(qd_\nu/(d);\,q)_n} Q'_{\mu\nu}, \tag{9.3.5}$$

where

$$Q_{\mu\nu} = (-1)^{(A-E-J)\,m+(C-G-J)\,n} \Pi(ej/a)^m \Pi(gj/c)^n\, a_\mu^{(A-E-J)\,m}$$

$$\times c_\nu^{(C-G-J)\,n}\, q^{(A-E)\frac{1}{2}m(m+1)+(C-G)\frac{1}{2}n(n+1)}\, q^{-\frac{1}{2}J(m+n)\,(m+n+1)}$$

and

$$Q'_{\mu\nu} = (-1)^{(B-F-K)\,m+(D-H-K)\,n} \Pi(fk/b)^m \Pi(hk/d)^n\, b_\mu^{(B-F-K)\,m}$$

$$\times d_\nu^{(D-H-K)\,n}\, q^{(B-F)\frac{1}{2}m(m+1)+(D-H)\frac{1}{2}n(n+1)}\, q^{-\frac{1}{2}K(m+n)\,(m+n+1)}.$$

The conditions for convergence are:

  (i)   $B+D > F+H+2K$, or $B+D = F+H+2K$ and
       $\text{Rl}\,\Pi\{fkh/(bd)\} > 0$,

  (ii)  $B > F+K$, or $B = F+K$ and $\text{Rl}\,\Pi(fk/b) > 0$,

  (iii) $D > H+K$, or $D = H+K$ and $\text{Rl}\,\Pi(hk/d) > 0$,

  (iv) $A+C > E+G+2J$, or $A+C = E+G+2J$ and
       $\text{Rl}\,\Pi\{egj/(ac)\} > 0$,

  (v)  $A > E+J$, or $A = E+J$ and $\text{Rl}\,\Pi(ej/a) > 0$,

  (vi) $C > G+J$, or $C = G+J$ and $\text{Rl}\,\Pi(ej/c) > 0$.

The first set of double series are all convergent if

(vii) $A > E + J$, or $C > G + J$, or $A = E$, $C = G$ and $J = 0$,

and the second set of double series are all convergent if

(viii) $B > K + F$, or $D > K + H$, or $D = H$, $B = F$ and $K = 0$.

If any of these convergence conditions are not satisfied, the equality sign in the theorem can be replaced by an asymptotic approximation.

As a special case, when $J = 0$ and $K = 0$, the theorem gives transformations between products of ordinary hypergeometric series. Again this method can be extended to triple integrals, and, in general, to multiple integrals, though again the results quickly become very complicated and they have not been investigated systematically in the literature.

### 9.3.1 Single basic integrals of Appell functions.

As we might expect, the general theorem of § 5.5.1 can be extended to provide a basic contour integral of each of the four basic Appell functions. These integrals are expressed in the integral

$$I_\nu = \frac{t}{2\pi i} \int_{-i\pi/t}^{i\pi/t} \Pi \begin{bmatrix} (a)\, q^s, (b)\, q^{-s}, z q^s, q^{1-s}/z; \\ (c)\, q^s, (d)\, q^{-s}; \end{bmatrix} \Phi^{(\nu)}[x, y]\, ds, \quad (9.3.1.1)$$

where $\Phi^{(\nu)}[x, y]$, $(\nu = 1, 2, 3, 4)$ are the four basic Appell functions.

### 9.3.2 Integrals of products of basic functions.

The general theorem of § 5.5.1 can also be extended to include products of basic functions in the integrand, by considering integrals of the general type

$$I = \frac{t}{2\pi i} \int_{-i\pi/t}^{i\pi/t} \Pi \begin{bmatrix} (a)\, q^s, (b)\, q^{-s}, z q^s, q^{1-s}/z; \\ (c)\, q^s, (d)\, q^{-s}; \end{bmatrix} q$$

$$\times {}_{C'+D'+E}\Phi_{A'+B'+F} \begin{bmatrix} (c')\, q^s, (d')\, q^{-s}, (e); \\ (a')\, q^s, (b')\, q^{-s}, (f); \end{bmatrix} q, x$$

$$\times {}_{C''+D''+G}\Phi_{A''+B''+H} \begin{bmatrix} (c'')\, q^s, (d'')\, q^{-s}, (g); \\ (a'')\, q^s, (b'')\, q^{-s}, (h); \end{bmatrix} q, y \end{bmatrix} ds, \quad (9.3.2.1)$$

where $\mathrm{Rl}\, x > 0$, $\mathrm{Rl}\, y > 0$ and $|q| < 1$.

Then (i) $\qquad\qquad\qquad I = \Sigma_C,$

when (a) $C > A$ or when

$$C = A \quad \text{and} \quad \mathrm{Rl}\{(a_1 a_2 \dots a_A z)/(c_1 c_2 \dots c_C)\} > 0,$$

(b) $C' > A'$ or

$$C' = A' \quad \text{and} \quad \mathrm{Rl}\{(a_1' a_2' \dots a_A')/(c_1' c_2' \dots c_C')\} > 0,$$

and (c) $C'' > A''$ or
$$C'' = A'' \quad \text{and} \quad \text{Rl}\{(a_1'' a_2'' \dots a_A'')/(c_1'' c_2'' \dots c_C'')\} > 0.$$
(ii) $I \sim \Sigma_C$ when   (a) $C < A$,   or   (b) $C' < A'$,   or   (c) $C'' < A''$.

Also (iii)          $I = \Sigma_D$, when   (a) $D > B$

or when
$$D = B \quad \text{and} \quad \text{Rl}\{(b_1 b_2 \dots b_B q)/(d_1 d_2 \dots d_D z)\} > 0,$$
    (b) $D' > B'$ or
$$D' = B' \quad \text{and} \quad \text{Rl}\{(b_1' b_2' \dots b_B')/(d_1' d_2' \dots d_D')\} > 0,$$
and (c) $D'' > B''$ or
$$D'' = B'' \quad \text{and} \quad \text{Rl}\{(b_1'' b_2'' \dots b_B'')/(d_1'' d_2'' \dots d_D'')\} > 0.$$
(iv) $I \sim \Sigma_D$ when   (a) $D < B$,   or   (b) $D' < B'$,   or   (c) $D'' < B''$.

The proof follows on the same lines as that of §5.5.1. As an example of this theorem, we quote the basic analogue of Meixner's integral (8.6.1.6), which is

$$\frac{t}{2\pi i} \int_{-i\pi/t}^{i\pi/t} \Pi[c, q;\ cq^s, q^{-s};\ q]\,{}_1\Phi_1[cq^s;\ a;\ q, x]\,{}_1\Phi_1[q^{-s};\ b;\ q, y]\, ds$$

$$= \sum_{p=0}^{\infty} \sum_{m=0}^{\infty} \frac{(c;\ q)_{p+m}\,(-1)^p\, q^{\frac{1}{2}p(p+1)}\, x^m}{(q;\ q)_p\,(a;\ q)_p\,(q;\ q)_m}\,{}_1\Phi_1[cq^{p+m};\ b;\ q, y], \quad (9.3.2.2)$$

where $\text{Rl}\, x > 0$ and $\text{Rl}\, y > 0$.

## 9.4 The general contour integral of basic Lauricella functions

We have had several generalizations and extensions of the theorem of §5.5.1; we shall now state the basic analogue of the theorem of §8.7 for the Lauricella functions. We shall write

$$I \equiv \frac{t}{2\pi i} \int_{-i\pi/t}^{i\pi/t} \sum_{m=0}^{\infty} \sum_{n=0}^{\infty} \Pi(q^s)\, A_m B_n\, ds, \quad (9.4.1)$$

where
$$\Pi(q^s) \equiv \Pi \begin{bmatrix} (a)\ q^{u_a m + v_a n + s},\ (b)\ q^{u_b m + v_b n - s},\ zq^s, q^{1-s}/z; \\ (c)\ q^{u_c m + v_c n + s},\ (d)\ q^{u_d m + v_d n - s}; \end{bmatrix} q \end{bmatrix},$$

$$A_m = \frac{((e);\ q)_m\, x^m}{(q;\ q)_m\,((f);\ q)_m} \quad \text{and} \quad B_n = \frac{((g);\ q)_n\, y^n}{(q;\ q)_n\,((h);\ q)_n}.$$

Also, let
$$\Sigma_C \equiv \sum_{\nu=1}^{C} \sum_{m=0}^{\infty} \sum_{n=0}^{\infty} \sum_{p=0}^{\infty} \frac{\Pi(q^{-u_\nu m - v_\nu n - p}/c_\nu)}{(q;\ q)_p}\, A_m B_n (-1)^p\, q^{\frac{1}{2}p(p+1)}$$

and
$$\Sigma_D \equiv \sum_{\mu=1}^{D} \sum_{m=0}^{\infty} \sum_{n=0}^{\infty} \sum_{p=0}^{\infty} \frac{\Pi(d_\mu q^{u_\mu m + v_\mu n + p})}{(q;\ q)_p}\, A_m B_n (-1)^p\, q^{\frac{1}{2}p(p+1)},$$

where all the $u$'s and $v$'s are integers or zero, and
$$|q| < 1,\, \text{Rl}\, x > 0,\, \text{Rl}\, y > 0.$$

Then $\qquad\qquad I = \Sigma_C \quad \text{or} \quad I \sim \Sigma_C.$

Also $\qquad\qquad I = \Sigma_D \quad \text{or} \quad I \sim \Sigma_D,$

under the various conditions for convergence or divergence, similar to those outlined in the previous sections.

In this way, we can apply the ideas of §5.5.1, to contour integrals with all the types of function which can occur in the integrand. These integrands can, therefore, involve double, triple or multiple summation of groups of products of the general type

$$\Pi((a)\, q^{um+vn+wp+\dots \pm s}),$$

where the $a$'s are complex numbers and the $u, v, w, \dots$, are integers or zero.

From contour integrals of this type, we can deduce integrals representing all the known types of functions which can properly be called basic hypergeometric. In every case, each integral will lead to transformations, analytic continuation formulae, or asymptotic representations of sums of such functions.

# APPENDIX I

## Relations between products of the type $(a)_n$

$$(a)_n = a(a+1)(a+2)(a+3)\ldots(a+n-1), \quad (a)_0 = 1. \tag{I.1}$$

$$(a)_n = \frac{\Gamma(a+n)}{\Gamma(a)}. \tag{I.2}$$

$$(a+n)_n = \frac{(a)_{2n}}{(a)_n}, \tag{I.3}$$

and, in general,

$$(a+kn)_n = \frac{(a)_{(k+1)n}}{(a)_{kn}}. \tag{I.4}$$

$$(a-n)_n = (-1)^n(1-a)_n, \tag{I.5}$$

and, in general,

$$(a-kn)_n = \frac{(-1)^n(1-a)_{kn}}{(1-a)_{(k-1)n}}. \tag{I.6}$$

$$(a)_{-n} = \frac{(-1)^n}{(1-a)_n}. \tag{I.7}$$

$$(a)_{-n} = \frac{\Gamma(a-n)}{\Gamma(a)}. \tag{I.8}$$

$$(a)_{N-n} = \frac{(-1)^n(a)_N}{(1-a-N)_n}. \tag{I.9}$$

$$(a+n)_{N-n} = \frac{(a)_N}{(a)_n}, \tag{I.10}$$

and, in general,

$$(a+kn)_{N-n} = \frac{(a)_N(a+N)_{(k-1)n}}{(a)_{kn}}. \tag{I.11}$$

$$(a-n)_{N-n} = \frac{(a)_N(-1)^n(1-a)_n}{(1-a-N)_{2n}}, \tag{I.12}$$

and, in general,

$$(a-kn)_{N-n} = \frac{(-1)^n(a)_N(1-a)_{kn}}{(1-a-N)_{(k+1)n}}. \tag{I.13}$$

$$(a+n)_N = \frac{(a)_N(a+N)_n}{(a)_n}. \tag{I.14}$$

$$(a-n)_N = \frac{(1-a)_n(a)_N}{(1-a-N)_n}. \tag{I.15}$$

$$(a+n)_{N-2n} = \frac{(-1)^n(a)_N}{(1-a-N)_n(a)_n}, \tag{I.16}$$

and, in general,

$$(a+kn)_{N-jn} = \frac{(a)_N\,(a+N)_{kn-jn}}{(a)_{kn}}. \tag{I.17}$$

$$(a-n)_{N-2n} = \frac{(a)_N\,(1-a)_n}{(1-a-N)_{3n}}, \tag{I.18}$$

and, in general,

$$(a-kn)_{N-jn} = \frac{(a)_N\,(-1)^{(k+j)\,n}\,(1-a)_{kn}}{(1-a-N)_{jn}}. \tag{I.19}$$

$$(a+n)_{2n} = \frac{(a)_{3n}}{(a)_n}, \tag{I.20}$$

and, in general,

$$(a+kn)_{jn} = \frac{(a)_{(k+j)n}}{(a)_{kn}}. \tag{I.21}$$

$$(a-n)_{2n} = (-1)^n\,(a)_n\,(1-a)_n, \tag{I.22}$$

and, in general,

$$(a-kn)_{jn} = (a)_{(j-k)n}\,(-1)^{kn}\,(1-a)_{kn}, \quad \text{if} \quad j \geqslant k, \tag{I.23}$$

$$= \frac{(1-a)_{kn}\,(-1)^{jn}}{(1-a)_{(k-j)\,n}}, \quad \text{if} \quad j < k. \tag{I.24}$$

$$(a)_{2n} = (\tfrac{1}{2}a)_n\,(\tfrac{1}{2}a+\tfrac{1}{2})_n\,2^{2n}, \tag{I.25}$$

and, in general,

$$(a)_{kn} = (a/k)_n\,((a+1)/k)_n\,((a+2)/k)_n \cdots ((a+k-1)/k)_n\,k^{kn}. \tag{I.26}$$

$$\Gamma(1-z) = \frac{\pi}{\sin(\pi z)\,\Gamma(z)}. \tag{I.27}$$

$$\Gamma(1+z) = z! = \int_0^\infty u^z e^{-u}\,du. \tag{I.28}$$

$$\Gamma(a-n) = \Gamma(a)/(a-n)_n. \tag{I.29}$$

$$\Gamma(a-n) = \Gamma(a)\,(-1)^n/(1-a)_n. \tag{I.30}$$

# APPENDIX II

## Relations between products of the type $(a; q)_n$

$$(a; q)_n = (1-a)(1-aq)(1-aq^2) \ldots (1-aq^{n-1}), \quad (a; q)_0 = 1. \quad \text{(II.1)}$$

$$(a; q)_n = \prod_{r=0}^{\infty} \frac{(1-aq^r)}{(1-aq^{n+r})} = \Pi \begin{bmatrix} a & ; \\ aq^n; & q \end{bmatrix}. \quad \text{(II.2)}$$

$$(aq^n; q)_n = \frac{(a; q)_{2n}}{(a; q)_n}, \quad \text{(II.3)}$$

and, in general,

$$(aq^{kn}; q)_n = \frac{(a; q)_{(k+1)n}}{(a; q)_{kn}}. \quad \text{(II.4)}$$

$$(aq^{-n}; q)_n = (-a)^n q^{-\frac{1}{2}(n+1)n} (q/a; q)_n, \quad \text{(II.5)}$$

and, in general,

$$(aq^{-kn}; q)_n = (-a)^n q^{\frac{1}{2}(n-1)n-kn} \frac{(q/a; q)_{kn}}{(q/a; q)_{(k-1)n}}. \quad \text{(II.6)}$$

$$(a; q)_{-n} = \frac{(-a)^{-n} q^{\frac{1}{2}n(n+1)}}{(q/a; q)_n}. \quad \text{(II.7)}$$

$$(a; q)_{-n} = \prod_{r=0}^{\infty} \frac{(1-aq^{-n+r})}{(1-aq^r)}. \quad \text{(II.8)}$$

$$(a; q)_{N-n} = \frac{(a; q)_N \, q^{\frac{1}{2}(n+1)n}}{(q^{1-N}/a; q)_n \, (-a)^n \, q^{Nn}}, \quad = \frac{(aq^{-n}; q)_N}{(a; q)_{-n}}. \quad \text{(II.9)}$$

$$(aq^n; q)_{N-n} = \frac{(a; q)_N}{(a; q)_n}, \quad \text{(II.10)}$$

and, in general,

$$(aq^{kn}; q)_{N-n} = \frac{(a; q)_N \, (aq^N; q)_{(k-1)n}}{(a; q)_{kn}}. \quad \text{(II.11)}$$

$$(aq^{-n}; q)_{N-n} = \frac{(a; q)_N \, (q/a; q)_n \, (-1)^n \, q^{\frac{1}{2}(3n+1)n-2Nn}}{(q^{1-N}/a; q)_{2n} \, a^n}. \quad \text{(II.12)}$$

$$(aq^{-n}; q)_N = \frac{(a; q)_N \, (q/a; q)_n}{(q^{1-N}/a; q)_n} \, q^{-Nn}. \quad \text{(II.13)}$$

$$(aq^n; q)_N = \frac{(a; q)_N \, (aq^N; q)_n}{(a; q)_n}. \quad \text{(II.14)}$$

$$(aq^n; q)_{N-2n} = \frac{(a; q)_N (-1)^n q^{\frac{1}{2}(n+1)n}}{(a; q)_n (q^{1-N}/a; q)_n a^n q^{Nn}}. \tag{II.15}$$

$$(a; q^2)_n = (\sqrt{a}; q)_n (-\sqrt{a}; q)_n. \tag{II.16}$$

$$(a; q)_{2n} = (a; q^2)_n (aq; q^2)_n. \tag{II.17}$$

$$(a; q)_{3n} = (a; q^3)_n (aq; q^3)_n (aq^2; q^3)_n. \tag{II.18}$$

$$\frac{q^{nr}(q^{-n}; q)_r}{(q; q)_n} = \frac{(-1)^r q^{\frac{1}{2}r(r-1)}}{(q; q)_{n-r}}. \tag{II.19}$$

# APPENDIX III

## Summation theorems for ordinary hypergeometric series

The binomial theorem,

$$_1F_0[a; \; ; z] = (1-z)^{-a}. \tag{III.1}$$

Saalschutz's theorem,

$$_3F_2[a, b, -n; c, d; 1] = \frac{(c-a)_n (c-b)_n}{(c)_n (c-a-b)_n}, \tag{III.2}$$

provided that $c+d = a+b-n+1$.
Gauss's theorem,

$$_2F_1[a, b; c; 1] = \Gamma\begin{bmatrix} c, c-a-b \\ c-a, c-b \end{bmatrix}, \tag{III.3}$$

or, when $b = -n$, Vandermonde's theorem,

$$_2F_1[a, -n; c; 1] = \frac{(c-a)_n}{(c)_n}. \tag{III.4}$$

Kummer's theorem,

$$_2F_1[a, b; 1+a-b; -1] = \Gamma\begin{bmatrix} 1+a-b, 1+\frac{1}{2}a \\ 1+a, 1+\frac{1}{2}a-b \end{bmatrix}. \tag{III.5}$$

Gauss's second theorem,

$$_2F_1[a, b; \tfrac{1}{2}+\tfrac{1}{2}a+\tfrac{1}{2}b; \tfrac{1}{2}] = \Gamma\begin{bmatrix} \frac{1}{2}, \frac{1}{2}+\frac{1}{2}a+\frac{1}{2}b \\ \frac{1}{2}+\frac{1}{2}a, \frac{1}{2}+\frac{1}{2}b \end{bmatrix}. \tag{III.6}$$

Bailey's theorem,

$$_2F_1[a, 1-a; c; \tfrac{1}{2}] = \Gamma\begin{bmatrix} \frac{1}{2}c, \frac{1}{2}+\frac{1}{2}c \\ \frac{1}{2}c+\frac{1}{2}a, \frac{1}{2}+\frac{1}{2}c-\frac{1}{2}a \end{bmatrix}. \tag{III.7}$$

Dixon's theorem,

$$_3F_2\begin{bmatrix} a, & b, & c; \\ & 1+a-b, 1+a-c; \end{bmatrix} 1$$
$$= \Gamma\begin{bmatrix} 1+\frac{1}{2}a, 1+a-b, 1+a-c, 1+\frac{1}{2}a-b-c \\ 1+a, 1+\frac{1}{2}a-b, 1+\frac{1}{2}a-c, 1+a-b-c \end{bmatrix}, \tag{III.8}$$

or, if $c = -n$,

$$_3F_2\begin{bmatrix} a, & b, & -n; \\ & 1+a-b, 1+a+n; \end{bmatrix} 1 = \frac{(1+a)_n (1+\frac{1}{2}a-b)_n}{(1+\frac{1}{2}a)_n (1+a-b)_n}. \tag{III.9}$$

$$_4F_3\begin{bmatrix} a, 1+\frac{1}{2}a, & b, & c; \\ \frac{1}{2}a, 1+a-b, 1+a-c; \end{bmatrix} -1 = \Gamma\begin{bmatrix} 1+a-b, 1+a-c \\ 1+a, 1+a-b-c \end{bmatrix}, \tag{III.10}$$

or, if $c = -n$,

$$_4F_3\left[\begin{matrix} a, 1+\tfrac{1}{2}a, & b, & -n; \\ \tfrac{1}{2}a, 1+a-b, 1+a+n; \end{matrix} -1\right] = \frac{(1+a)_n}{(1+a-b)_n}. \quad \text{(III. 11)}$$

$$_5F_4\left[\begin{matrix} a, 1+\tfrac{1}{2}a, & b, & c, & d; \\ \tfrac{1}{2}a, 1+a-b, 1+a-c, 1+a-d; \end{matrix} 1\right]$$

$$= \Gamma\left[\begin{matrix} 1+a-b, 1+a-c, 1+a-d, 1+a-b-c-d \\ 1+a, 1+a-b-c, 1+a-b-d, 1+a-c-d \end{matrix}\right], \quad \text{(III. 12)}$$

or, if $d = -n$,

$$_5F_4\left[\begin{matrix} a, 1+\tfrac{1}{2}a, & b, & c, & -n; \\ \tfrac{1}{2}a, 1+a-b, 1+a-c, 1+a+n; \end{matrix} 1\right] = \frac{(1+a)_n(1+a-b-c)_n}{(1+a-b)_n(1+a-c)_n}. \quad \text{(III. 13)}$$

Dougall's theorem,

$$_7F_6\left[\begin{matrix} a, 1+\tfrac{1}{2}a, & b, & c, & d, & e, & -n; \\ \tfrac{1}{2}a, 1+a-b, 1+a-c, 1+a-d, 1+a-e, 1+a+n; \end{matrix} 1\right]$$

$$= \frac{(1+a)_n(1+a-b-c)_n(1+a-b-d)_n(1+a-c-d)_n}{(1+a-b)_n(1+a-c)_n(1+a-d)_n(1+a-b-c-d)_n}, \quad \text{(III. 14)}$$

provided that $1 + 2a = b + c + d + e - n$.

Nearly-poised summation theorems,

$$_3F_2\left[\begin{matrix} a, 1+\tfrac{1}{2}a, -n; \\ \tfrac{1}{2}a, & b; \end{matrix} 1\right] = (b-a-1-n)\frac{(b-a)_{n-1}}{(b)_n}, \quad \text{(III. 15)}$$

$$_3F_2\left[\begin{matrix} a, & b, & -n; \\ 1+a-b, 1+2b-n; \end{matrix} 1\right] = \frac{(a-2b)_n(1+\tfrac{1}{2}a-b)_n(-b)_n}{(1+a-b)_n(\tfrac{1}{2}a-b)_n(-2b)_n}, \quad \text{(III. 16)}$$

$$_4F_3\left[\begin{matrix} a, 1+\tfrac{1}{2}a, & b, & -n; \\ \tfrac{1}{2}a, 1+a-b, 1+2b-n; \end{matrix} 1\right] = \frac{(a-2b)_n(-b)_n}{(1+a-b)_n(-2b)_n}, \quad \text{(III. 17)}$$

and

$$_4F_3\left[\begin{matrix} a, 1+\tfrac{1}{2}a, & b, & -n; \\ \tfrac{1}{2}a, 1+a-b, 2+2b-n; \end{matrix} 1\right]$$

$$= \frac{(a-2b-1)_n(\tfrac{1}{2}a+\tfrac{1}{2}-b)_n(-b-1)_n}{(1+a-b)_n(\tfrac{1}{2}a-\tfrac{1}{2}-b)_n(-2b-1)_n}. \quad \text{(III. 18)}$$

$$_7F_6\left[\begin{matrix} a, 1+\tfrac{1}{2}a, & \tfrac{1}{2}d, & \tfrac{1}{2}+\tfrac{1}{2}d, a-d, 1+2a-d+n, -n; \\ \tfrac{1}{2}a, 1+a-\tfrac{1}{2}d, a+\tfrac{1}{2}-\tfrac{1}{2}d, 1+d, d-a-n, 1+a+n; \end{matrix} 1\right]$$

$$= \frac{(1+a)_n(1+2a-2d)_n}{(1+a-d)_n(1+2a-d)_n}. \quad \text{(III. 19)}$$

$$_4F_3\left[\begin{array}{c}\tfrac{1}{2}a,\ \tfrac{1}{2}+\tfrac{1}{2}a,\ b+n,\ -n;\\ \tfrac{1}{2}b,\ \tfrac{1}{2}b+\tfrac{1}{2},\ 1+a;\end{array}1\right]=\frac{(b-a)_n}{(b)_n}.\qquad\text{(III.20)}$$

$$_3F_2\left[\begin{array}{c}a,\ 1+\tfrac{1}{2}a,\qquad b;\\ \tfrac{1}{2}a,\ 1+a-b;\end{array}-1\right]=\Gamma\left[\begin{array}{c}\tfrac{1}{2}+\tfrac{1}{2}a,\ 1+a-b\\ 1+a,\ \tfrac{1}{2}+\tfrac{1}{2}a-b\end{array}\right].\quad\text{(III.21)}$$

$$_4F_3\left[\begin{array}{c}a,\ 1+\tfrac{1}{2}a,\qquad b,\qquad c;\\ \tfrac{1}{2}a,\ 1+a-b,\ 1+a-c;\end{array}1\right]$$

$$=\Gamma\left[\begin{array}{c}\tfrac{1}{2}+\tfrac{1}{2}a,\ 1+a-b,\ 1+a-c,\ \tfrac{1}{2}+\tfrac{1}{2}a-b-c\\ 1+a,\ \tfrac{1}{2}+\tfrac{1}{2}a-b,\ \tfrac{1}{2}+\tfrac{1}{2}a-c,\ 1+a-b-c\end{array}\right].\quad\text{(III.22)}$$

Watson's theorem,

$$_3F_2\left[\begin{array}{c}a,\qquad b,\ c;\\ \tfrac{1}{2}+\tfrac{1}{2}a+\tfrac{1}{2}b,\ 2c;\end{array}1\right]=\Gamma\left[\begin{array}{c}\tfrac{1}{2},\ c+\tfrac{1}{2},\ \tfrac{1}{2}+\tfrac{1}{2}a+\tfrac{1}{2}b,\ \tfrac{1}{2}-\tfrac{1}{2}a-\tfrac{1}{2}b+c\\ \tfrac{1}{2}+\tfrac{1}{2}a,\ \tfrac{1}{2}+\tfrac{1}{2}b,\ \tfrac{1}{2}-\tfrac{1}{2}a+c,\ \tfrac{1}{2}-\tfrac{1}{2}b+c\end{array}\right].$$
$$\text{(III.23)}$$

Whipple's theorem,

$$_3F_2[a,b,c;\ d,e;\ 1]=\pi 2^{1-2c}\Gamma\left[\begin{array}{c}d,\ e\\ \tfrac{1}{2}a+\tfrac{1}{2}e,\ \tfrac{1}{2}a+\tfrac{1}{2}d,\ \tfrac{1}{2}d+\tfrac{1}{2}e,\ \tfrac{1}{2}b+\tfrac{1}{2}d\end{array}\right]$$
$$\text{(III.24)}$$

provided that $a+b=1$, and that $d+e=1+2c$.

$$_3F_2\left[\begin{array}{c}a,\ 1+\tfrac{1}{2}a,\qquad -n;\\ \tfrac{1}{2}a,\ 1+a+n;\end{array}-1\right]=\frac{(1+a)_n}{(\tfrac{1}{2}+\tfrac{1}{2}a)_n}.\qquad\text{(III.25)}$$

$$_4F_3\left[\begin{array}{c}a,\ 1+\tfrac{1}{2}a,\qquad b,\qquad -n;\\ \tfrac{1}{2}a,\ 1+a-b,\ 1+a+n;\end{array}1\right]=\frac{(1+a)_n\,(\tfrac{1}{2}+\tfrac{1}{2}a-b)_n}{(\tfrac{1}{2}+\tfrac{1}{2}a)_n\,(1+a-b)_n}.\quad\text{(III.26)}$$

$$_6F_5\left[\begin{array}{c}a,\ 1+\tfrac{1}{2}a,\qquad b,\qquad c,\qquad d,\qquad e;\\ \tfrac{1}{2}a,\ 1+a-b,\ 1+a-c,\ 1+a-d,\ 1+a-e;\end{array}-1\right]$$

$$=\Gamma\left[\begin{array}{l}1+a-b,\ 1+a-c,\ 1+a-d,\ 1+a-e,\\ a,\ 1+a,\ 1+a-c-d,\ a+c+e,\\ \qquad\qquad 1+\tfrac{1}{2}+\tfrac{1}{2}a-\tfrac{1}{2}b-\tfrac{1}{2}c,\ \tfrac{1}{2}+\tfrac{1}{2}a-\tfrac{1}{2}d-\tfrac{1}{2}e\\ \qquad\qquad 1+\tfrac{1}{2}a-\tfrac{1}{2}b-\tfrac{1}{2}d,\ 1+\tfrac{1}{2}a-\tfrac{1}{2}c-\tfrac{1}{2}e\end{array}\right],$$
$$\text{(III.27)}$$

where $1=b+c=d+e$.

Bilateral series,

$$_2H_2\left[\begin{array}{c}a,b;\\ c,d;\end{array}1\right]=\Gamma\left[\begin{array}{c}c,d,1-a,c+d-a-b-1\\ c-a,d-a,c-b,d-b\end{array}\right],\qquad\text{(III.28)}$$

$$_5H_5\left[\begin{array}{c}1+\tfrac{1}{2}a,\qquad b,\qquad c,\qquad d,\qquad e;\\ \tfrac{1}{2}a,\ 1+a-b,\ 1+a-c,\ 1+a-d,\ 1+a-e;\end{array}1\right]$$

$$=\Gamma\left[\begin{array}{l}1-b,1-c,1-d,1-e,1+2a-b-c-d-e,1+a-b,\\ 1+a,1-a,1+a-b-c,1+a-b-d,1+a-b-e,\\ \qquad\qquad 1+a-c,1+a-d,1+a-e\\ \qquad\qquad 1+a-c-d,1+a-c-e,1+a-d-e\end{array}\right],$$
$$\text{(III.29)}$$

and if $e = \frac{1}{2}a$,

$$_3H_3\left[\begin{matrix} b, & c, & d; \\ 1+a-b, 1+a-c, 1+a-d; \end{matrix} \, 1\right]$$

$$= \Gamma\left[\begin{matrix} 1-b, 1-c, 1-d, 1-\tfrac{1}{2}a, 1+\tfrac{3}{2}a-b-c-d, 1+a-b, \\ 1+a, 1-a, 1+a-b-c, 1+a-b-d, 1+\tfrac{1}{2}a-b, \end{matrix}\right.$$
$$\left.\begin{matrix} 1+a-c, 1+a-d, 1+\tfrac{1}{2}a \\ 1+\tfrac{1}{2}a-c, 1+\tfrac{1}{2}a-d, 1+a-c-d \end{matrix}\right].$$

$$\text{(III. 30)}$$

Saalschutz's theorem, in the non-terminating form,

$$_3F_2\left[\begin{matrix} e, & f, & g; \\ 1+a, e+f+g-a; \end{matrix} \, 1\right]$$

$$+\Gamma\left[\begin{matrix} 1+a, 1+a-e-f, 1+a-e-g, 1+a-f-g, e+f+g-a-1 \\ 1+a-e-f-g, 2+2a-e-f-g, e, f, g \end{matrix}\right]$$

$$\times\, _3F_2\left[\begin{matrix} 1+a-f-g, 1+a-e-g, 1+a-e-f; \\ 2+a-e-f-g, 2+2a-e-f-g; \end{matrix} \, 1\right]$$

$$= \Gamma\left[\begin{matrix} e, f, g, e+a, f+a, g+a \\ e+f-a, e+g-a, f+g-a \end{matrix}\right]. \quad \text{(III. 31)}$$

Dougall's theorem in the non-terminating form,

$$_7F_6\left[\begin{matrix} a, 1+\tfrac{1}{2}a, & b, & c, & d, & e, & f; \\ \tfrac{1}{2}a, 1+a-b, 1+a-c, 1+a-d, 1+a-e, 1+a-f; \end{matrix} \, 1\right]$$

$$+\Gamma\left[\begin{matrix} 1+2b-a, b+c-a, b+d-a, b+e-a, b+f-a, \\ 1+b-c, 1+b-d, 1+b-e, 1+b-f, b-a, \end{matrix}\right.$$
$$\left.\begin{matrix} a-b, 1+a-c, 1+a-d, 1+a-e, 1+a-f \\ 1+a, c, d, e, f \end{matrix}\right]$$

$$\times\, _7F_6\left[\begin{matrix} 2b-a, 1+b-\tfrac{1}{2}a, & b, b+c-a, b+d-a, b+e-a, b+f-a; \\ b-\tfrac{1}{2}a, 1+b-a, 1+b-c, 1+b-d, 1+b-e, 1+b-f; \end{matrix} \, 1\right]$$

$$= \Gamma\left[\begin{matrix} 1+a-c, 1+a-d, 1+a-e, 1+a-f, b+c-a, \\ 1+a, b-a, 1+a-d-e, 1+a-c-e, 1+a-c-d, \end{matrix}\right.$$
$$\left.\begin{matrix} b+d-a, b+e-a, b+f-a \\ 1+a-c-f, 1+a-d-f, 1+a-e-f \end{matrix}\right],$$

$$\text{(III. 32)}$$

where $1 + 2a = b + c + d + e + f$.

## APPENDIX IV

### Summation theorems for basic series

Vandermonde's analogue,

$$_2\Phi_1[a, q^{-n}; b; q, q] = \frac{(b/a; q)_n a^n}{(b; q)_n}. \tag{IV.1}$$

Gauss's analogue,

$$_2\Phi_1[a, b; c; q, c/ab] = \Pi\begin{bmatrix} c/a, c/b; \\ c, c/ab; \end{bmatrix} q \Big], \tag{IV.2}$$

or, when $b = q^{-n}$,

$$_2\Phi_1[a, q^{-n}; c; q, cq^n/a] = \frac{(c/a; q)_n}{(c; q)_n}. \tag{IV.3}$$

Saalschutz's analogue,

$$_3\Phi_2[a, b, q^{-n}; c, d; q, q] = \frac{(c/a; q)_n (c/b; q)_n}{(c; q)_n (c/ab; q)_n}, \tag{IV.4}$$

provided that $cd = abq^{1-n}$.

Dixon's analogue,

$$_4\Phi_3\begin{bmatrix} a, -q\sqrt{a}, & b, & q^{-n}; \\ -\sqrt{a}, aq/b, aq^{1+n}; \end{bmatrix} q, q^{1+n}\sqrt{a/b}\Big] = \frac{(aq; q)_n (\sqrt{a}q/b; q_n)}{(\sqrt{a}q; q)_n (aq/b; q)_n}, \tag{IV.5}$$

or for general $c$,

$$_4\Phi_3\begin{bmatrix} a, -q\sqrt{a}, & b, & c; \\ -\sqrt{a}, aq/b, aq/c; \end{bmatrix} q, q\sqrt{a}/bc\Big] = \Pi\begin{bmatrix} aq, \sqrt{a}q/b, \sqrt{a}q/c, aq/bc; \\ aq/b, aq/c, \sqrt{a}q, \sqrt{a}q/bc; \end{bmatrix} q \Big]. \tag{IV.6}$$

$$_6\Phi_5\begin{bmatrix} a, q\sqrt{a}, -q\sqrt{a}, & b, & c, & d; \\ \sqrt{a}, & -\sqrt{a}, aq/b, aq/c, aq/d; \end{bmatrix} q, aq/bcd\Big]$$

$$= \Pi\begin{bmatrix} aq, aq/cd, aq/bd, aq/bc; \\ aq/b, aq/c, aq/d, aq/bcd; \end{bmatrix} q \Big]. \tag{IV.7}$$

Jackson's theorem,

$$_8\Phi_7\begin{bmatrix} a, q\sqrt{a}, -q\sqrt{a}, & b, & c, & d, & e, & q^{-n}; \\ \sqrt{a}, & -\sqrt{a}, aq/b, aq/c, aq/d, aq/e, aq^{1+n}; \end{bmatrix} q, q\Big]$$

$$= \frac{(aq)_n (aq/bc)_n (aq/cd)_n (aq/bd)_n}{(aq/b)_n (aq/c)_n (aq/d)_n (aq/bcd)_n}, \tag{IV.8}$$

provided that

$$a^2q = bcdeq^{-n}.$$

$$_6\Phi_5\begin{bmatrix} a, q\sqrt{a}, -q\sqrt{a}, & b, & c, & q^{-N}; \\ \sqrt{a}, & -\sqrt{a}, aq/b, aq/c, aq^{1+N}; \end{bmatrix} q, aq^{1+N}/bc\Big]$$

$$= \frac{(aq; q)_N (\sqrt{a}q/b; q)_N}{(aq/b; q)_N (aq/c; q)_N}. \tag{IV.9}$$

Euler's theorem,

$$_0\Phi_0[\ ;\ ;q,z] = 1/\Pi(z;q). \tag{IV.10}$$

Heine's theorem,

$$_1\Phi_0[a;\ ;q,z] = \Pi\begin{bmatrix} az; \\ z; \end{bmatrix}q\,. \tag{IV.11}$$

Basic bilateral series, Jacobi's theorem,

$$_1\Psi_1[a;b;q,z] = \Pi\begin{bmatrix} b/a, az, q/az, q; \\ q/a, b/az, b, z; \end{bmatrix}q\,, \tag{IV.12}$$

$$_2\Psi_2\begin{bmatrix} b, & c; \\ aq/b, aq/c; \end{bmatrix}q, -aq/bc\,]$$

$$= \Pi\begin{bmatrix} aq/bc; \\ q/b, q/c, aq/b, aq/c, -aq/bc; \end{bmatrix}q\,]$$

$$\times \Pi[aq^2/b^2, aq^2/c^2, q^2, aq, q/a; q^2], \tag{IV.13}$$

$$_6\Psi_6\begin{bmatrix} q\sqrt{a}, -q\sqrt{a}, & b, & c, & d, & e; \\ \sqrt{a}, & -\sqrt{a}, aq/b, aq/c, aq/d, aq/e; \end{bmatrix}q, a^2q/bcde\,]$$

$$= \Pi\begin{bmatrix} aq, aq/bc, aq/bd, aq/be, aq/cd, aq/ce, aq/de, q, q/a; \\ q/b, q/c, q/d, q/e, aq/b, aq/c, aq/d, aq/e, a^2q/bcde; \end{bmatrix}q\,]. \tag{IV.14}$$

The non-terminating form of Jackson's theorem,

$$_8\Phi_7\begin{bmatrix} a, q\sqrt{a}, -q\sqrt{a}, & b, & c, & d, & e, & f; \\ \sqrt{a}, & -\sqrt{a}, aq/b, aq/c, aq/d, aq/e, aq/f; \end{bmatrix}q, q\,]$$

$$+ \Pi\begin{bmatrix} bq/c, bq/d, bq/e, bq/f, aq, c, d, e, f, b/a; \\ b^2q/a, bc/a, bd/a, be/a, bf/a, aq/c, aq/d, aq/e, aq/f, a/b; \end{bmatrix}q\,]$$

$$\times\,_8\Phi_7\begin{bmatrix} b^2/a, qb/\sqrt{a}, -qb/\sqrt{a}, & b, bc/a, bd/a, be/a, bf/a; \\ b/\sqrt{a}, & -b/\sqrt{a}, bq/a, bq/c, bq/d, bq/e, bq/f; \end{bmatrix}q, q\,]$$

$$= \Pi\begin{bmatrix} aq, b/a, aq/de, aq/ce, aq/cd, aq/cf, aq/df, aq/ef; \\ aq/c, aq/d, aq/e, aq/f, bc/a, bd/a, be/a, bf/a; \end{bmatrix}q\,], \tag{IV.15}$$

where $a^2q = bcdef$.

The basic analogue of the non-terminating form of Saalschutz's theorem,

$$\Pi\begin{bmatrix} a, b, c; \\ e, f; \end{bmatrix}q\,]\,_3\Phi_2\begin{bmatrix} a, b, c; \\ e, f; \end{bmatrix}q, q\,] - \frac{q}{e}\Pi\begin{bmatrix} qa/e, qb/e, qc/e; \\ q^2/e, qf/e; \end{bmatrix}q\,]$$

$$\times\,_3\Phi_2\begin{bmatrix} aq/e, bq/e, cq/e; \\ q^2/e, qf/e; \end{bmatrix}q, q\,]$$

$$= \Pi\begin{bmatrix} e, q/e, f/a, f/b, f/c; \\ a, b, c, aq/e, bq/e, cq/e; \end{bmatrix}q\,], \tag{IV.16}$$

where $ef = qabc$.

# APPENDIX V

## Table 1 $\quad 1 \Big/ \prod_{n=0}^{\infty} (1 - aq^n)$

| $q$ | $a = -0.90$ | $a = -0.85$ | $a = -0.80$ | $a = -0.75$ |
|---|---|---|---|---|
| 0·00 | 0·52631548 | 0·54054030 | 0·55555556 | 0·57142857 |
| 0·05 | 0·50246091 | 0·51734652 | 0·53306550 | 0·54968941 |
| 0·10 | 0·47807310 | 0·49352871 | 0·50986729 | 0·52716498 |
| 0·15 | 0·45289097 | 0·46883510 | 0·48571737 | 0·50361931 |
| 0·20 | 0·42668491 | 0·44303758 | 0·46038855 | 0·47882595 |
| 0·25 | 0·39925062 | 0·41592615 | 0·43366506 | 0·45256307 |
| 0·30 | 0·37040698 | 0·38730716 | 0·40534013 | 0·42461071 |
| 0·35 | 0·33999878 | 0·35700492 | 0·37521725 | 0·39475116 |
| 0·40 | 0·30905880 | 0·32487019 | 0·34311687 | 0·36277383 |
| 0·45 | 0·27406231 | 0·29079755 | 0·30889210 | 0·32848831 |
| 0·50 | 0·23849058 | 0·25475774 | 0·27245804 | 0·29175072 |
| 0·55 | 0·20135928 | 0·21685347 | 0·23384509 | 0·25251266 |
| 0·60 | 0·16307631 | 0·17741297 | 0·19329080 | 0·21090974 |
| 0·65 | 0·12443515 | 0·13714206 | 0·15139534 | 0·16741712 |
| 0·70 | 0·08682771 | 0·09735812 | 0·10937324 | 0·12311483 |
| 0·75 | 0·05250466 | 0·06030678 | 0·06942476 | 0·08010961 |
| 0·80 | 0·02471088 | 0·02942449 | 0·03513521 | 0·04207657 |
| 0·85 | 0·007041 | 0·008904 | 0·011301 | 0·01439873 |
| 0·90 | 0·0003 | 0·0006 | 0·0011 | 0·001674 |
| 0·95 | 0·0000 | 0·0000 | 0·0000 | 0·0002 |
| 1·00 | 0·00000000 | 0·00000000 | 0·00000000 | 0·00000000 |

| $q$ | $a = -0.70$ | $a = -0.65$ | $a = -0.60$ | $a = -0.55$ |
|------|-------------|-------------|-------------|-------------|
| 0·00 | 0·58823517 | 0·60606061 | 0·62500000 | 0·64516121 |
| 0·05 | 0·56729803 | 0·58598112 | 0·60583936 | 0·62698658 |
| 0·10 | 0·54550666 | 0·56498774 | 0·58571542 | 0·60781075 |
| 0·15 | 0·52263204 | 0·54285780 | 0·56441165 | 0·58742344 |
| 0·20 | 0·49844819 | 0·51936579 | 0·54170319 | 0·56560093 |
| 0·25 | 0·47272728 | 0·49427815 | 0·51735132 | 0·54210017 |
| 0·30 | 0·44523637 | 0·46734921 | 0·49109839 | 0·51665276 |
| 0·35 | 0·41573630 | 0·43831883 | 0·46266399 | 0·48895936 |
| 0·40 | 0·38398496 | 0·40691322 | 0·43174321 | 0·45868507 |
| 0·45 | 0·34974717 | 0·37285141 | 0·39800908 | 0·42545784 |
| 0·50 | 0·31281674 | 0·33586256 | 0·36112437 | 0·38887346 |
| 0·55 | 0·27306043 | 0·29572291 | 0·32077030 | 0·34851487 |
| 0·60 | 0·23050050 | 0·25233036 | 0·27671005 | 0·30400173 |
| 0·65 | 0·18546683 | 0·20584866 | 0·22892021 | 0·25510367 |
| 0·70 | 0·13886993 | 0·15698049 | 0·17785536 | 0·20198523 |
| 0·75 | 0·09266613 | 0·10746615 | 0·12496486 | 0·14572203 |
| 0·80 | 0·05054242 | 0·06090435 | 0·07363421 | 0·08933409 |
| 0·85 | 0·01841635 | 0·02365363 | 0·03051320 | 0·03954215 |
| 0·90 | 0·002446 | 0·003569 | 0·005243 | 0·00775262 |
| 0·95 | 0·000 | 0·000 | 0·000 | 0·000 |
| 1·00 | 0·00000000 | 0·00000000 | 0·00000000 | 0·00000000 |

| $q$ | $a = -0.50$ | $a = -0.45$ | $a = -0.40$ | $a = -0.35$ |
|------|-------------|-------------|-------------|-------------|
| 0·00 | 0·66666667 | 0·68965510 | 0·71428571 | 0·74074074 |
| 0·05 | 0·64955170 | 0·67368149 | 0·69954366 | 0·72733073 |
| 0·10 | 0·63141095 | 0·65667219 | 0·68377291 | 0·71291804 |
| 0·15 | 0·61204046 | 0·63843029 | 0·66678457 | 0·69732333 |
| 0·20 | 0·59121844 | 0·61873727 | 0·64836507 | 0·68034032 |
| 0·25 | 0·56869888 | 0·59734593 | 0·62826846 | 0·66172771 |
| 0·30 | 0·54420434 | 0·57397219 | 0·60620743 | 0·64119909 |
| 0·35 | 0·51741855 | 0·54828588 | 0·58184180 | 0·61840992 |
| 0·40 | 0·48797892 | 0·51989999 | 0·55476526 | 0·59294138 |
| 0·45 | 0·45547020 | 0·48835980 | 0·52448923 | 0·56427938 |
| 0·50 | 0·41942238 | 0·45313276 | 0·49042475 | 0·53178860 |
| 0·55 | 0·37931891 | 0·41360450 | 0·45186531 | 0·49468133 |
| 0·60 | 0·33462960 | 0·36909223 | 0·40797807 | 0·45198500 |
| 0·65 | 0·28489938 | 0·31890271 | 0·35782528 | 0·40252158 |
| 0·70 | 0·22996164 | 0·26250056 | 0·30047280 | 0·34494253 |
| 0·75 | 0·17042903 | 0·19994385 | 0·23533651 | 0·27794846 |
| 0·80 | 0·10877617 | 0·13295599 | 0·16316431 | 0·20108438 |
| 0·85 | 0·05148837 | 0·06738085 | 0·08864451 | 0·11726684 |
| 0·90 | 0·01154307 | 0·01731472 | 0·02617512 | 0·03989475 |
| 0·95 | 0·0 | 0·0 | 0·0 | 0·0 |
| 1·00 | 0·00000000 | 0·00000000 | 0·00000000 | 0·00000000 |

| q | a = − 0·30 | a = − 0·25 | a = − 0·20 | a = − 0·15 |
|------|------------|------------|------------|------------|
| 0·00 | 0·76923072 | 0·80000000 | 0·83333333 | 0·86956517 |
| 0·05 | 0·75726492 | 0·78960392 | 0·82464844 | 0·86275143 |
| 0·10 | 0·74434403 | 0·77832520 | 0·81518154 | 0·85528877 |
| 0·15 | 0·73030049 | 0·76601072 | 0·80479782 | 0·84706540 |
| 0·20 | 0·71493845 | 0·75247911 | 0·79333544 | 0·83794560 |
| 0·25 | 0·69802547 | 0·73751221 | 0·78059740 | 0·82776220 |
| 0·30 | 0·67928159 | 0·72084379 | 0·76634045 | 0·81630658 |
| 0·35 | 0·65836524 | 0·70214465 | 0·75025992 | 0·80331428 |
| 0·40 | 0·63485434 | 0·68100186 | 0·73196844 | 0·78844481 |
| 0·45 | 0·60822126 | 0·65689051 | 0·71096577 | 0·77125185 |
| 0·50 | 0·57779912 | 0·62913362 | 0·68659468 | 0·75113887 |
| 0·55 | 0·54273731 | 0·59684563 | 0·65797571 | 0·72729084 |
| 0·60 | 0·50194448 | 0·55885248 | 0·62390843 | 0·69856562 |
| 0·65 | 0·45402284 | 0·51358015 | 0·58272002 | 0·66331639 |
| 0·70 | 0·39721711 | 0·45891157 | 0·53203284 | 0·61909079 |
| 0·75 | 0·32947115 | 0·39204961 | 0·46842183 | 0·56210621 |
| 0·80 | 0·24892520 | 0·30960504 | 0·38700631 | 0·48633319 |
| 0·85 | 0·15604030 | 0·20891974 | 0·28155273 | 0·38207636 |
| 0·90 | 0·06133182 | 0·09514910 | 0·14903849 | 0·23583887 |
| 0·95 | 0·0 | 0·0 | 0·0 | 0·0 |
| 1·00 | 0·00000000 | 0·00000000 | 0·00000000 | 0·00000000 |

| q | a = − 0·10 | a = − 0·05 | a = 0 | a = 0·05 | a = 0·10 |
|------|------------|------------|-------|------------|------------|
| 0·00 | 0·90909091 | 0·95238093 | 1·00 | 1·05263156 | 1·11111111 |
| 0·05 | 0·90433005 | 0·94988093 | 1·00 | 1·05540860 | 1·11698849 |
| 0·10 | 0·89909088 | 0·94711650 | 1·00 | 1·05850920 | 1·12358272 |
| 0·15 | 0·89329063 | 0·94404156 | 1·00 | 1·06199128 | 1·13102440 |
| 0·20 | 0·88682773 | 0·94059892 | 1·00 | 1·06592791 | 1·13947947 |
| 0·25 | 0·87957590 | 0·93671690 | 1·00 | 1·07041234 | 1·14916150 |
| 0·30 | 0·87137583 | 0·93230406 | 1·00 | 1·07556558 | 1·16034989 |
| 0·35 | 0·86202334 | 0·92724208 | 1·00 | 1·08154733 | 1·17341697 |
| 0·40 | 0·85125236 | 0·92137492 | 1·00 | 1·08857292 | 1·18886968 |
| 0·45 | 0·83870923 | 0·91449252 | 1·00 | 1·09693948 | 1·20741631 |
| 0·50 | 0·82391363 | 0·90630501 | 1·00 | 1·10706901 | 1·23007668 |
| 0·55 | 0·80619624 | 0·89640107 | 1·00 | 1·11958151 | 1·25837390 |
| 0·60 | 0·78459628 | 0·88417672 | 1·00 | 1·13542623 | 1·29468667 |
| 0·65 | 0·75768502 | 0·86870795 | 1·00 | 1·15613326 | 1·34294464 |
| 0·70 | 0·72324581 | 0·84850682 | 1·00 | 1·18433706 | 1·41012744 |
| 0·75 | 0·67765707 | 0·82101593 | 1·00 | 1·22498863 | 1·50990168 |
| 0·80 | 0·61461780 | 0·78144489 | 1·00 | 1·28860510 | 1·67299843 |
| 0·85 | 0·52232588 | 0·71969073 | 1·00 | 1·40206924 | 1·98497175 |
| 0·90 | 0·37725038 | 0·61044293 | 1·00 | 1·65986682 | 2·79441768 |
| 0·95 | 0·2 | 0·4 | 1·00 | 2·75418435 | 7·7971904 |
| 1·00 | 0·00000000 | 0·00000000 | 1·00 | ∞ | ∞ |

| $q$ | $a = 0 \cdot 15$ | $a = 0 \cdot 20$ | $a = 0 \cdot 25$ | $a = 0 \cdot 30$ |
|---|---|---|---|---|
| 0·00 | 1·17647054 | 1·25000000 | 1·33333333 | 1·42857135 |
| 0·05 | 1·18582882 | 1·26329108 | 1·35109975 | 1·45147209 |
| 0·10 | 1·19638000 | 1·27835036 | 1·37132958 | 1·47767806 |
| 0·15 | 1·20834580 | 1·29551361 | 1·39450168 | 1·50784791 |
| 0·20 | 1·22200984 | 1·31521350 | 1·42123718 | 1·54284156 |
| 0·25 | 1·23774016 | 1·33801539 | 1·45235357 | 1·58379870 |
| 0·30 | 1·25602195 | 1·36467082 | 1·48894609 | 1·63225817 |
| 0·35 | 1·27750766 | 1·39619929 | 1·53251355 | 1·69034429 |
| 0·40 | 1·30309495 | 1·43401748 | 1·58515956 | 1·76106811 |
| 0·45 | 1·33405243 | 1·48015087 | 1·64992600 | 1·84883368 |
| 0·50 | 1·37223194 | 1·53759583 | 1·73137321 | 1·96033131 |
| 0·55 | 1·42044386 | 1·61097247 | 1·83664396 | 2·10620481 |
| 0·60 | 1·48316334 | 1·70777906 | 1·97755111 | 2·30439117 |
| 0·65 | 1·56796160 | 1·84100204 | 2·17503302 | 2·58742424 |
| 0·70 | 1·68869897 | 2·03512853 | 2·46974116 | 3·02033108 |
| 0·75 | 1·87361155 | 2·34201447 | 2·95111913 | 3·75174241 |
| 0·80 | 2·18976707 | 2·89159989 | 3·85546653 | 5·1956037 |
| 0·85 | 2·83985074 | 4·10944635 | 6·0209939 | 8·9428294 |
| 0·90 | 4·77680298 | 8·30161101 | 14·6892657 | 26·5085132 |
| 0·95 | 22·7378273 | 68·464270 | 213·43837 | 691·11111 |
| 1·00 | $\infty$ | $\infty$ | $\infty$ | $\infty$ |

| $q$ | $a = 0 \cdot 35$ | $a = 0 \cdot 40$ | $a = 0 \cdot 45$ | $a = 0 \cdot 50$ |
|---|---|---|---|---|
| 0·00 | 1·53846144 | 1·66666667 | 1·81818182 | 2·00000000 |
| 0·05 | 1·56730757 | 1·70247214 | 1·86223752 | 2·05398421 |
| 0·10 | 1·60048246 | 1·74385833 | 1·91341791 | 2·11701817 |
| 0·15 | 1·63887133 | 1·79199738 | 1·97326120 | 2·19111187 |
| 0·20 | 1·68363740 | 1·84843972 | 2·04381535 | 2·27895874 |
| 0·25 | 1·73633319 | 1·91526911 | 2·12785250 | 2·38423079 |
| 0·30 | 1·79907128 | 1·99534310 | 2·22920393 | 2·51204374 |
| 0·35 | 1·87479348 | 2·09267608 | 2·35329983 | 2·66971356 |
| 0·40 | 1·96771097 | 2·21307203 | 2·50807075 | 2·86803484 |
| 0·45 | 2·08405404 | 2·36521658 | 2·70552286 | 3·12354330 |
| 0·50 | 2·23341299 | 2·56265968 | 2·96464089 | 3·46274644 |
| 0·55 | 2·43128658 | 2·82764608 | 3·31709223 | 3·93058166 |
| 0·60 | 2·70428863 | 3·19910371 | 3·81936477 | 4·60878229 |
| 0·65 | 3·10181750 | 3·75098459 | 4·58132189 | 5·6601578 |
| 0·70 | 3·72552471 | 4·63999709 | 5·8427943 | 7·4510494 |
| 0·75 | 4·81679978 | 6·2529040 | 8·2193356 | 10·9600859 |
| 0·80 | 7·0845879 | 9·7883809 | 13·7262705 | 19·5767617 |
| 0·85 | 13·4840856 | 20·6749449 | 32·3023422 | 51·5567364 |
| 0·90 | 48·8854119 | 92·342659 | 179·176364 | 358·352727 |
| 0·95 | 2332·8737 | 8244·9779 | 30668·661 | 120818·95 |
| 1·00 | $\infty$ | $\infty$ | $\infty$ | $\infty$ |

| $q$ | $a = 0.55$ | $a = 0.60$ | $a = 0.65$ | $a = 0.70$ |
|------|------------|------------|------------|------------|
| 0·00 | 2·22222222 | 2·50000000 | 2·85714242 | 3·33333333 |
| 0·05 | 2·28837309 | 2·58139473 | 2·95817842 | 3·46060495 |
| 0·10 | 2·36600853 | 2·67741270 | 3·07798255 | 3·61230453 |
| 0·15 | 2·45775443 | 2·79149606 | 3·22110646 | 3·79453477 |
| 0·20 | 2·56715207 | 2·92831947 | 3·39377254 | 4·01569876 |
| 0·25 | 2·69906491 | 3·09435012 | 3·60465580 | 4·28760380 |
| 0·30 | 2·86032388 | 3·29874731 | 3·86614724 | 4·62726326 |
| 0·35 | 3·06078757 | 3·55485492 | 4·19647372 | 5·0599538 |
| 0·40 | 3·31515505 | 3·88278186 | 4·62340686 | 5·6246274 |
| 0·45 | 3·64621480 | 4·31409013 | 5·1910899 | 6·3840135 |
| 0·50 | 4·09101255 | 4·90082825 | 5·9734008 | 7·4447096 |
| 0·55 | 4·71339665 | 5·7342342 | 7·1021246 | 9·0003320 |
| 0·60 | 5·6318263 | 6·9870710 | 8·8321041 | 11·4335502 |
| 0·65 | 7·0881073 | 9·0209660 | 11·7108091 | 15·5889209 |
| 0·70 | 9·6434756 | 12·7025101 | 17·0943101 | 23·6317798 |
| 0·75 | 14·8600725 | 20·5483390 | 29·0982994 | 42·4442405 |
| 0·80 | 28·4718544 | 42·3698870 | 64·811264 | 102·558239 |
| 0·85 | 84·328123 | 141·926010 | 247·11411 | 448·448464 |
| 0·90 | 741·91428 | 1598·74491 | 3611·5281 | 8635·9006 |
| 0·95 | — | — | — | — |
| 1·00 | ∞ | ∞ | ∞ | ∞ |

| $q$ | $a = 0.75$ | $a = 0.80$ | $a = 0.85$ | $a = 0.90$ |
|------|------------|------------|------------|------------|
| 0·00 | 4·00000000 | 5·00000000 | 6·66666667 | 10·00000000 |
| 0·05 | 4·16406084 | 5·2193187 | 6·9781807 | 10·4960538 |
| 0·10 | 4·36063428 | 5·4834836 | 7·3553796 | 11·0999020 |
| 0·15 | 4·5980905 | 5·8043936 | 7·8162251 | 11·8419271 |
| 0·20 | 4·8880383 | 6·1986754 | 8·3860104 | 12·7652500 |
| 0·25 | 5·2469214 | 6·6900756 | 9·1011700 | 13·9325418 |
| 0·30 | 5·6986501 | 7·3134498 | 10·0157070 | 15·4376338 |
| 0·35 | 6·2791135 | 8·1216759 | 11·2124569 | 17·4261209 |
| 0·40 | 7·0442714 | 9·1982146 | 12·8238355 | 20·1338528 |
| 0·45 | 8·0854737 | 10·6812440 | 15·0723280 | 23·9632830 |
| 0·50 | 9·5604302 | 12·8132974 | 18·3552751 | 29·6464039 |
| 0·55 | 11·7610326 | 16·0521758 | 23·4383707 | 38·6250463 |
| 0·60 | 15·2774586 | 21·3458726 | 31·9476659 | 54·0443888 |
| 0·65 | 21·4495129 | 30·9113793 | 47·8076344 | 83·755752 |
| 0·70 | 33·8388121 | 50·8718967 | 82·308702 | 151·362397 |
| 0·75 | 64·329199 | 102·741685 | 177·383329 | 349·975351 |
| 0·80 | 169·479541 | 296·91158 | 566·44822 | 1246·4538 |
| 0·85 | 857·44976 | 1756·8741 | 3973·0984 | 10545·708 |
| 0·90 | 22164·017 | 62359·823 | 199343·62 | — |
| 0·95 | — | — | — | — |
| 1·00 | ∞ | ∞ | ∞ | ∞ |

# APPENDIX VI

## Table 2 $\quad 1\Big/\prod_{n=1}^{\infty}(1-q^n)$

| q | | q | | q | | q | |
|---|---|---|---|---|---|---|---|
| 0·000 | 1·00000000 | 0·225 | 1·37934675 | 0·450 | 2·70552286 | 0·675 | 16·1657587 |
| 0·005 | 1·00505036 | 0·230 | 1·39318901 | 0·455 | 2·76680960 | 0·680 | 17·3538181 |
| 0·010 | 1·01020304 | 0·235 | 1·40739756 | 0·460 | 2·83080798 | 0·685 | 18·6737604 |
| 0·015 | 1·01546036 | 0·240 | 1·42198535 | 0·465 | 2·89768016 | 0·690 | 20·1445401 |
| 0·020 | 1·02082481 | 0·245 | 1·43696598 | 0·470 | 2·96760068 | 0·695 | 21·7884436 |
| 0·025 | 1·02629888 | 0·250 | 1·45235357 | 0·475 | 3·04075722 | 0·700 | 23·6317798 |
| 0·030 | 1·03188522 | 0·255 | 1·46816304 | 0·480 | 3·11735207 | 0·705 | 25·7057391 |
| 0·035 | 1·03758649 | 0·260 | 1·48440987 | 0·485 | 3·19760323 | 0·710 | 28·0474623 |
| 0·040 | 1·04340554 | 0·265 | 1·50111034 | 0·490 | 3·28174626 | 0·715 | 30·7013834 |
| 0·045 | 1·04934525 | 0·270 | 1·51828156 | 0·495 | 3·37003560 | 0·720 | 33·7209194 |
| 0·050 | 1·05540860 | 0·275 | 1·53594131 | 0·500 | 3·46274645 | 0·725 | 37·1706139 |
| 0·055 | 1·06159870 | 0·280 | 1·55410838 | 0·505 | 3·56017687 | 0·730 | 41·1288727 |
| 0·060 | 1·06791878 | 0·285 | 1·57280248 | 0·510 | 3·66265026 | 0·735 | 45·6914768 |
| 0·065 | 1·07437215 | 0·290 | 1·59204415 | 0·515 | 3·77051755 | 0·740 | 50·9761207 |
| 0·070 | 1·08096222 | 0·295 | 1·61185513 | 0·520 | 3·88416028 | 0·745 | 57·1283279 |
| 0·075 | 1·08769259 | 0·300 | 1·63225817 | 0·525 | 4·00399359 | 0·750 | 64·329201 |
| 0·080 | 1·09456695 | 0·305 | 1·65327718 | 0·530 | 4·13046989 | 0·755 | 72·805690 |
| 0·085 | 1·10158909 | 0·310 | 1·67493735 | 0·535 | 4·26408291 | 0·760 | 82·844246 |
| 0·090 | 1·10876302 | 0·315 | 1·69726512 | 0·540 | 4·40537215 | 0·765 | 94·809207 |
| 0·095 | 1·11609282 | 0·320 | 1·72028840 | 0·545 | 4·55492779 | 0·770 | 109·167748 |
| 0·100 | 1·12358272 | 0·325 | 1·74403653 | 0·550 | 4·71339669 | 0·775 | 126·524054 |
| 0·105 | 1·13123716 | 0·330 | 1·76854041 | 0·555 | 4·88148860 | 0·780 | 147·666645 |
| 0·110 | 1·13906071 | 0·335 | 1·79383267 | 0·560 | 5·05998364 | 0·785 | 173·634578 |
| 0·115 | 1·14705812 | 0·340 | 1·81994774 | 0·565 | 5·24974053 | 0·790 | 205·811192 |
| 0·120 | 1·15523429 | 0·345 | 1·84692189 | 0·570 | 5·45170606 | 0·795 | 246·058392 |
| 0·125 | 1·16359437 | 0·350 | 1·87479348 | 0·575 | 5·6669261 | 0·800 | 296·91159 |
| 0·130 | 1·17214360 | 0·355 | 1·90360298 | 0·580 | 5·9865572 | 0·805 | 361·86657 |
| 0·135 | 1·18088754 | 0·360 | 1·93339334 | 0·585 | 6·1418817 | 0·810 | 445·80793 |
| 0·140 | 1·18983189 | 0·365 | 1·96420976 | 0·590 | 6·4043230 | 0·815 | 555·65889 |
| 0·145 | 1·19898258 | 0·370 | 1·99610032 | 0·595 | 6·6854644 | 0·820 | 701·38435 |
| 0·150 | 1·20834580 | 0·375 | 2·02911579 | 0·600 | 6·9870710 | 0·825 | 897·56537 |
| 0·155 | 1·21792792 | 0·380 | 2·06331004 | 0·605 | 7·3111135 | 0·830 | 1165·9218 |
| 0·160 | 1·22773568 | 0·385 | 2·09874025 | 0·610 | 7·6597976 | 0·835 | 1539·4369 |
| 0·165 | 1·23777595 | 0·390 | 2·13546701 | 0·615 | 8·0355970 | 0·840 | 2069·2555 |
| 0·170 | 1·24805597 | 0·395 | 2·17355479 | 0·620 | 8·4412919 | 0·845 | 2836·5053 |
| 0·175 | 1·25858323 | 0·400 | 2·21307203 | 0·625 | 8·8800146 | 0·850 | 3973·0986 |
| 0·180 | 1·26936557 | 0·405 | 2·25409152 | 0·630 | 9·3553018 | 0·855 | 5699·4012 |
| 0·185 | 1·28041112 | 0·410 | 2·29669077 | 0·635 | 9·8711576 | 0·860 | 8394·6331 |
| 0·190 | 1·29172835 | 0·415 | 2·34095233 | 0·640 | 10·432125 | 0·865 | 12733·016 |
| 0·195 | 1·30332608 | 0·420 | 2·38696419 | 0·645 | 11·043376 | 0·870 | 19957·261 |
| 0·200 | 1·31521351 | 0·425 | 2·43482019 | 0·650 | 11·710809 | 0·875 | 32451·276 |
| 0·205 | 1·32740021 | 0·430 | 2·48462065 | 0·655 | 12·441174 | 0·880 | 54996·208 |
| 0·210 | 1·33989624 | 0·435 | 2·53647252 | 0·660 | 13·242219 | 0·885 | 97670·57 |
| 0·215 | 1·35271201 | 0·440 | 2·59049039 | 0·665 | 14·122861 | 0·890 | 182942·54 |
| 0·220 | 1·36585838 | 0·445 | 2·64679675 | 0·670 | 15·093397 | 0·895 | — |

# BIBLIOGRAPHY

*This list contains most of the works on hypergeometric functions published since 1934, together with earlier works of importance referred to in the text.*

## I. BOOKS

*All these works contain extensive lists of references.*

APPELL, P. & KAMPÉ DE FÉRIET, J. (1926). *Fonctions hypergeometriques et hypersphériques.* Paris: Gauthier Villars.

BAILEY, W. N. (1935). (Second edition 1964) *Generalized hypergeometric series.* Cambridge Mathematical Tract No. 32. Cambridge University Press.

ERDÉLYI, A., MAGNUS, W., OBERHETTINGER, F. & TRICOMI, G. F. (1953). *Higher transcendental functions.* Vols. I, II and III. New York: McGraw Hill.

ERDÉLYI, A., MAGNUS, W., OBERHETTINGER, F. & TRICOMI, G. F. (1954). *Tables of integral transforms.* Vols. I and II. New York: McGraw Hill.

HEINE, E. (1898). *Handbuch die Kugelfunctionen, Theorie und Anwendung.* Vols. I and II. Berlin: Springer Verlag.

KLEIN, F. (1933). *Vorlesungen uber die hypergeometrische Funktion.* Haupt edition. Berlin: Springer Verlag.

## II. PAPERS, AND OTHER WORKS

AGARWAL, NIRMALA (1959). Certain basic hypergeometric identities of the Cayley–Orr type. *J. London Math. Soc.* **34**, 37–46.

AGARWAL, R. P. (1950). On self-reciprocal functions involving complex variables. *Ganita*, **1**, 17–25.

AGARWAL, R. P. (1950). Sur une généralisation de la transformation de Hankel. *Annales de la Société scientifiques de Bruxelles*, 164–8.

AGARWAL, R. P. (1953). On integral analogues of certain transformations of well-poised basic hypergeometric series. *Quart. J. Math.* (Oxford), (2), **4**, 161–7.

AGARWAL, R. P. (1954). On the partial sums of series of hypergeometric type. *Proc. Camb. Phil. Soc.* **49**, 441–5.

ALDER, H. L. (1948). The nonexistence of certain identities in the theory of partitions and compositions. *Bull. Amer. Math. Soc.* **54**, 712–22.

ALDER, H. L. (1954). Generalizations of the Rogers–Ramanujan identities. *Pacific J. Math.* **4**, 161–8.

AL-SALEM, W. A. (1959). Some functions related to Bessel Polynomials. *Duke Math. J.* **26**, 519–39.

BAILEY, W. N. (1928). Products of generalized hypergeometric series. *Proc. London Math. Soc.* (2), **28**, 242–54.

BAILEY, W. N. (1929a). Transformations of generalized hypergeometric series. *Proc. London Math. Soc.* (2), **29**, 495–502.

BAILEY, W. N. (1929b). Some identities involving generalized hypergeometric series. *Proc. London Math. Soc.* (2), **29**, 503–16.

BAILEY, W. N. (1929c). An identity involving Heine's basic hypergeometric series. *J. London Math. Soc.* **4**, 254–7.

BAILEY, W. N. (1930a). An extension of Whipple's theorem on well-poised hypergeometric series. *Proc. London Math. Soc.* (2), **31**, 505–11.

BAILEY, W. N. (1930b). A generalization of an integral due to Ramanujan. *J. London Math. Soc.* 5, 200–2.

BAILEY, W. N. (1931a). An elementary proof of Saalschutz's theorem on generalized hypergeometric series. *Math. Gazette*, 15, 299–300.

BAILEY, W. N. (1931b). The partial sum of the coefficients of the hypergeometric series. *J. London Math. Soc.* 6, 40–1.

BAILEY, W. N. (1931c). A note on an integral due to Ramanujan. *J. London Math. Soc.* 6, 216–7.

BAILEY, W. N. (1932a). On one of Ramanujan's theorems. *J. London Math. Soc.* 7, 34–6.

BAILEY, W. N. (1932b). Some transformations of generalized hypergeometric series, and contour integrals of Barnes's type. *Quart. J. Math.* (Oxford), 3, 168–82.

BAILEY, W. N. (1933a). On certain relations between hypergeometric series of higher order. *J. London Math. Soc.* 8, 100–7.

BAILEY, W. N. (1933b). A reducible case of the fourth type of Appell's hypergeometric functions of two variables. *Quart. J. Math.* (Oxford), 4, 305–8.

BAILEY, W. N. (1934a). Transformations of well-poised hypergeometric series. *Proc. London Math. Soc.* (2), 36, 235–40.

BAILEY, W. N. (1934b). On the reducibility of Appell's function $F_4$. *Quart. J. Math.* (Oxford), 5, 291–2.

BAILEY, W. N. (1935a). Some theorems concerning products of hypergeometric series. *Proc. London Math. Soc.* (2), 38, 377–84.

BAILEY, W. N. (1935b). Some expansions in Bessel functions involving Appell's function $F_4$. *Quart. J. Math.* (Oxford), 6, 233–8.

BAILEY, W. N. (1936). Series of hypergeometric type which are infinite in both directions. *Quart. J. Math.* (Oxford), 7, 105–15.

BAILEY, W. N. (1937a). A new proof of Dixon's theorem on hypergeometric series. *Quart. J. Math.* (Oxford), 8, 113–4.

BAILEY, W. N. (1937b). Associated hypergeometric series. *Quart. J. Math.* (Oxford), 8, 115–8.

BAILEY, W. N. (1941a). On the double integral representation of Appell's function $F_4$. *Quart. J. Math.* (Oxford), 12, 12–4.

BAILEY, W. N. (1941b). A note on certain $q$-identities. *Quart. J. Math.* (Oxford), 12, 173–5.

BAILEY, W. N. (1947a). Some identities in combinatory analysis. *Proc. London Math. Soc.* (2) 49, 421–35.

BAILEY, W. N. (1947b). Well-poised basic hypergeometric series. *Quart. J. Math.* (Oxford), 18, 157–66.

BAILEY, W. N. (1947c). A transformation of nearly-poised basic hypergeometric series. *J. London Math. Soc.* 22, 237–40.

BAILEY, W. N. (1949). Identities of the Rogers–Ramanujan type. *Proc. London Math. Soc.* (2) 50, 1–10.

BAILEY, W. N. (1950a). On the basic bilateral hypergeometric series $_2\Psi_2$. *Quart. J. Math.* (Oxford), (2), 1, 194–8.

BAILEY, W. N. (1950b). On the analogue of Dixon's theorem for basic bilateral series. *Quart. J. Math.* (Oxford), (2), 1, 318–20.

BAILEY, W. N. (1951). On the simplification of some identities of the Rogers–Ramanujan type. *Proc. London Math. Soc.* (3), 1, 217–22.

BAILEY, W. N. (1952a). A note on two of Ramanujan's formulae. *Quart. J. Math.* (Oxford), (2), 3, 29–31.

BAILEY, W. N. (1952*b*). A further note on two of Ramanujan's formulae. *Quart. J. Math.* (Oxford), (2), **3**, 158–60.

BAILEY, W. N. (1953*a*). An expression for $\vartheta_1(nz)/\vartheta_1(z)$. *Proc. American Math. Soc.* **4**, 569–72.

BAILEY, W. N. (1953*b*). On the sum of a terminating $_3F_2(1)$. *Quart. J. Math.* (Oxford), (2), **4**, 237–40.

BAILEY, W. N. (1954*a*). Contiguous hypergeometric functions of the type $_3F_2(1)$. *Proc. Glasgow Math. Assoc.* **2**, 62–5.

BAILEY, W. N. (1954*b*). Ernest William Barnes, (obituary). *J. London Math. Soc.* **29**, 498–503.

BAILEY, W. N. (1956). A note on a $_3F_2$. *Math. Gazette*, **40**, 277–8.

BAILEY, W. N. (1959*a*). On the sum of a particular bilateral hypergeometric series $_3H_3$. *Quart. J. Math.* (Oxford), (2), **10**, 92–4.

BAILEY, W. N. (1959*b*). On two manuscripts by Bishop Barnes. *Quart. J. Math.* (Oxford), (2), **10**, 236–40.

*Note.* A complete list of all Professor Bailey's publications will be found in his obituary notice (Slater 1963).

BARNES, E. W. (1907*a*). The asymptotic expansion of integral functions defined by generalized hypergeometric series. *Proc. London Math. Soc.* (2), **5**, 59–116.

BARNES, E. W. (1907*b*). A new development of the theory of hypergeometric functions. *Proc. London Math. Soc.* (2), **6**, 141–77.

BARNES, E. W. (1907*c*). On functions defined by simple types of hypergeometric series. *Trans. Camb. Phil. Soc.* **20**, 253–79.

BARNES, E. W. (1910). A transformation of generalized hypergeometric series. *Quart. J. Math.* **41**, 136–40.

BROMWICH, T. J. I'A. (1909). An asymptotic formula for the generalized hypergeometric series. *Proc. London Math. Soc.* (2), **7**, 101–6.

BURCHNALL, J. L. (1932). A relation between hypergeometric series. *Quart. J. Math.* (Oxford), **3**, 318–20.

BURCHNALL, J. L. (1939). The differential equations of Appell's function $F_4$. *Quart. J. Math.* (Oxford), **10**, 145–50.

BURCHNALL, J. L. & CHAUNDY, T. W. (1940). Expansions of Appell's double hypergeometric functions, I. *Quart. J. Math.* (Oxford), **11**, 249–70.

BURCHNALL, J. L. & CHAUNDY, T. W. (1941*a*). Expansions of Appell's double hypergeometric functions, II. *Quart. J. Math.* (Oxford), **12**, 112–28.

BURCHNALL, J. L. (1941*b*). A note on the polynomials of Hermite. *Quart. J. Math.* (Oxford), **12**, 9–11.

BURCHNALL, J. L. (1942). Differential equations associated with hypergeometric functions. *Quart. J. Math.* (Oxford), **13**, 90–106.

BURCHNALL, J. L. & CHAUNDY, T. W. (1944). The hypergeometric identities of Cayley, Orr and Bailey. *Proc. London Math. Soc.* (2), **50**, 56–74.

BURCHNALL, J. L. (1948). On the well-poised $_3F_2$. *J. London Math. Soc.* **23**, 253–7.

BURCHNALL, J. L. & LAKIN, A. (1950). The theorems of Saalschutz and Dougall. *Quart. J. Math.* (Oxford), (2), **1**, 161–4.

BURCHNALL, J. L. (1951). An algebraic property of the classical polynomials. *Proc. London Math. Soc.* (3) **1**, 232–40.

BURWELL, W. R. (1924). Asymptotic expansions of generalized hypergeometric functions. *Proc. London Math. Soc.* **22**, 57–72.

BUSBRIDGE, I. W. (1939). The evaluation of certain integrals involving products of Hermite polynomials. *J. London Math. Soc.* **14**, 93–7.

BUSBRIDGE, I. W. (1948). Some integrals involving Hermite polynomials. *J. London Math. Soc.* **23**, 135–41.

BUSBRIDGE, I. W. (1950). On the integro-exponential function and the evaluation of some integrals involving it. *Quart. J. Math.* (Oxford), (2), **1**, 176–84.

CARLITZ, L. (1959). Some q-polynomials in two variables. *Math. Nachr.* **17**, 224–38.

CARLSON, F. (1914). Sur une classe de séries de Taylor. *Dissertation*, Upsala.

CAYLEY, A. (1858). On a theorem relating to hypergeometric series. *Phil. Mag.* (4), **16**, 356–7, reprinted as *Collected papers*, Vol. 3, 268–9.

CHAPMAN, S. (1933). Some ratios of infinite determinents. *J. London Math. Soc.* **8**, 266–72.

CHAUNDY, T. W. (1935). Hypergeometric partial differential equations, I. *Quart. J. Math.* (Oxford), **6**, 288–303.

CHAUNDY, T. W. (1936). Hypergeometric partial differential equations, II. *Quart. J. Math.* (Oxford), **7**, 306–15.

CHAUNDY, T. W. (1938). Linear partial differential equations, I. *Quart. J. Math.* (Oxford), **9**, 234–40.

CHAUNDY, T. W. (1939). Hypergeometric partial differential equations, III. *Quart. J. Math.* (Oxford), **10**, 219–40.

CHAUNDY, T. W. (1941). Singular solutions of differential equations, I. *Quart. J. Math.* (Oxford), **12**, 129–46.

CHAUNDY, T. W. (1942). Expansions of hypergeometric functions. *Quart. J. Math.* (Oxford), **13**, 159–71.

CHAUNDY, T. W. (1943). An extension of hypergeometric functions, I. *Quart. J. Math.* (Oxford), **14**, 55–78.

CHAUNDY, T. W. (1949). Differential equations with polynomial solutions. *Quart. J. Math.* (Oxford), **20**, 105–20.

CHAUNDY, T. W. (1951). Some hypergeometric identities. *J. London Math. Soc.* **26**, 42–4.

CHAUNDY, T. W. (1953). Second-order linear differential equations with polynomial solutions. *Quart. J. Math.* (Oxford), (2), **4**, 81–95.

CHAUNDY, T. W. (1962). F. H. Jackson, (obituary). *J. London Math. Soc.* **37**, 126–8.

CLAUSEN, T. (1828). Ueber die Fälle wenn die Reihe $y = 1 + \frac{\alpha \cdot \beta}{1 \cdot \gamma} x + \dots$ . *J. für Math.* **3**, 89–95.

DARLING, H. B. C. (1930). On a proof of one of Ramanujan's theorems. *J. London Math. Soc.* **5**, 8–9.

DARLING, H. B. C. (1932). On certain relations between hypergeometric series of higher orders. *Proc. London Math. Soc.* (2), **34**, 323–39.

DARLING, H. B. C. (1935). On the differential equation satisfied by the hypergeometric series of the second order. *J. London Math. Soc.* **10**, 63–70.

DIXON, A. C. (1891). On the sum of the cubes of the coefficients in a certain expansion by the binomial theorem. *Mess. Math.* **20**, 79–80.

DIXON, A. C. (1903). Summation of a certain series. *Proc. London Math. Soc.* **35**, 285–9.

DIXON, A. C. (1905). On a certain double integral. *Proc. London Math. Soc.* (2), **2**, 8–15.

DOUGALL, J. (1907). On Vandermonde's theorem and some more general expansions. *Proc. Edin. Math. Soc.* **25**, 114–32.

DYSON, F. (1943). Three identities in combinatory analysis. *J. London Math. Soc.* **18**, 35–9.

EDWARDES, D. (1923). An expansion in factorials similar to Vandermonde's theorem, and some applications. *Mess. Math.* **52**, 129–36.

ERBER, T. (1960). Inequalities for hypergeometric functions. *Arch. Rational Mech. Anal.* **4**, 341–51.

ERDÉLYI, A. (1939a). The transformation of Eulerian hypergeometric integrals. *Quart. J. Math.* (Oxford), **10**, 129–34.

ERDÉLYI, A. (1939b). Transformation of hypergeometric integrals by means of fractional integration by parts. *Quart. J. Math.* (Oxford), **10**, 176–89.

ERDÉLYI, A. (1940). A class of hypergeometric functions. *J. London Math. Soc.* **15**, 209–12.

ERDÉLYI, A. (1941). Integration of the differential equation of Appell's function $F_4$. *Quart. J. Math.* (Oxford), **12**, 68–77.

ERDÉLYI, A. (1950). The general form of hypergeometric series of two variables. *Proc. International Congr. Math.* Vol. 1.

ERDÉLYI, A. (1951). On some functional transformations. *Rendiconti del Seminario Matematico*, **10**, 217–34.

EULER, L. (1748). *Introductio in Analysis Infinitorum.* Lausanne. Vol. i.

FOX, C. (1927). The expansion of hypergeometric series in terms of similar series. *Proc. Lond Math. Soc.* (2), **26**, 201–10.

FOX, C. (1928). The asymptotic expansion of generalized hypergeometric functions. *Proc. London Math. Soc.* (2), **27**, 380–400.

GAUSS, C. F. (1812). Disquisitiones generales circa seriem infinitam, *Thesis*, GOTTINGEN; published in *Ges. Werke* Gottingen (1866). Vol. ii, 437–45; iii, 123–63; iii, 207–29; iii, 446–60.

GOULD, H. W. (1959). Dixon's theorem expressed as a convolution. *Nordisk. Mat. Tidskr.* **7**, 73–6.

GUINAND, A. P. (1945). Gauss sums and primative characters. *Quart. J. Math.* (Oxford), **16**, 59–63.

HAHN, W. (1949a). Über orthogonalpolynome, die $q$-Differenzengleichungenugen. *Math. Nachr.* **2**, 4–34.

HAHN, W. (1949b). Über polynome, die gleichzeitig zwei verschiedenen Orthogonalsystemen angehoren. *Math. Nachr.* **2**, 263–78.

HAHN, W. (1949c). Beitrage zur theorie der Heineschen Reihen, Die 24 integrale der hypergeometrischen $q$-Differenzengleichung, Das $q$-Analogen der Laplace Transformation. *Math. Nachr.* **2**, 340–79.

HAHN, W. (1950). Über die hoheren Heineschen Reihen und eine einheitliche Theorie der sogenannten speziellen Funktionen. *Math. Nachr.* **3**, 257–94.

HAHN, W. (1951). Über die Reduzibilitat einer speziellen geometrischen Differenzengleichung. *Math. Nachr.* **5**, 347–54.

HAHN, W. (1952). Über uneigentliche Losungen linearer geometrischer Differenzengleichungen. *Math. Annalen*, **125**, 67–81.

HAHN, W. (1953). Die mechanische Deutung einer geometrischen Differenzengleichung. *Zeit. f. ang. Math. u. Mech.* **33**, 1–3.

HAHN, W. (1954). Über einige Grenzwertbeziehungen bei unendlichen Produkten. *Math. Zeitschr.* **60**, 488–94.

HAHN, W. (1955). Über analytische Losungen linearer Differential-Differenzengleichungen. *Math. Zeitschr.* **63**, 313–19.

HAHN, W. (1956). Eine Bemerkung zur zweiten Methode von Liapunov. *Math. Nachr.* **14**, 349–54.

HALL, N. A. (1936). An algebraic identity. *J. London Math. Soc.* **11**, 276.

HARDY, G. H. (1920). On two theorems of F. Carlson and S. Wigert. *Acta. Math.* **42**, 327–39.

HARDY, G. H. (1923a). A chapter from Ramanujan's notebook. *Proc. Camb. Phil. Soc.* **21**, 492–503.

HARDY, G. H. (1923b). Some formulae of Ramanujan. *Proc. London Math. Soc.* (2), **22**, xii-xiii, (Records for 14th December 1922).

HILL, M. J. M. (1908). On a formula for the sum of a finite number of terms of the hypergeometric series when the fourth element is equal to unity. *Proc. London Math. Soc.* (2), **5**, 335–41, and **6**, 339–48.

HILL, M. J. M. & WHIPPLE, F. J. W. (1910). A reciprocal relation between generalized hypergeometric series. *Quart. J. Math.* **41**, 128–35.

HILL, M. J. M. (1918). On the continuation of the hypergeometric series. *Proc. London Math. Soc.* (2), **17**, 320–33.

HODGKINSON, J. (1918). An application of conformal representation to certain hypergeometric series. *Proc. London Math. Soc.* (2), **17**, 17–24.

HODGKINSON, J. (1931). A note on one of Ramanujan's theorems. *J. London Math. Soc.* **6**, 42–3.

HORN, J. (1937). Hypergeometrische Funktionen zweier Veranderlichen im Schnittpunkt dreier Singularitaten. *Math. Annalen*, **115**, 435–55.

INCE, E. L. (1920). On continued fractions connected with the hypergeometric equation. *Proc. London Math. Soc.* (2), **18**, 236–48.

JACKSON, F. H. (1897). Certain expansions of $x^n$ in hypergeometric series. *Proc. Edin. Math. Soc.* **15**, 90–6.

JACKSON, F. H. (1905a). Some properties of a generalized hypergeometric function. *Amer. J. Math.* **27**, 1–6.

JACKSON, F. H. (1905b). Pseudo-periodic functions analogous to the circular functions. *Mess. Math.* **34**, 32–9.

JACKSON, F. H. (1905c). The application of basic numbers to Bessel's and Legendre's functions. *Proc. London Math. Soc.* (2), **2**, 192–220, and **3**, 1–20.

JACKSON, F. H. (1905d). The basic gamma function and the elliptic functions. *Proc. Roy. Soc.* A, **76**, 127–44.

JACKSON, F. H. (1908a). On a formula relating to hypergeometric series. *Mess. Math.* **37**, 123–6.

JACKSON, F. H. (1908b). On $q$-functions and a certain difference operator. *Trans. Roy. Soc. Edin.* **46**, 253–81.

JACKSON, F. H. (1909). $q$-Form of Taylor's theorem. *Mess. Math.* **38**, 57–61.

JACKSON, F. H. (1910a). $q$-Difference equations. *Amer. J. Math.* **32**, 305–14.

JACKSON, F. H. (1910b). Transformation of $q$-series. *Mess. Math.* **39**, 145–51.

JACKSON, F. H. (1910c). Borel's integral and $q$-series. *Proc. Roy. Soc. Edin.* **30**, 378–85.

JACKSON, F. H. (1910d). On $q$-definite integrals. *Quart. J. Pure and Appl. Math.* **41**, 193–203.

JACKSON, F. H. (1910e). A $q$-generalization of Abel's series. *Rend. di Palermo*, **29**, 1–7.

JACKSON, F. H. (1911). The products of $q$-hypergeometric functions. *Mess. Math.* **40**, 92–100.

JACKSON, F. H. (1917). The $q$-integral analogous to Borel's integral. *Mess. Math.* **47**, 57–64.

JACKSON, F. H. (1921a). Summation of $q$-hypergeometric series. *Mess. Math.* **50**, 101–12.

JACKSON, F. H. (1921b). Examples of a generalization of Euler's transformation for power series. *Mess. Math.* **50**, 169–87.

JACKSON, F. H. (1921c). A new transformation of Heinean series. *Mess. Math.* **50**, 377–84.

JACKSON, F. H. (1941a). Certain $q$-identities. *Quart. J. Math.* (Oxford), **12**, 167–72.

JACKSON, F. H. (1941b). Hypergeometric series and 'set' numbers. *Quart. J. Math.* (Oxford), **12**, 201–10.

JACKSON, F. H. (1941c). $q^\theta$-equations and Fibonacci numbers. *Quart. J. Math.* (Oxford), **12**, 211–5.

JACKSON, F. H. (1942). On basic double hypergeometric functions. *Quart. J. Math.* (Oxford), **13**, 69–82.

JACKSON, F. H. (1944). Basic double hypergeometric functions. *Quart. J. Math.* (Oxford), **15**, 49–61.

JACKSON, F. H. (1946). Basic functions and polynomial sequences. *Quart. J. Math.* (Oxford), **17**, 99–110.

JACKSON, F. H. (1951). Basic integration. *Quart. J. Math.* (Oxford), (2), **2**, 1–16.

*Note.* For a complete list of Rev. Jackson's publications, see his obituary notice (Chaundy 1962).

JACKSON, M. (1949a). On some formulae in partition theory and bilateral basic hypergeometric series. *J. London Math. Soc.* **24**, 233–7.

JACKSON, M. (1949b). A generalization of the theorems of Watson and Whipple on the sum of the series $_3F_2$. *J. London Math. Soc.* **24**, 238–40.

JACKSON, M. (1950a). On well-poised bilateral hypergeometric series of the type $_8\Psi_8$. *Quart. J. Math.* (Oxford), (2), **1**, 63–8.

JACKSON, M. (1950b). On Lerch's transcendent and the basic bilateral hypergeometric series $_2\Psi_2$. *J. London Math. Soc.* **25**, 189–95.

JACKSON, M. (1952a). Transformations of series of the type $_3H_3$ with unit argument. *J. London Math. Soc.* **27**, 116–23.

JACKSON, M. (1952b). A note on the sum of a particular well-poised $_6H_6$ with argument $-1$. *J. London Math. Soc.* **27**, 124–6.

JACKSON, M. (1954). Transformations of series of the type $_3\Psi_3$. *Pacific J. Math.* **4**, 557–62.

JACKSON, M. (1958). A note on the reducibility of the bilateral hypergeometric series $_3H_3$. *J. London Math. Soc.* **33**, 475–6.

JACOBI, C. G. J. (1829). Fundamenta nova. *Ges. Werke*, **1**, 497–538.

JAEGER, J. C. (1938). A continuation formula for Appell's function $F_3$. *J. London Math. Soc.* **13**, 254.

KNOPP, K. (1947). *Theorie und Andendungen der uneinhlichen Reihen.* Berlin: Springer-Verlag.

KNOTTNERUS, U. J. (1960). *Approximation formulae for generalized hypergeometric functions.* Groningen: Wolters.

KUMMER, E. E. (1836). Uber die hypergeometrische Reihe $F(\alpha, \beta, x)$. *J. für Math.* **15**, 39–83 and 127–72.

LAKIN, A. (1952). A hypergeometric identity related to Dougall's theorem. *J. London Math. Soc.* **27**, 229–34.

LAMBE, C. G. & WARD, D. R. (1934). Some differential equations and associated integral equations. *Quart. J. Math.* (Oxford), **5**, 81–97.

LAURICELLA, G. (1893). Sulle funzioni ipergeometriche a piu variabili. *Rendiconti del Circolo Matematico di Palermo*, **7**, 111–3.

LERCH, M. (1892). Some transcendents. *Rozprany České Akademie, Cisaře, Františka Josefa*, 1, 2 and 3, summary in *Jahrbuch für Math.* (1893), 442–5.

LEVELT, A. H. M. (1960). On a formula of C. S. Meijer. *Proc. Kon. Akad. v. Wetensch.* **63**, 102–5.

LEVELT, A. H. M. (1961). *Hypergeometric functions.* Amsterdam: Doctorial thesis.

LITTLEWOOD, D. E. & RICHARDSON, A. R. (1935). Some special *S*-functions and *q*-series. *Quart. J. Math.* (Oxford), **6**, 184–98.

MACFARLANE, G. G. (1949). The application of Mellin transforms to the summation of slowly convergent series. *Phil. Mag.* **37**, 188–98.

MACMAHON, P. A. (1902). The sums of the powers of the binomial coefficients. *Quart. J. Math.* **33**, 274–88.

MACMAHON, P. A. (1921). Permutations, lattice permutations, and the hypergeometric series. *Proc. London Math. Soc.* (2), **19**, 216–27.

MACROBERT, T. M. (1938). Proofs of some formulae for the generalized hypergeometric function and certain related functions. *Phil. Mag.* **26**, 82–93.

MACROBERT, T. M. (1939). Solution in multiple series of a type of generalized hypergeometric equation. *Proc. Roy. Soc. Edin.* **59**, 49–54.

MACROBERT, T. M. (1942). Some integrals involving *E*-functions and confluent hypergeometric functions. *Quart. J. Math.* (Oxford), **13**, 65–8.

MACROBERT, T. M. (1943). Associated Legendre functions of the first kind. *Quart. J. Math.* (Oxford), **14**, 1–4.

MACROBERT, T. M. (1947). *Functions of a complex variable.* Third edition. London: Macmillan.

MACROBERT, T. M. (1959). Multiplication formulae for the *E*-functions regarded as functions of their parameters. *Pacific J. Math.* **9**, 759–61.

MAUNSELL, F. G. (1930). Some notes on extended continued fractions. *Proc. London Math. Soc.* (2), **30**, 127–32.

MEIJER, C. S. (1934). Über Whittakersche bezw. Besselsche Funktionen und deren Produkte. *Niew. Arch. voor Wisk.* (2), **18**, 10–39.

MEIJER, C. S. (1935). Einge Integraldarstellungen fur Produkte von Whittakerschen Funktionen. *Quart. J. Math.* (Oxford), **6**, 241–8.

MEIJER, C. S. (1936). Neue Integraldarstellungen aus der Theorie der Whittakerschen und Hankelschen Funktionen. *Math. Ann.* **112**, 469–89.

MEIJER, C. S. (1939). Über Produkte von Legendreschen Funktionen. *Proc. Kon. Akad. v. Wetensch.* (Amsterdam), **42**, 930–7.

MEIJER, C. S. (1940a, b). Über Besselsche, Struvesche, und Lommelsche Funktionen. *Proc. Kon. Akad. v. Wetensch.* (Amsterdam), **43**. (1940a), 198–210; (1940b), 366–378.

MEIJER, C. S. (1941a). Multiplikationstheoreme für die Funktion $G^{m,\,n}_{p,\,q}(z)$. *Proc. Kon. Akad. v. Wetensch.* (Amsterdam), **44**, 1062–70.

MEIJER, C. S. (1941b–f). Neue Integraldarstellungen für Whittakersche Funktionen. *Proc. Kon. Akad. v. Wetensch.* (Amsterdam), **44**. (1941b), 81–92; (1941c), 186–94; (1941d), 298–307; (1941e), 442–51; (1941f), 590–8.

MEIJER, C. S. (1946a–h). On the *G*-function. *Proc. Kon. Akad. v. Wetensch.* (Amsterdam), **49**. (1946a), 227–37; (1946b), 344–56; (1947c), 457–69; (1947d), 632–41; (1947e), 765–72; (1947f), 936–43; (1947g), 1063–72; (1947h), 1165–75.

MEIJER, C. S. (1952a–1956). Expansion theorems for the *G*-function. *Proc. Kon. Akad. v. Wetensch.* (Amsterdam), **55**, (1952a), 369–79; (1952b), 483–87; **56**, (1953a), 43–9; (1953b), 187–93; (1953c), 349–57; **57**, (1954a), 77–91; (1954b), 273–9; **58**, (1955a), 243–51; (1955b), 309–14; **59**, (1956), 70–82.

MEIJER, C. S. (1955c). Outwickelungen wan gegeneraliseerde hypergeometrische functiones. *Simon Stevin*, **31**, 117–39.

NANDA, V. S. (1951). Partition theory and the thermodynamics of multi-dimensional oscillator assemblies. *Proc. Camb. Phil. Soc.* **45**, 591–601.

NÖRLUND, N. S. (1956). Sur les fonctions hypergéometriques d'ordre supérieur. *Mat. Fys. Skr. Dan. Vid. Selsk.* **1**, 1–47.

OLVER, F. W. J. (1953). Note on the asymptotic expansion of generalized hypergeometric series. *J. London Math. Soc.* **28**, 462–3.

OLVER, F. W. J. (1954). The asymptotic solution of linear differential equations of the second order for large values of a parameter. *Phil. Trans. Roy. Soc.* (London), A, **247**, 307–68.

OLVER, F. W. J. (1956). The asymptotic solution of linear differential equations of the second order in a domain containing one transition point. *Phil. Trans. Roy. Soc.* (London), A, **249**, 65–97.

ORR, W. McF. (1899). Theorems relating to the products of two hypergeometric series. *Trans. Camb. Phil. Soc.* **17**, 1–15.

PAPPERITZ, E. (1885). Über verwandte *s*-Funktionen. *Math. Ann.* **25**, 212–21.

PERRON, O. (1917). Über das Verhalten der hypergeometrische Reihe. *Ak. der Wiss. Math. nat. wiss. Klasse*, VIII A, 1–69.

PIDDUCK, F. B. (1946). Lommel's functions of small argument. *Quart. J. Math.* (Oxford), **97**, 193–6.

POCHAMMER, L. (1870). Über hypergeometrische Funktionen *n*-ter Ordnung. *J. für Math.* (Crelle), **71**, 316–40.

POOLE, E. G. C. (1935). On associated hypergeometric series. *Quart. J. Math.* (Oxford), **6**, 214–6.

POOLE, E. G. C. (1938). A transformation of Eulerian hypergeometric integrals. *Quart. J. Math.* (Oxford), **9**, 230–3.

PREECE, C. T. (1923). Dougall's theorem on hypergeometric functions. *Proc. Camb. Phil. Soc.* **21**, 595–8.

PREECE, C. T. (1924). The product of two generalized hypergeometric functions. *Proc. London Math. Soc.* (2), **22**, 370–80.

RAGAB, F. (1952). Recurrence formulae for the $E$-functions. *Proc. Math. and Phys. Soc. Egypt*, **4**, 127–36.

RAGAB, F. (1953$a$). Integrals of $E$-functions expressed in terms of $E$-functions. *Proc. Glasgow Math. Assoc.* **1**, 192–5.

RAGAB, F. (1953$b$). A linear relation between $E$-functions. *Proc. Glasgow Math. Assoc.* **1**, 185–6.

RAGAB, F. (1954$a$). Integrals involving $E$-functions and modified Bessel functions of the second kind. *Proc. Glasgow Math. Assoc.* **2**, 52–6.

RAGAB, F. (1954$b$). Further integrals involving $E$-functions. *Proc. Glasgow Math. Assoc.* **2**, 77–84.

RAGAB, F. (1954$c$). An integral involving a product of two modified Bessel functions. *Proc. Glasgow Math. Assoc.* **2**, 85–8.

RAGAB, F. (1954$d$). Integrals involving $E$-functions and Bessel functions. *Proc. Kon. Akad. v. Wetensch.* (Amsterdam), A 57, 414–23.

RAGAB, F. (1954$e$). Linear relations between $E$-functions and Bessel functions. *Acta Math.* **92**, 1–11.

RAGAB, F. (1955$a$). New integrals involving Bessel functions. *Math. Zeitschr.* **61**, 386–90.

RAGAB, F. (1955$b$). A sum of two $E$-functions expressed as a sum of two $E$-functions. *Proc. Glasgow Math. Assoc.* **2**, 124–6.

RAINVILLE, E. D. (1945). The contiguous function relations for ${}_pF_q$ with applications. *Bull. Amer. Math. Soc.* **51**, 714–23.

RAMANUJAN, S. & ROGERS, L. J. (1919). Proof of certain identities in combinatory analysis. *Proc. Camb. Phil. Soc.* **19**, 211–6.

RAMANUJAN, S. (1935). *Collected papers.* Cambridge University Press.

RICHMOND, H. W. (1892). The sum of the cubes of the coefficients in $(1-x)^{2n}$. *Mess. Math.* **21**, 77–8.

RIEMANN, G. F. B. (1857). *P*-Funktionen. *Ges. Math. Werke*, Gottingen (republished Leipzig, 1892), 67–84.

RIESZ, M. (1921). Sur le principé de Phragmén-Lindelöf. *Proc. Camb. Phil. Soc.* **20**, 205–7, and **21**, 6–7.

ROGERS, L. J. (1894). On the expansion of some infinite products. *Proc. London Math. Soc.* **25**, 318–43.

ROGERS, L. J. (1916). On two theorems of combinatory analysis and some allied identities. *Proc. London Math. Soc.* (2), **16**, 315–36.

RUSHFORTH, J. M. (1952). Congruence properties of the partition function and associated functions. *Proc. Camb. Phil. Soc.* **48**, 402–13.

SAALSCHUTZ, L. (1890). Eine summationsformel. *Zeitschr. für Math. und Physik.* **35**, 186–8.

SAALSCHUTZ, L. (1891). Über einen Sezialfall der hypergeometrischen Reihe dritter Ordnung. *Zeitschr. für Math. und Physik.* **36**, 278–95 and 321–7.

SANDHAM, H. F. (1952). Five series of partitions. *J. London Math. Soc.* **27**, 107–15.

SANDHAM, H. F. (1954). A square as the sum of 9, 11 and 13 squares. *J. London Math. Soc.* **29**, 31–7.

SANDHAM, H. F. (1956). A square and a product of hypergeometric functions. *Quart. J. Math.* (Oxford), (2), **7**, 153–4.

SARAN, S. (1955a). Integrals associated with hypergeometric functions of three variables. *Proc. Nat. Inst. Sci.* (India), A **21**, 83–90.

SARAN, S. (1955b). Transformations of hypergeometric functions of three variables. *Ganita*, **6**, 9–23.

SAWER, W. W. (1949). Differential equations with polynomial solutions. *Quart. J. Math.* (Oxford), **20**, 22–30.

SCHUR, I. (1917). Ein Beitrag zur additiven Zahlentheorie und zur Theorie der Kettenbruche. *Berliner Sitgungs.* **23**, 301–21.

SEARLE, J. H. C. (1909). The summation of certain series. *Mess. Math.* **38**, 138–44.

SEARS, D. B. (1950). Transformation theory of hypergeometric functions. *Proc. London Math. Soc.* (2), **52**, 14–35.

SEARS, D. B. (1951a). Transformations of basic hypergeometric functions of special type. *Proc. London Math. Soc.* (2), **52**, 467–83.

SEARS, D. B. (1951b). On the transformation theory of hypergeometric functions and cognate trigonometrical series. *Proc. London Math. Soc.* (2), **53**, 138–57.

SEARS, D. B. (1951c). On the transformation theory of basic hypergeometric functions. *Proc. London Math. Soc.* (2), **53**, 158–80.

SEARS, D. B. (1951d). Transformations of basic hypergeometric functions of any order. *Proc. London Math. Soc.* (2), **53**, 181–91.

SEARS, D. B. (1952). Two identities of Bailey. *J. London Math. Soc.* **27**, 510–1.

SHANKER, H. (1946). On the Hankel transform of generalized hypergeometric functions. *J. London Math. Soc.* **21**, 194–8.

SHEPPARD, W. F. (1912). Summation of the coefficients of some terminating hypergeometric series. *Proc. London Math. Soc.* (2), **10**, 469–78.

SHUCKLA, H. S. (1956). Certain transformations of nearly-poised bilateral hypergeometric series. *Ganita*, **7**, 113–21.

SHUCKLA, H. S. (1957a). Certain transformations of nearly-poised bilateral hypergeometric series of special type. *Ganita*, **8**, 195–201.

SHUCKLA, H. S. (1957b). On certain relations between products of bilateral hypergeometric series. *Proc. Glasgow Math. Assoc.* **3**, 141–44.

SHUCKLA, H. S. (1958a). On certain transformations of nearly-poised basic bilateral hypergeometric series of the type $_M\Psi_M$. *Math. Zeitschr.* **69**, 195–201.

SHUCKLA, H. S. (1958b). Certain transformations of nearly-poised basic bilateral hypergeometric series of special type. *Canad. J. Math.* **10**, 195–201.

SHUCKLA, H. S. (1959). Certain theorems of Cayley and Orr type for bilateral hypergeometric series. *Quart. J. Math.* (Oxford), (2), **10**, 48–59.

SHUCKLA, H. S. (1960). A note on the sums of certain bilateral hypergeometric series. *Proc. Camb. Phil. Soc.* **55**, 262–6.

SINGH, V. N. (1956). A note on the partial sums of certain basic bilateral hypergeometric series. *Proc. Camb. Phil. Soc.* **52**, 756–8.

SINGH, V. N. (1957). Certain generalized hypergeometric identities of the Rogers–Ramanujan type. *Pacific J. Math.* **7**, 1011–4.

SLATER, L. J. (1951). A new proof of Rogers's transformations of infinite series. *Proc. London Math. Soc.* (2), **53**, 460–75.

SLATER, L. J. (1952a). Further identities of the Rogers–Ramanujan type. *Proc. London Math. Soc.* (2), **54**, 147–67.

SLATER, L. J. (1952b). General transformations of bilateral series. *Quart. J. Math.* (Oxford), (2), **3**, 72–80.

SLATER, L. J. (1952c). Integrals representing general hypergeometric transformations. *Quart. J. Math.* (Oxford), (2), **3**, 206–16.

SLATER, L. J. (1952d). An integral of hypergeometric type. *Proc. Camb. Phil. Soc.* **48**, 578–82.

SLATER, L. J. (1953). Two double hypergeometric integrals. *Quart. J. Math.* (Oxford), (2), **4**, 127–31.

SLATER, L. J. (1954a). A note on equivalent product theorems. *Math. Gazette*, **38**, 127–8.

SLATER, L. J. (1954b). Some new results on equivalent products. *Proc. Camb. Phil. Soc.* **50**, 394–403.

SLATER, L. J. (1954c). The evaluation of the basic confluent hypergeometric functions. *Proc. Camb. Phil. Soc.* **50**, 404–13.

SLATER, L. J. (1954d). Expansions of generalized Whittaker functions. *Proc. Camb. Phil. Soc.* **50**, 628–31.

SLATER, L. J. (1955a). A note on the partial sum of a certain hypergeometric series. *Math. Gazette*, **39**, 217–18.

SLATER, L. J. (1955b). Integrals for asymptotic expansions of hypergeometric functions. *Proc. Amer. Math. Soc.* **6**, 226–31.

SLATER, L. J. (1955c). The integration of hypergeometric functions. *Proc. Camb. Phil. Soc.* **51**, 288–96.

SLATER, L. J. (1955d). Some basic hypergeometric transforms. *J. London Math. Soc.* **30**, 351–60.

SLATER, L. J. (1955e). Hypergeometric Mellin transforms. *Proc. Camb. Phil. Soc.* **51**, 577–89.

SLATER, L. J. & LAKIN, A. (1956). Two proofs of the $_6\Psi_6$ summation theorem. *Proc. Edin. Math. Soc.* 116–121.

SLATER, L. J. (1960). *Confluent hypergeometric functions.* Cambridge University Press.

SLATER, L. J. (1963). W. N. Bailey, (obituary). *J. London Math. Soc.* **37**, 504–12.

TAYLOR, P. R. (1940). The functional equation for Epstein's zeta function. *Quart. J. Math.* (Oxford), **11**, 177–82.

THOMAE, J. (1879). Über die Funktionen welche durch Reihen von der Form dargestellt werden: $1 + \dfrac{p\,p'\,p''}{1\,q'\,q''} + \ldots$ *J. für Math.* (Crelle), **87**, 26–73.

TITCHMARSH, E. C. (1948*a*). *Fourier integrals.* Oxford: University Press.

TITCHMARSH, E. C. (1948*b*). Some integrals involving Hermite polynomials. *J. London Math. Soc.* **23**, 15–6.

TRICOMI, F. G. (1951). Expansion of hypergeometric functions in series of confluent functions and applications to Jacobi polynomials. *Math. Helvetici*, **25**, 10–21.

TRICOMI, F. G. (1954). Introduction to elliptic functions with applications. *Bull. Amer. Math. Soc.* **60**, 587.

VARMA, R. S. (1937*a*). An infinite integral involving Bessel functions and parabolic cylinder functions. *Proc. Camb. Phil. Soc.* **33**, 210–1.

VARMA, R. S. (1937*b*). Summation of some infinite series of Weber's parabolic cylinder functions. *J. London Math. Soc.* **12**, 25–7.

WALLIS, J. (1655). *Arithmetica Infinitorum.* London.

WATSON, G. N. (1906). The general solution of Laplace's equation in *n* dimensions. *Mess. Math.* **36**, 98–106.

WATSON, G. N. (1907). The expansions of products of hypergeometric functions. *Quart. J. Pure and Appl. Math.* **39**, 27–51.

WATSON, G. N. (1908). A series for the square of the hypergeometric function. *Quart. J. Pure and Appl. Math.* **40**, 46–57.

WATSON, G. N. (1909). The cubic transformation of the hypergeometric function. *Quart. J. Pure and Appl. Math.* **41**, 70–9.

WATSON, G. N. (1910*a*). The continuation of functions defined by generalized hypergeometric series. *Trans. Camb. Phil. Soc.* **11**, 281–99.

WATSON, G. N. (1910*b*). The singularities of functions defined by Taylor's series. *Quart. J. Pure and Appl. Math.* **42**, 41–53.

WATSON, G. N. (1911*a*). A theory of asymptotic series. *Phil. Trans. Roy. Soc.* **A 211**, 279–313.

WATSON, G. N. (1911*b*). The characteristics of asymptotic series. *Quart. J. Pure and Appl. Math.* **43**, 152–60.

WATSON, G. N. (1918). Asymptotic expansions of hypergeometric functions. *Trans. Camb. Phil. Soc.* **14**, 277–308.

WATSON, G. N. (1920). The product of two hypergeometric functions. *Proc. London Math. Soc.* (2), **20**, 191–5.

WATSON, G. N. (1924*a*). Dixon's theorem on generalized hypergeometric functions. *Proc. London Math. Soc.* (2), **23**, xxxi–iii.

WATSON, G. N. (1924*b*). The theorems of Clausen and Cayley on products of hypergeometric functions. *Proc. London Math. Soc.* (2), **22**, 163–70.

WATSON, G. N. (1925). A note on generalized hypergeometric series. *Proc. London Math. Soc.* (2), **23**, xiii–v.

WATSON, G. N. (1930*a*–1931*c*). Theorems stated by Ramanujan.

— (1930*a*). *J. London Math. Soc.* **3**, 216–25.

— (1930*b*). *J. London Math. Soc.* **3**, 282–9.

— (1930*c*). *Proc. London Math. Soc.* (2), **29**, 293–308.

— (1930*d*). *J. London Math. Soc.* **4**, 39–48.

— (1930e). *J. London Math. Soc.* **4**, 82–6.

— (1930f). *J. London Math. Soc.* **4**, 231–7.

— (1931a). *J. London Math. Soc.* **6**, 59–65.

— (1931b). *J. London Math. Soc.* **6**, 65–70.

— (1931c). *J. London Math. Soc.* **6**, 126–32.

WATSON, G. N. (1930g). A new proof of the Rogers–Ramanujan identities. *J. London Math. Soc.* **4**, 4–9.

WATSON, G. N. (1931d). Ramanujan's note books. *J. London Math. Soc.* **6**, 137–53.

WATSON, G. N. (1932a–1937b). Some singular moduli, (I)–(VI).

— (1932a). *Quart. J. Math.* (Oxford), **3**, 81–98,

— (1932b). *Quart. J. Math.* (Oxford), **3**, 189–212.

— (1936a). *Proc. London Math. Soc.* (2), **40**, 83–142.

— (1936b). *Acta Arithmetica*, (2), **1**, 284–323.

— (1937a). *Proc. London Math. Soc.* (2), **42**, 377–97.

— (1937b). *Proc. London Math. Soc.* (2), **42**, 398–409.

WATSON, G. N. (1933a). General transforms. *Proc. London Math. Soc.* (2), **35**, 156–99.

WATSON, G. N. (1933b). Proof of certain identities in combinatory analysis. *J. Ind. Math. Soc.* **20**, 57–69.

WATSON, G. N. (1933c–f). Notes on generating functions of polynomials. *J. London Math. Soc.* **8**, 189–92, 194–99, 289–92 and **9**, 22–8.

WATSON, G. N. (1934). Ramanujan's continued fraction. *Proc. Camb. Phil. Soc.* **31**, 7–17.

WATSON, G. N. (1936c, d). The final problem, an account of the mock theta functions. *J. London Math. Soc.* **11**, 55–80 and *Proc. London Math. Soc.* (2), **42**, 377–97.

WATSON, G. N. (1936e). Ramanujan's integrals and Gauss's sums. *Quart. J. Math.* (Oxford), **7**, 175–83.

WATSON, G. N. (1937c). A note on Lerch's functions. *Quart. J. Math.* (Oxford), **8**, 43–7.

WATSON, G. N. (1939). Three triple integrals. *Quart. J. Math.* (Oxford), **10**, 266–76.

WATSON, G. N. (1948a). *Bessel functions*. Cambridge University Press.

WATSON, G. N. (1948b). A table of Ramanujan's function $\tau(n)$. *Proc. London Math. Soc.* (2), **51**, 1–13.

WATSON, G. N. (1952). Periodic sigma functions. *Proc. London Math. Soc.* (3), **2**, 129–49.

WATSON, G. N. (1956). A bilinear transformation. *Edin. Math. notes*, **40**, 7–14.

WATSON, G. N. (1959). A theorem on continued fractions. *Proc. Edin. Math. Soc.* **11**, 167–74.

WHIPPLE, F. J. W. & HILL, M. J. M. (1910). A reciprocal relation between generalized hypergeometric series. *Quart. J. Pure and Appl. Math.* **41**, 128–35.

WHIPPLE, F. J. W. (1925). A group of generalized hypergeometric series; relations between 120 allied series of the type $F(a,b,c; e,f)$. *Proc. London Math. Soc.* (2), **23**, 104–14.

WHIPPLE, F. J. W. (1926a). On well-poised series, generalized hypergeometric series having parameters in pairs, each pair with the same sum. *Proc. London Math. Soc.* (2), **24**, 247–63.

WHIPPLE, F. J. W. (1926b). Well-poised series and other generalized hypergeometric series. *Proc. London Math. Soc.* (2), **25**, 525–44.

WHIPPLE, F. J. W. (1926c). A fundamental relation between hypergeometric series. *J. London Math. Soc.* **1**, 138–45.

WHIPPLE, F. J. W. (1927a). Some transformations of generalized hypergeometric series. *Proc. London Math. Soc.* (2), **26**, 257–72.

WHIPPLE, F. J. W. (1927b). Algebraic proofs of the theorems of Cayley and Orr concerning the products of certain hypergeometric series. *J. London Math. Soc.* **2**, 85–90.

WHIPPLE, F. J. W. (1928). On a theorem due to F. S. Macaulay, concerning the enumeration of power products. *Proc. London Math. Soc.* (2), **28**, 431–7.

WHIPPLE, F. J. W. (1929). On a formula implied in Orr's theorems concerning the products of hypergeometric series. *J. London Math. Soc.* **4**, 48–50.

WHIPPLE, F. J. W. (1930a). On series allied to the hypergeometric series with argument $-1$. *Proc. London Math. Soc.* (2), **30**, 81–94.

WHIPPLE, F. J. W. (1930b). The sum of the coefficients of a hypergeometric series. *J. London Math. Soc.* **5**, 192–3.

WHIPPLE, F. J. W. (1934). On transformations of terminating well-poised hypergeometric series of the type ${}_9F_8$. *J. London Math. Soc.* **9**, 137–40.

WHIPPLE, F. J. W. (1936). Relations between well-poised hypergeometric series of the type ${}_7F_6$. *Proc. London Math. Soc.* (2), **40**, 336–44.

WHIPPLE, F. J. W. (1937). Well-poised hypergeometric series and cognate trigonometric series. *Proc. London Math. Soc.* (2), **42**, 410–21.

WHITTAKER, E. T. & WATSON, G. H. (1947). *Modern analysis*. Cambridge University Press.

WRIGHT, E. M. (1933a). On the coefficients of power series having exponential singularities. *J. London Math. Soc.* **8**, 71–9.

WRIGHT, E. M. (1933b). The asymptotic expansion of the generalized Bessel function. *Proc. London Math. Soc.* (2), **38**, 257–70.

WRIGHT, E. M. (1934). Asymptotic partition formulae; partitions into $k$-th powers. *Acta Math.* **63**, 112–20.

WRIGHT, E. M. (1935). The asymptotic expansion of the generalized hypergeometric function. *J. London Math. Soc.* **10**, 286–93.

WRIGHT, E. M. (1940a). The asymptotic expansion of the generalized hypergeometric function, II. *Proc. London Math. Soc.* (2), **46**, 389–408.

WRIGHT, E. M. (1940b). The generalized Bessel function of order greater than one. *Quart. J. Math.* (Oxford), **11**, 36–48.

WRIGHT, E. M. (1940c). The asymptotic expansion of integral functions defined by Taylor series. *Phil. Trans. Roy. Soc.* (London), **A 238**, 423–51 and **A 239**, 217–32.

WRIGHT, E. M. (1948). The asymptotic expansion of integral functions and of the coefficients in their Taylor series. *Trans. Amer. Math. Soc.* **64**, 409–38.

WRIGHT, E. M. (1949a). The Taylor coefficients of integral functions. *J. London Math. Soc.* **24**, 40–3.

WRIGHT, E. M. (1949b). On the coefficients of power series having exponential singularities, II. *J. London Math. Soc.* **24**, 304–9.

WRIGHT, E. M. & YATES, B. G. (1950). The asymptotic expansion of a certain integral. *Quart. J. Math.* (Oxford), (2), **1**, 41–53.

WRIGHT, E. M. (1951). Corrigendum. *Proc. London Math. Soc.* (2), **54**, 254.

WRIGHT, E. M. (1952). The asymptotic expansion of the generalized hypergeometric function, a correction. *J. London Math. Soc.* **27**, 256.

WRINCH, D. M. (1924). The hypergeometric function with $k$ denominators. *Proc. London Math. Soc.* (2), **22**, xxii–xxiii.

# SYMBOLIC INDEX

**271**

# GENERAL INDEX